Power Systems

Electrical power has been the technological foundation of industrial societies for many years. Although the systems designed to provide and apply electrical energy have reached a high degree of maturity, unforeseen problems are constantly encountered, necessitating the design of more efficient and reliable systems based on novel technologies. The book series Power Systems is aimed at providing detailed, accurate and sound technical information about these new developments in electrical power engineering. It includes topics on power generation, storage and transmission as well as electrical machines. The monographs and advanced textbooks in this series address researchers, lecturers, industrial engineers and senior students in electrical engineering.

Power Systems is indexed in Scopus

Hui Wang

Advanced Matrix Converters

Topology, Modulation, and Control

Springer

Hui Wang
School of Automation
Central South University
Changsha, China

ISSN 1612-1287 ISSN 1860-4676 (electronic)
Power Systems
ISBN 978-3-031-69901-6 ISBN 978-3-031-69902-3 (eBook)
https://doi.org/10.1007/978-3-031-69902-3

This Springer imprint is published by the registered company Springer Nature Switzerland AG
The registered company address is: Gewerbestrasse 11, 6330 Cham, Switzerland

Preface

Three-phase AC-AC power converter is a key element for power generation, transmission, and distribution systems that enables efficient electric energy conversion and has elicited considerable research interest. It acts as an essential interface between the utility grid, the power source, and the load, and has been widely applied to AC-AC power conversion occasions, such as adjustable speed drives (ASDs), uninterrupted power supplies (UPS), wind energy conversion system (WECS), and flexible AC transmission system (FACTS). Among the various AC-AC power converters, the three-phase passive front-end converters including diode rectifier and thyristor rectifier front-end converters show some advantages concerning high efficiency, low electro-magnetic interference (EMI) noise, low cost, simple structure, and high reliability. The undesirable features of the passive front-end converters are the unidirectional power flow and the greatly distorted line currents. Voltage-source-type back-to-back PWM converters (VBBC) and current-source-type back-to-back PWM converters (CBBC) are two typical alternatives. However, the heavy and bulky passive energy storage elements hinder the compactness design and affect the reliability of the above systems.

As an "all-silicon" solution, the matrix converters (MCs) have attracted an increasing attention in recent years because the bulky passive components are eliminated and then, enhanced power density, conversion efficiency, and reliability of the converters are achieved. They are especially suitable for volume/weight-critical, efficiency-critical, and lifetime-critical applications. In the past decades, although extensive researches have been done on MCs and lots of achievements have been made, there are still many problems that need to be solved including the nonlinearity resulting from strict synchronization, complicated main circuit and commutations, relatively high switching losses and EMI noise, and limited input reactive power capability. Besides, with the development demand on higher efficiency, higher input/output power quality, enhanced control performance, and lower costs for emerging fields, advanced MCs technology with high performance topology, modulation, and control is urgently in need for research.

This book gives a comprehensive and in-depth introduction into the development of advanced MCs technology. For each topic, topology configuration, operating principles, modulation strategy, control scheme, system design, and specific application are backed by very concrete simulations and experiments that enable the reader to not only understand the topic but also develop and apply the related technology.

With this book, I intend to enable the reader to: (1) get the latest knowledge on the concept and operation of the advanced direct AC-AC conversion technology from the fundamental to the whole picture and (2) properly select advanced direct AC-AC topologies, modulation strategies, and control schemes according to the specific application.

The main research results of this book have been originally taken from the author who carried out the related research together for almost 16 years, which is a comprehensive summary of the author's latest research results. This book is likely to be of interest to university researchers, R&D engineers, and graduate students in electrical engineering who wish to learn the topologies, modulations, controls, and applications of advanced MCs.

Changsha, China Hui Wang

Acknowledgments

The research covered in the book is supported by the Nature Science Foundation of China (NSFC) under Grant 62173351, the Science and Technology Innovation Program of Hunan Province under Grant 2021RC3010, the JieBang Headed Program of Hunan Province under Grant 2021GK1120, and the JieBang Headed Program of Changsha under Grant kq2102006. The author would like to thank postgraduate students Rouxi Chen, Xiang Yan, Shilin Yuan, Zhengmei Lu, and Shimiao Chen for their contributions and proofreading. Special thanks for the academic support and contributions go to Prof. Mei Su, Prof. Yao Sun, Prof. Yonglu Liu, and Prof. Guanguan Zhang for Chap. 2, Prof. Hanbing Dan for Chap. 11, Xida Chen for Chaps. 1, 6, 8, 9, 11, Xin Li for Chaps. 4, 10, Prof. Patrick Wheeler for Chaps. 6, 11, Prof. Frede Blaabjerg for Chap. 9, and Xiaocan Zhou for Chap. 8. Finally, the author would like to thank the long-term support and encouragement from their families.

The author would also like to thank the staff at Springer, in particular Michael McCabe, for their help and support.

Contents

List of Symbols

S_{ij}	The switching function of a single switch
U_{max}	The maximum value among the three-phase input voltages
U_{mid}	The medium value among the three-phase input voltages
U_{min}	The minimum value among the three-phase input voltages
P^*	The reference active power
P	The active power
Q	The reactive power
i_o	The output current
i_{ra}	The input current of phase a
i^*_{ra}	The desired fundamental current of phase a
$\bar{i}_{ra}$	The average input current of phase a
Δi_r	The harmonic current
m	The modulation index
q	The voltage transfer ratio
ω_i	The angular frequency of the input voltages
ω_o	The angular frequency of the output voltages
ϕ	Initial phase of the output voltages
γ	Ripple index
f_s	The switching frequency
f_n	Mains frequency
i_y	The third-harmonic current
i^*_y	The third-harmonic injection current reference
C_i	The input capacitor
L_{m1}	The magnetizing inductance of the transformer
i_{mm1}	The magnetizing current
T_s	The switching period
i_{ref}	The current reference
φ	The input power factor angle
G	The equivalent input conductance
D	The steady-state duty cycle of the upper switch
D'	The steady-state duty cycle of the lower switch

N	The number of output phases
ϕ_L	The load power factor angle
d	The duty ratio
U_{cmmax}	The amplitude of common-mode voltage
θ_{mid}	The phase corresponding to the input phase with the medium instantaneous voltage
δ	The voltage ripple index
i_{rmid}	The input current before filtering corresponding to the input phase with the medium instantaneous voltage
θ_i	The input phase voltage of the filtering capacitor of phase a
θ_{ca}	The phase of the filtering capacitor voltage of phase a
P_p	The number of pole pairs of the PW
P_c	The number of pole pairs of the CW
M_p	The PW excitation inductance
M_c	The CW excitation inductance
$L_{F(f)}$	The input filter inductance
$C_{F(f)}$	The input filter capacitor
u_{aO}	The a phase-to-neutral voltage
u_{bO}	The b phase-to-neutral voltage
u_{cO}	The c phase-to-neutral voltage
θ_{sa}	The phase of the input phase voltage u_{sa}
u_{pO}	The upper DC source voltage
u_{On}	The lower DC source voltage
u_{rs}	The output line-line voltage
u^*_{max}	The maximum value among the modulation signals
u^*_{mid}	The medium value among the modulation signals
u^*_{min}	The minimum value among the modulation signals
u_0	The zero-sequence voltage
I_{pm}	The amplitude of the active components of the input current
I_{pm}	The amplitude of the reactive components of the input current
$\bar{u}^*_{jp}$	The normalized positive modulation signal
$\bar{u}^*_{jn}$	The normalized negative modulation signal

List of Abbreviations

2LPWM	two-level PWM method
3LPWM	three-level PWM
3HIMC	third-harmonic injection indirect matrix converter
3MC	three-level-output-stage matrix converter
ASDs	adjustable speed drives
ADC	analog-to-digital converter
BW	bandwidth
BPF	band-pass filter
BDFIM	brushless doubly-fed induction machine
CBBC	current back-to-back converter
CC	cycloconverter
CW	control winding
CSI	current-source type inverter
CCM	continuous conduction mode
CSR	current-source rectifier
CBPWM	carrier-based pulse width modulation
CF	common output frequency
DFIM	doubly-fed induction machine
DTC	direct torque control
DM	differential-mode
DSVM	direct space vector modulation
DF	different output frequency
DMC	direct matrix converter
DSPWM	double-signal pulse width modulation method
DPWM	discontinuous pulse-width modulation
DSP	digital signal processor
EMI	electromagnetic interference
EMC	electro-magnetic compatibility
ESR	equivalent series resistance
FFT	fast Fourier transform
FPGA	field-programmable gate array

FCCC	forced commutated cycloconverter
FACTS	flexible AC transmission system
IM	induction motor
IO	input output
ISVM	indirect space vector modulation
IMC	indirect matrix converter
IVS	input voltage selector
IGBT	insulated gate bipolar transistor
I3SMC	indirect three-level sparse matrix converter
LPF	low-pass filter
LISN	line impedance stabilization network
MC	matrix converter
MPP	maximum power point
MEA	more electric aircraft
MPPT	maximum power point tracking
MPC	model predictive control
MOSFET	metal oxide semiconductor field-effect transistor
MT-BBC	matrix-type back-to-back converter
NPC	neutral-point clamped
NSM	new scalar method
OVM	optimized Venturini's method
P&O	perturbation and observation
PI	proportional-integration
PR	proportional resonant
PW	power winding
PV	photovoltaic
PF	power factors
PWM	pulse width modulation
PFC	power factor correction
PMSG	permanent magnet synchronous generator
PMSM	permanent magnet synchronous motor
PLL	phased locked loop
PI-SMC	proportional-integral second-order sliding mode control
RW	rotor winding
RMS	root mean square
SMC	sliding mode control
SVM	space vector modulation
SPWM	sinusoidal pulse width modulation
STATCOM	static synchronous compensator
THD	total harmonic distortion
TSMC	two-stage matrix converter
TSPWM	third-harmonic added sinusoidal pulse width modulation
T^3HIMC	three-level T-type indirect matrix converter
T^2HIMC	three-level third-harmonic injection IMC
T3LIMC	T-type three-level indirect matrix converter

USMC	ultra-sparse matrix converter
UPS	uninterrupted power supplies
VSI	voltage-source inverter
VTR	voltage transfer ratio
VBBC	voltage back-to-back converter
VSMC	very-sparse matrix converter
WECS	wind energy conversion system
ZCS	zero current switching

Chapter 1
Introduction

Three-phase AC-AC converters, for power conversion from a three-phase mains system to a three-phase load with an arbitrary voltage amplitude and frequency, have been widely used in many fields such as ASDs, marine electric propulsion, WECS, FACTS, and so on [1]. Desirable features in such systems mainly include simple and compact structure, sinusoidal input and output currents, bidirectional power flow, operation with unity power factor for any load, long-term reliability, and regeneration capability. Different from traditional back-to-back converters, MCs, as an array of controlled semiconductor switches, can inherently satisfy the above characteristics. Thereby, the emerging topology that achieves direct AC-AC power conversion without DC-link energy storage elements has gradually gained prominence [2]. In this chapter, the traditional three-phase AC-AC conversion systems are initially introduced in Sect. 1.1. Subsequently, their issues are analyzed in Sect. 1.2. The matrix converters and their key concerns are presented in Sects. 1.3 and 1.4. Then, the hybrid third-harmonic injection technique for optimizing MCs is introduced in Sect. 1.5. Section 1.6 provides a brief overview of the contents and organization of the entire book.

1.1 Traditional Three-Phase AC-AC Conversion Systems

The three-phase AC-AC converter with the DC-link is the scheme widely used in the ADSs field, which is composed of a rectifier, a DC-link energy storage element, and an inverter. In the simplest case, the mains coupling can be achieved via a diode bridge. A pulse-controlled braking resistor must be placed on the DC-link, or an antiparallel thyristor bridge must be inserted on the mains side to allow the braking operation of the load. The low-complexity scheme has the shortcomings of severe mains distortion and high reactive power requirements. Furthermore, in chemical plants and other factories where dangerous goods are produced, large-capacity braking resistors may cause fire, which is a potential security threat.

H. Wang, *Advanced Matrix Converters*, Power Systems,
https://doi.org/10.1007/978-3-031-69902-3_1

By coupling the DC-link of a PWM rectifier and a PWM inverter, the mains-friendly AC-AC converter with bidirectional power flow can be realized. According to the DC-link, there are two types: the VBBC with a DC-link capacitor (shown in Fig. 1.1a) and the CBBC with a DC-link inductor (shown in Fig. 1.1b). A PWM rectifier is controlled in such a way that it draws a sinusoidal supply current that is in phase or antiphase with the corresponding supply voltage. Meanwhile, power

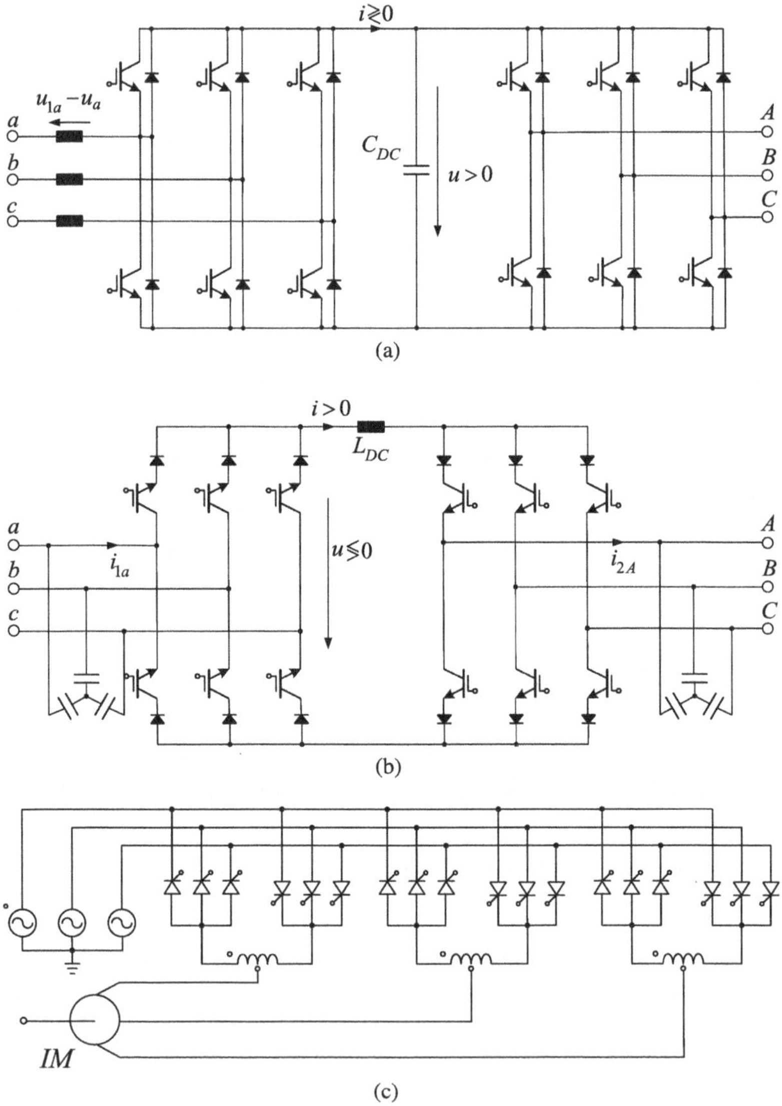

Fig. 1.1 Topologies of traditional three-phase AC–AC converter. (**a**) VBBC. (**b**) CBBC. (**c**) CC

factor correction can be done. The implementation of VBBC and CBBC requires 12 transistors and 12 diodes, or VBBC requires 12 reverse conduction IGBTs and CBBC requires 12 reverse blocking IGBTs. Due to the DC-link energy storage elements, there is an obvious advantage that the control of the two converter stages for the typical energy storage size is largely decoupled. Moreover, the PWM inverter stage has a constant input that is independent of the mains, which underlies the high power utilization.

As shown in Fig. 1.1c, the traditional three-phase AC-AC converter without the DC-link is called the CC, which typically consists of a thyristor network that connects each input line and each output line bidirectionally. The output voltage waveform can be modulated by continuously changing the trigger angle of the thyristor according to a specific rule. Due to the natural commutation of bidirectional switches, a multiphase power supply with the small phase displacement between voltage phasors is usually used to synthesize the better output voltage waveform. This is usually achieved by phase-shifting transformers. It is mainly used in the fields of high voltage, high power, and low frequency ADSs, such as rolling mills and mine hoists. The following are some typical applications of three-phase AC-AC converters.

(1) ASDs for AC Motors:
ADSs are the main application of three-phase AC-AC converters. The normal input voltage with fixed frequency and amplitude is converted by a three-phase AC-AC converter into an output voltage with continuously variable frequency and amplitude, which is applied to the AC motor to control its speed. When the input is always lagging, the three-phase AC-AC converter can provide lagging, leading, or unity power factor loads. Thereby, AC motors can draw any power factor current from the three-phase AC-AC converter. When an AC motor is driven by a three-phase AC-AC converter, the leakage inductance of the motor filters out most of the high-frequency harmonics and reduces the magnitudes of the low-order harmonics.

(2) Wind Power Generation
The wind turbine generator, such as the PMSG, is connected to the input of the converter, and the output of the converter is connected to the grid through the filter, which constitutes the wind energy conversion system. Furthermore, wind energy utilization can be realized by using the motor control strategy that includes an MPPT algorithm.

(3) AC System Interties
Some three-phase AC-AC converters are used to connect two independent AC power systems of equal or different frequencies to control the power flow between them. The basic idea is to use two converters in series, each with an input terminal connected to a common high-frequency base (LC circuit). The output terminals of the two converters are connected to the two systems to be tied. In this way, power flows from one system to another, from the output of the first converter to the input terminal, and then from the input of the second

converter to the output terminal. Meanwhile, there is a growing demand for controlled reactive energy sources used by industry and power systems. Some three-phase AC-AC converter can provide controllable reactive power to the AC system and achieve the reactive compensation.

1.2 Issues of Traditional Three-Phase AC-AC Converters

The VBBC and CBBC have been productization and practicality, but an underlying issue remains apparent. The DC-link energy storage elements possess a relatively large physical volume compared to the total volume of the converter, which does not cater to manufacturers and users. Moreover, the service life of the converter may be reduced when electrolytic capacitors are used for the DC-link of VBBC.

The CC removes the DC-link energy storage elements to ameliorate the above issues to some extent. However, there are still some inherent issues. The assembly requires many devices, and it becomes bulkier as the transformer's design power increases. Meanwhile, the input power factor is severely limited due to the high content of harmonics and subharmonics in the input current. Moreover, to suppress the circulating current when the CC is working in circulating current mode, the design of the circulating current reactor should be deliberate. For naturally commutated cycloconverters, the output frequency is usually limited to 40% of the input frequency. Then, the FCCC is developed to solve the problem [3]. However, the volume is further increased due to the use of external auxiliary commutation circuits.

To improve the above problems and pursue higher power density, an all-silicon "green" converter called MC came into being.

1.3 Matrix Converters

The MC is used to achieve direct AC-AC power conversion without DC-link energy storage elements and consists of an array of bidirectional switches. The highly attractive characteristics are as follows: (1) a simple and compact power circuit; (2) generation of load voltage with arbitrary amplitude and frequency; (3) sinusoidal input and output currents; (4) operation with unity power factor; and (5) regeneration capability.

1.3.1 Development

In 1967, L. Gyugyi and his colleagues at Westinghouse Electric proposed a modulation method for regulating the frequency and amplitude of the output voltage in FCCC, introducing the duty cycle to address the limitations of uncontrollable output voltage amplitude in the original FCCC. This marked the early development of MC technology. In his doctoral thesis published in 1970, L. Gyugyi first introduced the concept of MC as we understand it today. Subsequently, L. Gyugyi and B. Pelly put forward the idea of an MC circuit with bidirectional switches in 1976. However, due to limited advancements in power semiconductor devices at that time, research on MC remained primarily theoretical and did not gain widespread attention in the industry. It was not until 1980 when Italian scholars A. Alesina and M. Venturini successfully built the bidirectional switch using diodes and power MOSFETs, effectively implementing MC technology in engineering [4]. Then, they further proposed a nine-switch topology for MCs along with its direct switching function modulation method in 1981, but the maximum voltage transfer ratio was only 0.5. To overcome the limitation, M. Venturini introduced a third-harmonic injection technique that expanded the voltage transfer ratio range from 0.5 to 0.866 in 1989 [5]. Since then, the global academic community has shown great interest in MCs.

1.3.1.1 Topology

As shown in Fig. 1.2, the MC system mainly consists of the supply, an input *LC* filter, a switching array, and a load. Among them, the input of the switching array is a voltage source, while the output is a current source, and it must satisfy the constraints of no short circuits at the input and no open circuits at the output.

According to the switching structure of the MC, generally it can be divided into two categories of the DMC and the IMC, of which the most representative of the two topologies are shown in Figs. 1.3 and 1.4, respectively. The two converters have the consistency of the function, but there are differences in the modulation and commutation.

DMC, also known as a single-stage matrix converter, consists of nine bidirectional switches to form a 3×3 matrix, and each output phase is connected to the three-phase input through three bidirectional switches. Among them, each bidirectional switch has the ability of bidirectional conduction and shutoff, to meet the constraints that the input cannot be short-circuited and the output cannot be

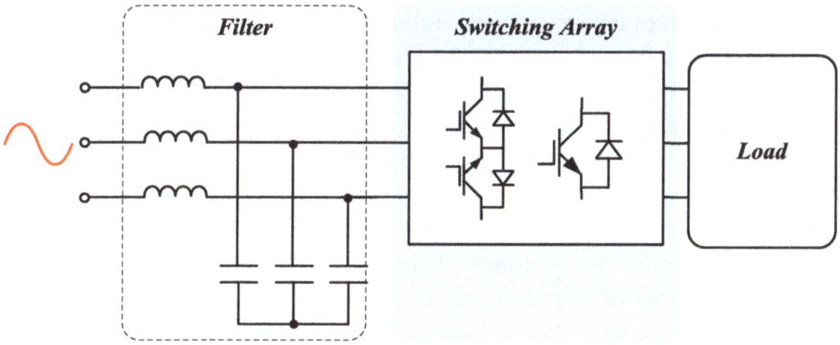

Fig. 1.2 Structure diagram of MCs

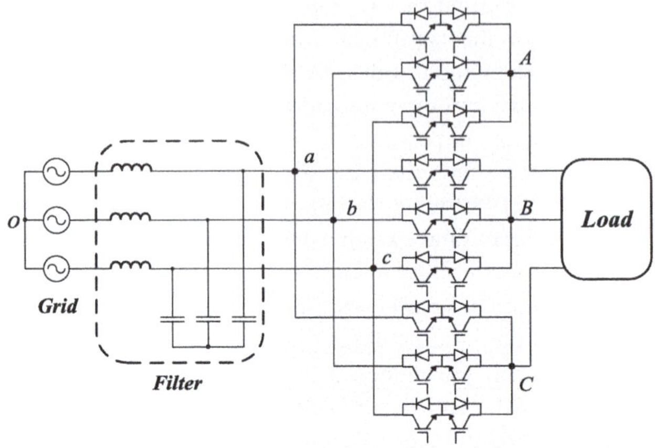

Fig. 1.3 Basic topology of a DMC

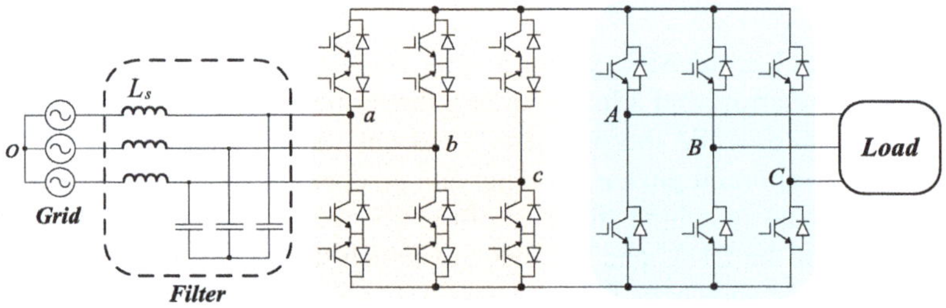

Fig. 1.4 Basic topology of an IMC

open-circuited. It is necessary to use a suitable multistep commutation technique to ensure the safe operation of the system.

IMC and VBBC in the structure are very similar to the existence of a physical DC-link but do not require intermediate energy storage devices. Due to the use of a two-stage structure, it is also known as the TSMC. The IMC can be structurally divided into two parts: the current-source rectifier and the voltage-source inverter, both of which can adopt the existing modulation strategy. The AC-AC conversion is realized through coordinated control of the two stages. The zero-current commutation of the rectifier stage is realized by controlling the switching state of the inverter stage without the complex commutation of a single-stage matrix converter, which reduces the control complexity and improves the reliability.

J. W. Kolar and other scholars have proposed a series of derived topologies to reduce the number of switches based on the IMC topology [6], such as 15-switches sparse matrix converter, 12-switches VSMC, and USMC with 9 switches. The reduction in the number of switching devices reduces the cost, but it can limit some of the functions of the converter, such as the USMC cannot realize the bidirectional flow of energy, and the load impedance angle is not allowed to exceed $\pm 30°$.

In addition, for some special requirements, some improved topologies are developed, such as the IMC with the expansion of the input reactive power capability [7], the four-legged MC for the unbalanced load [8], the AC-DC MC for the V2G field [9], and so on [10].

1.3.1.2 Modulation

The modulation strategy directly affects the performance of the converter such as the input/output waveform quality, the efficiency, the common-mode voltage, the input reactive power compensation capability, and the loss distribution. The modulation strategies of matrix converter mainly include the direct switching function method proposed by M. Venturini, dual-voltage synthesis modulation strategy proposed by A. Ishiguro, direct space vector modulation proposed by D. Casadei [11], indirect space vector modulation proposed by L. Huber [12], carrier-based modulation strategy proposed by Y. Yoon [13], mathematical construction modulation strategy proposed by Y. Sun, and the predictive control [14].

The direct transfer function method is first proposed with a maximum voltage transfer ratio of 0.5. Then, the voltage transfer ratio is increased to the maximum limit of 0.866 at linear modulation by injecting the third harmonic. The direct switching function method is based on the relationship between the input and output voltages or currents, and combined with physical constraints, the direct method of solving the switching function matrix. It is easy to understand by purely mathematical calculations but is computationally intensive and only possesses one solution.

The dual-voltage synthesis modulation strategy proposed by A. Ishiguro, which synthesizes the output voltage based on two input line-line voltages with larger amplitude, has the greatest feature of resisting the influence of unbalanced supply voltage. Thus, the strategy has received wide attention. X. Chen and X. Mu have carried out in-depth research on it.

M. Su proves the consistency between the dual-voltage synthesis modulation strategy and the space vector modulation strategy of MCs. There is no essential difference between the two strategies, and the results can be identical through reasonable arrangement of the switching sequence.

D. Casadei proposes a general modulation technique based on the space vector, whereby a general solution to the switching function matrix of a matrix converter can be found. It is shown that many of the known modulation strategies are a particular solution to this general solution, and the importance of choosing the appropriate zero vector is geometrically illustrated.

The indirect space vector modulation proposed by L. Huber decomposes the matrix converter into a virtual current-source rectifier and a virtual voltage-source inverter with proven space vector modulation, respectively. Then the corresponding switching states are equated to a DMC, whereas the modulation strategy can be directly applied to the IMC.

Y. Yoon proposed a carrier-based modulation strategy, which applies the idea of the carrier to realize the modulation of the DMC, and the key lies in the generation of specific carriers and the selection of offset voltages. It is essentially equivalent to the space vectors, and the difference is reflected in the aspects of the specific implementation and the calculation method.

The modulation strategy based on mathematical construction transforms the modulation problem into the problem of constructing the solution of the system with physical constraints. It has the characteristics of a large solution space and a low computation burden and enriches the connotation of the direct switching function modulation strategy.

The modulation strategy based on the idea of predictive control is first proposed by S. Muller, in which the switching states of a single-stage matrix converter are used as the control inputs, and the input power factor error and the output current error are weighted to form the cost function, and all the possible and reasonable combinations of the switching states are traversed to find out a switching state that minimizes the cost function [15]. This method transforms the modulation problem into a pure optimal control problem. The advantage is that there is no need to understand the basic connotations of the matrix converter, and the disadvantage is the lack of a fast optimization algorithm for solving the optimal solution, usually using an ergodic optimization method, which increases the computation burden of the microcontroller, and thus more difficult to implement.

The matrix converter modulation method based on hysteresis loop current control also transforms the modulation problem into a control problem by combining the ideas of space vector modulation and control, using space vector modulation to realize the unit power factor of the rectifier stage, and transforming the modulation of the inverter stage into the hysteresis loop control of the output current. The advantage is the good dynamic performance, while the disadvantage is the reduced quality of the input current and the larger output current ripple for the same equivalent switching frequency.

A general modulation strategy based on singular value decomposition is proposed by H. Hojabri, which is also a general strategy that can cover many off-the-shelf classical modulation strategies with appropriate parameter selection, such as direct switching function strategy, space vector modulation, and so on. Moreover, a new relationship between voltage transfer ratio and reactive power is revealed by the strategy, which improves the reactive power control capability of MCs.

Based on the above modulation strategies, a variety of modulation strategies have been derived to improve a certain performance of the matrix converter, such as improving the quality of the output waveform, reducing the common-mode voltage, improving the efficiency, expanding the reach of the input reactive power control, compensating the unbalanced input voltages, and so on.

1.3.1.3 Commutation

To satisfy the physical constraints that the input side of the matrix converter cannot be short-circuited, and the output side is not open-circuited, it is necessary to adopt a suitable commutation strategy for the bidirectional switches. The four-step commutation strategy for bidirectional switches is first proposed by N. Burány, and the method effectively solves the problem of commutation by means of strict logic control, which has become the basic method of bidirectional switching commutation.

P. Wheeler and J. Clare used intelligent modules to achieve the current detection, protection, and commutation of bidirectional switches and proposed a three-step commutation and one-step commutation strategy based on the direction of load current and obtained high-quality output currents.

W. Hofmann and others propose a seminatural two-step commutation strategy based on the polarity of the input line-line voltage and combined with the direction of the load current to propose a one-step commutation strategy. J. Mahlein and others propose a high speed digital logic device based on the full digital realization of the commutation strategy to be carried out, the integrated input voltage polarity, and the direction of the output current to obtain a complete natural four-step commutation [16]. The strategy limits the actual switching moment to the third step, which further improves the commutation reliability and the quality of output currents.

K. Sun proposes a current commutation method based on collector-emitter voltage polarity and designed a driver circuit for inverse resistance IGBTs, which integrates the protection and current detection functions and experimentally verifies the effectiveness of the method.

Based on the traditional voltage-based two-step commutation method, a safe commutation scheme using a four-step commutation in the transition interval is proposed by F. Zhang. When the voltage of the two commutating phases is close, three safe freewheel states are inserted to avoid the problem of the small current detection.

X. Ma proposes a variable step size commutation method. The safety zone and danger zone are divided according to the voltage magnitude and current direction, and then different strategies are used in different areas. The method not only fully ensures the reliability of the commutation, but also reduces the commutation time and improves the quality of the output waveforms.

X. Wang proposes a two-step commutation strategy based on the critical region in response to the problem that the voltage-based commutation strategy may lead to a short circuit due to the measurement error near the zero-crossing point of the input line-line voltage. The short-circuit phenomenon due to inaccurate judgment of the voltage in the critical region can be avoided by defining new states and adjusting the zero vector of each sample period and the commutation sequence.

1.3.1.4 Industrial Application

The power device is the key factor limiting the industrial application of MCs. With the increasing maturity of the theory research, some semiconductor device manufacturers such as Infineon, Semikron, Dynex, Fuji, IXYS, etc., begin the development of bidirectional switches and launch a number of power device modules suitable for MCs, laying the foundation for the industrial applications. Among them, Infineon, Semikron, and Dynex have launched commercial bidirectional switching modules, and Fuji is the launch of a 3×3 bidirectional switching module. Mainstream electric drive manufacturers attach great importance to the MC, including ABB, Alstom, Siemens, Fuji, Rockwell, Yaskawa, and so on. In 2005, Yaskawa developed the first commercial MC, named "Varispeed AC." The device is based on a DMC topology, where bidirectional switches are the RB-IGBT modules developed by Fuji. In practical applications, the MC has confirmed the low harmonic distortion and the bidirectional power flow. In 2006, Fuji launched the "FRENIC-Mx" series of MCs (30 kW~45 kW) for the Japanese market and updated the concept of the ASD system. In 2009, Yaskawa launched the "FSDrive-MX1S" MCs (3.3 kV, 200~3000 kVA and 6 kV, 400~6000 kVA) for medium and high voltage high power applications with a multimodular topology. The products possess sustained low-speed operation, power feedback, and good input currents and are suitable for direct-drive wind turbine generators and offshore large wind turbines.

1.3.2 Fundamentals

The MC represents a class of buck-type AC-AC converters with sinusoidal input and output currents, controlled input power factor, and strongly coupled input and output, where the desired output voltage and input current are directly synthesized by suitable modulation of the switching array. The MC is connected to a voltage source on the input side and a current source on the output side, and its switching action must satisfy the physical constraints of no short circuits at the input and no open circuits at the output.

In general, the MC can be divided into two main categories, single-stage matrix converters and two-stage matrix converters, or DMCs and IMCs, as shown in Fig. 1.5. Among them, the DMC consists of nine bidirectional switches to form a 3×3 matrix, each phase output is connected to the three-phase input through three bidirectional switches, and each bidirectional switch has bidirectional conduction and bidirectional blocking capability. To satisfy the constraints, a suitable commutation technique is required to ensure the safe operation of the system. The IMC is a two-stage AC-AC structure consisting of a CSR and a VSI. Among them, the CSR uses bidirectional switching to ensure bidirectional power flow, while the VSI is a traditional three-phase half-bridge topology. The zero-current commutation of the CSR is carried out when the VSI is at zero vector. Thereby, the IMC generally does not need a complex commutation strategy. To prevent the overvoltage generated by the load inductive energy during the shutoff, the MC needs to be equipped with the corresponding clamping circuit.

As shown in Fig. 1.5, $u_{a,b,c}$, $u_{A,B,C}$, $i_{a,b,c}$, and $i_{A,B,C}$ represent input phase voltages, output phase voltages, input currents, and output currents of the MC, respectively.

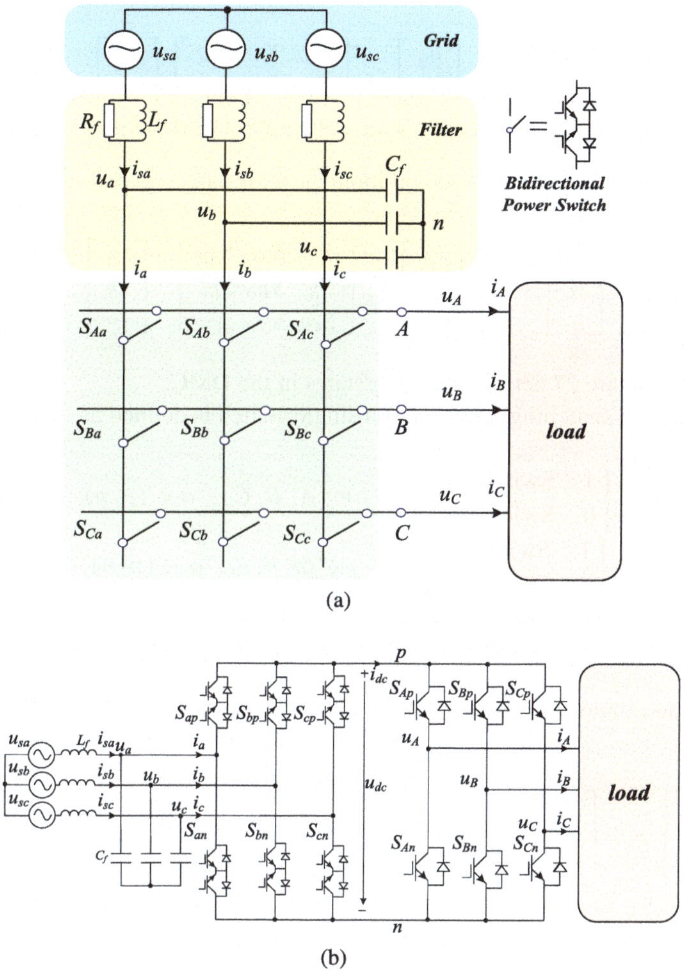

Fig. 1.5 Schematics of MCs. (a) DMC. (b) IMC

For DMC, the switching function S_{ij} of a single switch is defined as

$$S_{ij} = \begin{cases} 1, & \text{Switch } S_{ij} \text{ on} \\ 0, & \text{Switch } S_{ij} \text{ off} \end{cases} \quad i \in (A, B, C), j \in (a, b, c) \tag{1.1}$$

To avoid short circuits on the input side and open circuits on the output side, S_{ij} needs to satisfy

$$S_{ia} + S_{ib} + S_{ic} = 1 \tag{1.2}$$

The relationship between load and input phase voltages can be expressed as

$$\begin{bmatrix} u_A \\ u_B \\ u_C \end{bmatrix} = S \begin{bmatrix} u_a \\ u_b \\ u_c \end{bmatrix} = \begin{bmatrix} S_{Aa} & S_{Ab} & S_{Ac} \\ S_{Ba} & S_{Bb} & S_{Bc} \\ S_{Ca} & S_{Cb} & S_{Cc} \end{bmatrix} \begin{bmatrix} u_a \\ u_b \\ u_c \end{bmatrix} \tag{1.3}$$

The relationship between load and input currents can be expressed as

$$\begin{bmatrix} i_a \\ i_b \\ i_c \end{bmatrix} = S^T \begin{bmatrix} i_A \\ i_B \\ i_C \end{bmatrix} = \begin{bmatrix} S_{Aa} & S_{Ab} & S_{Ac} \\ S_{Ba} & S_{Bb} & S_{Bc} \\ S_{Ca} & S_{Cb} & S_{Cc} \end{bmatrix}^T \begin{bmatrix} i_A \\ i_B \\ i_C \end{bmatrix} \tag{1.4}$$

Hence, there are 27 active switching states in the DMC.

For IMC, the switching function of a single switch is defined as

$$\begin{aligned} S_{i\alpha} &= \begin{cases} 1, & \text{Switch } S_{i\alpha} \text{ on} \\ 0, & \text{Switch } \ S_{i\alpha} \text{ off} \end{cases} \quad i \in (A, B, C), \ \alpha \in (p, n) \\ S_{j\alpha} &= \begin{cases} 1, & \text{Switch } S_{j\alpha} \text{ on} \\ 0, & \text{Switch } \ S_{j\alpha} \text{ off} \end{cases} \quad j \in (a, b, c), \ \alpha \in (p, n) \end{aligned} \tag{1.5}$$

Note that the IMC needs to meet the switching function constraints of both CSR and VSI.

Then, the relationship between load and input phase voltages can be expressed as

$$\begin{aligned} \begin{bmatrix} u_A \\ u_B \\ u_C \end{bmatrix} &= S \begin{bmatrix} u_a \\ u_b \\ u_c \end{bmatrix} \\ &= \begin{bmatrix} S_{ap}S_{Ap} + S_{an}S_{An} & S_{bp}S_{Ap} + S_{bn}S_{An} & S_{cp}S_{Ap} + S_{cn}S_{An} \\ S_{ap}S_{Bp} + S_{an}S_{Bn} & S_{bp}S_{Bp} + S_{bn}S_{Bn} & S_{cp}S_{Bp} + S_{cn}S_{Bn} \\ S_{ap}S_{Cp} + S_{an}S_{Cn} & S_{bp}S_{Cp} + S_{bn}S_{Cn} & S_{cp}S_{Cp} + S_{cn}S_{Cn} \end{bmatrix} \begin{bmatrix} u_a \\ u_b \\ u_c \end{bmatrix} \end{aligned} \tag{1.6}$$

The relationship between load and input currents can be expressed as

$$\begin{aligned} \begin{bmatrix} i_a \\ i_b \\ i_c \end{bmatrix} &= S^T \begin{bmatrix} i_A \\ i_B \\ i_C \end{bmatrix} \\ &= \begin{bmatrix} S_{ap}S_{Ap} + S_{an}S_{An} & S_{bp}S_{Ap} + S_{bn}S_{An} & S_{cp}S_{Ap} + S_{cn}S_{An} \\ S_{ap}S_{Bp} + S_{an}S_{Bn} & S_{bp}S_{Bp} + S_{bn}S_{Bn} & S_{cp}S_{Bp} + S_{cn}S_{Bn} \\ S_{ap}S_{Cp} + S_{an}S_{Cn} & S_{bp}S_{Cp} + S_{bn}S_{Cn} & S_{cp}S_{Cp} + S_{cn}S_{Cn} \end{bmatrix}^T \begin{bmatrix} i_A \\ i_B \\ i_C \end{bmatrix} \end{aligned} \tag{1.7}$$

Hence, for IMC, the active switching states of the rectifier stage and inverter stage are $3^2 = 9$ and $2^3 = 8$, respectively.

According to (1.6), $S_{jp}S_{ip} + S_{jn}S_{in} = 1$ represents the input phase j and output phase i are connected, which is similar to $S_{ij} = 1$ in the DMC, and vice versa. Therefore, any switching state in the IMC can find the corresponding equivalent switching state in the DMC, i.e., the two converters have basically the same conversion function.

1.4 Key Issues of MCs

Over the last few decades, a large amount of research work has brought MCs closer to industrial applications. However, there are still some challenges that need to be overcome:

(1) Lower Voltage Transfer Ratio:

Neglecting the effect of the input filter, the VTR q that is the ratio between the fundamental component of the output voltage of the MC and the fundamental component of the supply voltage is defined as

$$q = \frac{U_{o(1)}}{U_{s(1)}} \tag{1.8}$$

The well-known maximum VTR of MCs is 0.866, which is analyzed by taking the classical SVM strategy as an example.

The principle of space vector modulation of MC is shown in Fig. 1.6. α_o and β_i represent the displacement angles in the sector, respectively. According to the SVM strategy, if the reference output voltage vector and input current vector are

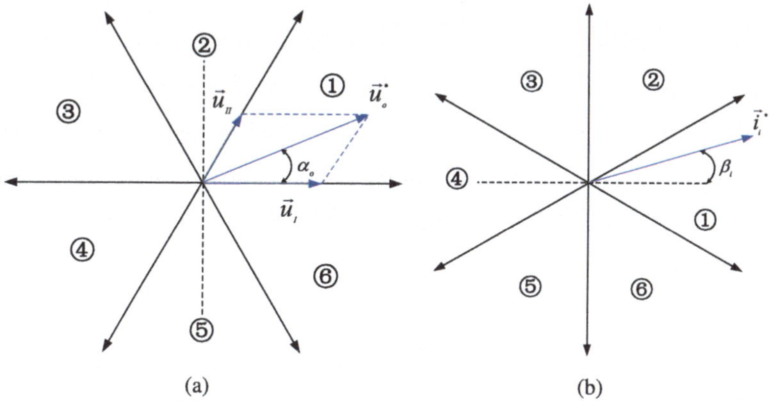

Fig. 1.6 Diagram of space vectors of (**a**) output voltages and (**b**) input currents

both in sector 1, the duty cycles corresponding to the four active vectors at linear modulation can be calculated as

$$\begin{cases} d_1 = (-1)^{n_v+n_i}\frac{2}{\sqrt{3}}q\frac{\sin\alpha_o\cos(\beta_i-\frac{\pi}{3})}{\cos\varphi_i} \\ d_2 = (-1)^{n_v+n_i+1}\frac{2}{\sqrt{3}}q\frac{\sin\alpha_o\cos(\beta_i+\frac{\pi}{3})}{\cos\varphi_i} \\ d_3 = (-1)^{n_v+n_i+1}\frac{2}{\sqrt{3}}q\frac{\sin(\frac{\pi}{3}-\alpha_o)\cos(\beta_i-\frac{\pi}{3})}{\cos\varphi_i} \\ d_4 = (-1)^{n_v+n_i}\frac{2}{\sqrt{3}}q\frac{\sin(\frac{\pi}{3}-\alpha_o)\cos(\beta_i+\frac{\pi}{3})}{\cos\varphi_i} \end{cases} \quad (1.9)$$

where n_v, n_i, and ϕ_i represent the output voltage sector, the input current sector, and the input power factor angle, respectively.

For the feasibility of the modulation strategy, the sum of the absolute values of the four duty cycles must satisfy

$$|d_1| + |d_2| + |d_3| + |d_4| \leq 1 \quad (1.10)$$

Then, the zero vectors are applied to complete the cycle period. Introducing (1.9) into (1.10), q must satisfy

$$q \leq \frac{\sqrt{3}}{2}\frac{|\cos\varphi_i|}{\cos\beta_i\sin(\alpha_o+\frac{\pi}{3})} \quad (1.11)$$

It is obvious that the VTR of the MC is compromised by the input power factor and needs to be less than 0.866 at linear modulation.

(2) Limited Input Reactive Power Control Range:

To ensure the normal operation of MCs, the adjustment range of the input power factor angle should be limited to [-π/6, π/6]. Thereby, the theoretical maximum control range of the input reactive power is

$$Q = P\tan\varphi_i \in \left[-\frac{P}{\sqrt{3}}, \frac{P}{\sqrt{3}}\right] \quad (1.12)$$

According to (1.11), note that the input reactive power control range may be smaller to ensure the appropriate VTR in industrial applications.

(3) Incidental Narrow Pulse:

Generally, the switching frequency of a switch is limited by its turn-on/turn-off delay and rise/fall time, i.e., there is a minimum turn-on time called the minimum pulse width, and all narrow pulses below the minimum pulse width must be limited to the minimum pulse width to ensure the achievability of the modulation and

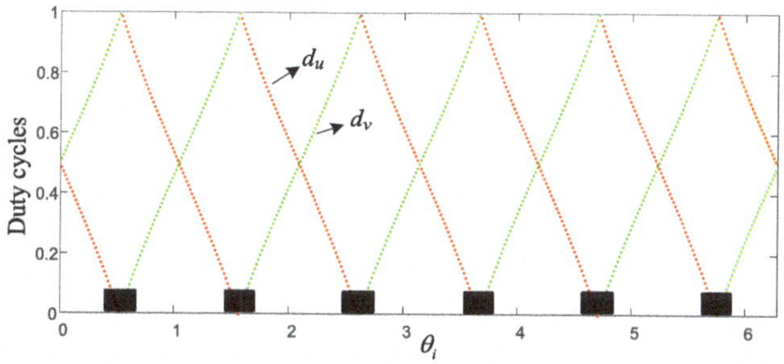

Fig. 1.7 Duty cycles of rectifier stage

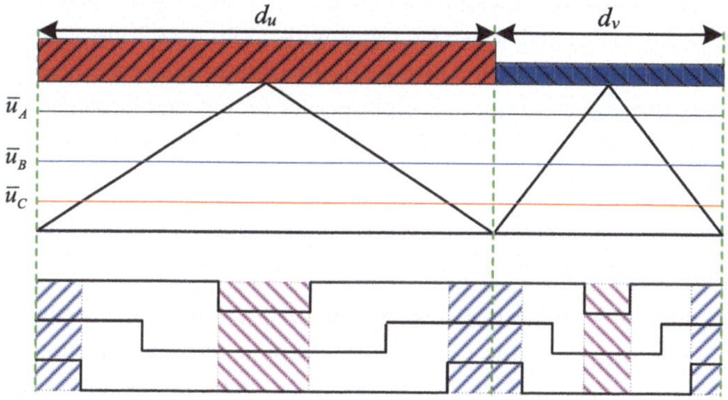

Fig. 1.8 Modulation strategy for IMC

the safety of the device. For an IMC, narrow pulses may also affect the safe commutation because the inverter stage must have enough zero vectors to ensure zero-current commutation in the rectifier stage.

The narrow pulse is particularly acute in IMCs due to the requirement for synchronous modulation strategies of the rectifier and inverter stage. The change rule of rectifier stage duty cycles d_u and d_v is shown in Fig. 1.7. the duration of a line-line voltage is extremely short in the edge of sectors (shown as black areas in Fig. 1.7). As shown in Fig. 1.8, assuming d_v is very small, modulation will not be realized without limited minimum pulse width, while the measure will inevitably damage waveforms.

Moreover, to ensure better input current quality and smaller input filter size, the rectifier stage of the IMC needs to adopt a higher switching frequency, which makes the IMC have a larger switching loss than the passive front-end converter, and at the same time increases the difficulty in the design of the EMI filter. In addition, the adoption of a large number of full-controlled switches significantly increases the cost and modulation complexity of the IMC.

1.5 Hybrid Third-Harmonic Injection Concept

The hybrid third-harmonic injection technique is a favorable choice to meet the above challenges of the traditional MCs.

A hybrid front-end converter is formed by integrating the front-end rectifier of a passive front-end converter and a portion of an active switch. The hybrid front-end converter inherits the advantages of the passive front-end converter, such as low switching losses and low cost, and at the same time is a compromise solution due to the control of the active switches, which results in a significant improvement of the input current quality and power factor. Hybrid front-end converters are generally characterized by the following features: (1) the presence of both natural and forced commutation; (2) the improvement of the input characteristics by switching the active switches; (3) passive devices such as diodes that bear the main current stress and power transfer tasks, and active switches with smaller volt-ampere capacity; and (4) the incomplete controllability that leads to limitations in certain performance aspects such as input current quality, input power factor, and power flow direction.

To improve the performance of hybrid front-end converters, the hybrid third-harmonic injection technique is developed. M. Jantsch and C.W. G. Verhoeve first came up with the concept to increase the lifetime of a three-phase grid-connected photovoltaic inverter without electrolytic capacitors in 1997. Subsequently, this concept is applied to the traditional diode rectifier-fed inverter that has a small DC-link capacitor for reducing the input current harmonics. Although the quality of the input current is improved greatly, the overvoltage problem may occur when feeding low output power factor loads or in regenerative operation mode due to the lack of energy storage components in the DC-link and the inherent characteristic of unidirectional power flow. In 2013, J. W. Kolar presents the design and implementation of the unidirectional active third-harmonic injection circuit (shown in Fig. 1.9) feeding a buck converter for DC distribution systems. Meanwhile, the derivation of the bidirectional AC-AC converter based on active third-harmonic injection circuits is also investigated. Based on the above research, detailed studies of the bidirectional

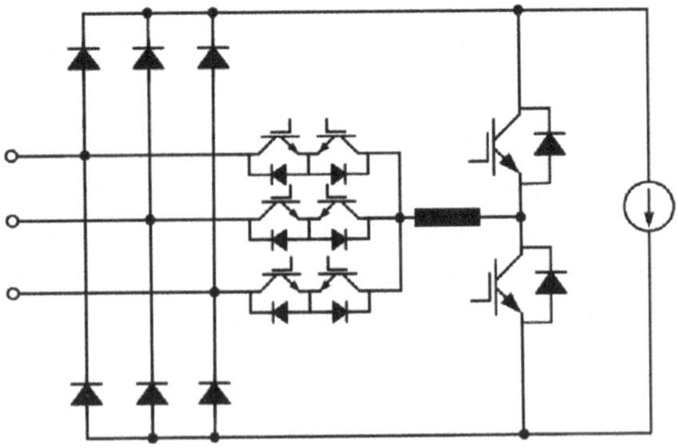

Fig. 1.9 Buck-type active third-harmonic injection converter

third-harmonic injection-based AC-AC converter are given in the book, including the input reactive power control range, voltage transfer ratio, and the experimental verification of the full function of the topology.

1.6 Organization of the Book

The central purpose of this book is to give a comprehensive and in-depth introduction to the topologies, modulation strategies, and control methods of advanced MCs. With this book, we intend to help researchers, engineers, and designers to properly select topologies and control methods of MCs according to the specific application.

Chapter 2 introduces the basic third-harmonic injection MC and demonstrates that the DC-link capacitor can be removed by bidirectional implementation and proper control of the rectifier. Meanwhile, an algorithm enhancing the input reactive power capability is developed. Thus, sinusoidal input currents are achieved, and the reactive power control range is extended significantly. Moreover, the design criteria of the third-harmonic injection inductor are discussed. Finally, the proposed topology and methods are verified by simulation and experimental results.

Chapter 3 presents an active 3HIMC topology with dual three-phase outputs. The presented 3HIMC has the advantages of sinusoidal input and output currents, bidirectional power flow, and can provide two sets of three-phase outputs, whose frequencies and amplitudes can be regulated independently. Compared with the conventional IMC, the input reactive power control range is extended significantly, and the converter can even work as a static compensator. Moreover, a maximal voltage transfer ratio of 0.866 is achievable under all operation conditions. Furthermore, the operations of the rectifier and inverters are independent. Thus, for a 3HIMC system containing multiple VSIs, modulation and control tasks can be implemented flexibly by a separate controller for each VSI, while it is difficult for conventional IMCs to achieve such a goal. All the features above make 3HIMC competitive for applications that require multiple three-phase outputs such as multiple drives. After describing the topology structure, operating principles, and control algorithm in detail, simulation and experimental results are presented to demonstrate the validity and effectiveness of the proposed topology.

Chapter 4 proposes a T^3HIMC topology as well as a carrier-based modulation method. In addition to inheriting the advantages of the third-harmonic injection IMC such as extended input reactive power control range and no need for synchronization in modulation, T^3HIMC could provide an improved output power quality. Besides, a closed-loop control algorithm for balancing the neutral-point voltage is developed. The presented modulation strategy and control algorithm are generalized and can be extended to other three-level IMCs. Finally, the proposed topology and method are verified by simulation and experimental results.

Chapter 5 aims to introduce a T^2HIMC topology, and a neutral-point voltage balancing algorithm. The topology consists of a line-commutated input voltage selector, a three-level inverter, and a third-harmonic injection circuit, where the split DC source voltage of the inverter is formed by two input line-line voltages. Furthermore, by adopting the active power filtering technique and utilizing the third-harmonic injection circuit to compensate the neutral-point current, the neutral-point potential is balanced without extra control effort. In addition, except the advantages such as bidirectional power flow, sinusoidal input/output currents, and multilevel output voltages, the synchronization in modulations is eliminated, and the control range of the input reactive power is extended significantly. Finally, the functionality and effectiveness of the proposed topology and methods are verified by simulations and experimental results.

Chapter 6 discusses the T3LIMC topology and its modulation strategy. The presented topology is composed of an *LC* input filter, a current-source rectifier, and a T-type three-level inverter, where the split DC source voltage of the inverter is formed by two input line-line voltages and the neutral-point voltage. A multi-carrier-based modulation strategy for the inverter and a specific zero-sequence voltage is also developed, and the desired neutral-point current of the inverter is constructed to compensate for the input currents. Consequently, the balanced neutral-point potential and symmetric sinusoidal input currents are achieved without additional control effort. In addition to advantages such as bidirectional power flow, sinusoidal input-output currents, controllable input power factor, no DC-link energy storage elements, and compact structure, the T3LIMC also improves the output waveform quality by generating the three-level output voltages. Meanwhile, the ZCS operation of the rectifier helps to improve efficiency. Finally, the functionality and effectiveness of the presented topology and modulation strategy are verified by experimental results.

Chapter 7 presents a three-phase micro-inverter without energy storage capacitors, which consists of a flyback stage, a third-harmonic injection circuit, and a line-commutated current-source-type inverter. The flyback stage realizes maximum power point tracking, while the third-harmonic injection circuit is responsible for output power factor correction. In addition to having the potential of achieving a compact structure, long lifetime, and high efficiency, the developed micro-inverter provides extended output reactive power control range and three-phase balanced output. Finally, the proposed topology and method are verified by simulations and experimental results.

Chapter 8 presents an MT-BBC with single-arm current injection, which consists of a three-phase rectifier with a small film DC-link capacitor and a three-phase inverter. A clamped modulation strategy is applied to the rectifier, which is featured by only one high-frequency switching bridge arm, reduced switching losses, and enhanced input reactive power control range. By controlling and injecting the quasi-third-harmonic current into the phase with a high-frequency switching bridge arm of the rectifier, sinusoidal input currents and controllable input power factor are achieved. A carrier modulation strategy with a variable carrier is applied to the inverter to compensate for the time-varying six-pulse DC-link voltage. The presented converter inherits the advantages of both the conventional VBBC and the MC including simple main circuit structure, sinusoidal inputs and outputs, bidirectional power flow capability, potentially high power density, high reliability, high efficiency, and direct AC-AC conversion. In addition, the input reactive power control range is extended significantly. The effectiveness of the proposed methods is verified by an experimental prototype.

Chapter 9 introduces a three-level output strategy for conventional $3\times N$ ($N \geq 3$) DMCs, where three-level output voltages are achieved without any changes in the topology. In this developed strategy, the DMC is treated as a two-stage converter that consists of a fictitious rectifier and an NPC three-level inverter. By applying a phase-shifted carrier modulation strategy to the rectifier and a modified carrier-based double-signal pulse width modulation strategy to the NPC inverter, sinusoidal input and output, reduced switching losses, and controllable input power factors are achieved in conventional $3\times N$ DMCs. The developed three-level output strategy improves output waveform quality while sacrificing input performance to some extent. Finally, the effectiveness of the developed strategy is verified by both simulations and experimental results.

Chapter 10 analyzes the VTR characteristic of 3HIMC and proposes a control method to enhance the linear VTR. With the developed method, the maximum linear VTR can break through the intrinsic limit and VTR >0.866 can be achieved by regulating the input reactive power under arbitrary output conditions theoretically. Finally, simulations and experimental results verify the correctness of the theoretical analysis, and the comparison of 3HIMC and other TSMCs with improved VTR demonstrates the effectiveness of the proposed method.

Chapter 11 presents a cascade PI-SMC method for a matrix converter-fed BDFIM. The PI-SMC method consists of a PI speed controller with an inner-loop second-order SMC. The PI controller provides a reference to the inner-loop second-order SMC with constraints according to the system requirements in terms of maximum current and speed. As the discontinuous sign function in the traditional SMC is replaced by the continuous super-twisting function, the chattering problem is eliminated for inner-loop second-order SMC. Moreover, the inner-loop second-order SMC has robustness and interference immunity against nonlinear disturbances of the BDFIM. Simulation and experimental results demonstrate the excellent robust tracking performance of the proposed PI-SMC method.

Chapter A gives conclusions and perspectives for future work.

References

1. J.W. Kolar, T. Friedli, J. Rodriguez, P.W. Wheeler, Review of three-phase PWM AC-AC converter topologies. IEEE Trans. Ind. Electron. **58**(11), 4988–5006 (2011)
2. P.W. Wheeler, J. Rodriguez, J.C. Clare, L. Empringham, A.Weinstein, Matrix converters: a technology review. IEEE Trans. Ind. Electron. **49**(2), 276–288 (2002)
3. A. Ishiguro, T. Furuhashi, S. Okuma, A novel control method for forced commutated cycloconverters using instantaneous values of input line-to-line voltages. IEEE Trans. Ind. Electron. **38**(3), 166–172 (1991)
4. M. Venturini, A new sine wave in sine wave out, conversion technique which eliminates reactive elements. Proc. Powercon **7**, E3/1–E3/15 (1980)

5. A. Alesina, M. Venturini, Analysis and design of optimum-amplitude nine-switch direct AC-AC converters. IEEE Trans. Power Electron. **4**(1), 101–112 (1989)
6. J.W. Kolar, F. Schafmeister, S.D. Round, H. Ert, Novel three-phase AC-AC sparse matrix converters. IEEE Trans. Power Electron. **22**(5), 1649–1661 (2007)
7. Y. Sun, X. Li, M. Su, H. Wang, H. Dan, W. Xiong, Indirect matrix converter-based topology and modulation schemes for enhancing input reactive power capability. IEEE Trans. Power Electron. **30**(9), 4669–4681 (2015)
8. Y. Sun, M. Su, X. Li, H. Wang, W. Gui, Indirect four-leg matrix converter based on robust adaptive back-stepping control. IEEE Trans. Ind. Electron. **58**(9), 4288–4298 (2011)
9. M. Su, H. Wang, Y. Sun, J. Yang, W. Xiong, Y. Liu, AC/DC matrix converter with an optimized modulation strategy for V2G applications. IEEE Trans. Power Electron. **28**(12), 5736–5745 (2013)
10. H. Wang, M. Su, Y. Sun, J. Yang, G. Zhang, W. Gui, J. Feng, Two-stage matrix converter based on third-harmonic injection technique. IEEE Trans. Power Electron. **31**(1), 533–547 (2016)
11. D. Casadei, G. Serra, A. Tani, L. Zarri, Matrix converter modulation strategies: a new general approach based on space-vector representation of the switch state. IEEE Trans. Ind. Electron. **49**(2), 370–381 (2002)
12. L. Huber, D. Borojevic, Space vector modulated three-phase to three-phase matrix converter with input power factor correction. IEEE Trans. Ind. Appl. **31**(6), 1234–1246 (1995)
13. Y.D. Yoon, S.K. Sul, Carrier-based modulation technique for matrix converter. IEEE Trans. Power Electron. **21**(6), 1691–1703 (2006)
14. G. Zhang, J. Yang, Y. Sun, M. Su, Q. Zhu, F. Blaabjerg, A predictive-control-based over-modulation method for conventional matrix converters. IEEE Trans. Power Electron. **33**(4), 3631–3643 (2018)
15. S. Muller, U. Ammann, S. Rees, New time-discrete modulation scheme for matrix converters. IEEE Trans. Ind. Electron. **52**(6), 1607–1615 (2005)
16. J. Mahlein, J. Igney, J. Weigold, M. Braun, O. Simon, Matrix converter commutation strategies with and without explicit input voltage sign measurement. IEEE Trans. Ind. Electron. **49**(2), 407–414 (2002)

Chapter 2
Basic Third-Harmonic Injection Matrix Converter

2.1 Introduction

Due to the attractive characteristics such as low cost, simple, and reliable structure, high efficiency, and low EMI noise, three-phase diode rectifier front-end converters have been widely applied to AC-AC power conversion applications such as ASDs and UPS. However, one of the undesirable features of the diode rectifier front-end converters is the greatly distorted line current, which may cause harmonic pollution in the power system [1]. Usually, there are two solutions to solve this problem: One is using the "green" active PFC converters to replace the diode rectifier front-end converters [2]; the other is applying the active power filter technique to compensate for the harmonic current generated by the diode rectifiers.

For the first solution, PWM-VBBC and PWM-CBBC are two typical active PFC converters. Both of them carry advantageous features such as bidirectional power flow, sinusoidal input currents, and controllable input power factor. However, the heavy and bulky energy storage elements in the DC-link hinder the compactness design and system integration of converters [3–5]. An alternative is the MCs, which have attracted an increasing attention in recent years due to the absence of energy storage elements [6–11]. MCs can be classified into two types: DMC and IMC. The IMC topology illustrated in Fig. 2.1 provides zero-current commutation that is not available in DMC. Besides, the clamp circuit for overvoltage protection is simpler compared to DMC. Due to the features above, IMC has been extensively investigated recently [8–11]. In most applications, a unity input power factor operating condition is preferred for MCs. However, when MCs are applied in a WECS [12] or a FACTS [13], there is a requirement of generating input reactive power. For IMCs employing the conventional modulation [11], the input reactive power can be regulated to some extent by changing the input displacement angle. Unfortunately, the maximum voltage transfer ratio achievable decreases with the decreasing of the input power factor [14, 15], which is of great disadvantage in ASD applications since the output capability is restricted. Moreover, strict

H. Wang, *Advanced Matrix Converters*, Power Systems,
https://doi.org/10.1007/978-3-031-69902-3_2

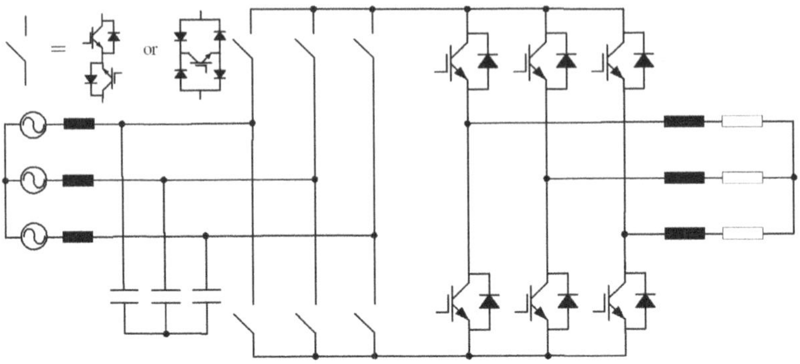

Fig. 2.1 Conventional IMC topology

Fig. 2.2 Diode rectifier front-end converter with paralleled active power filter

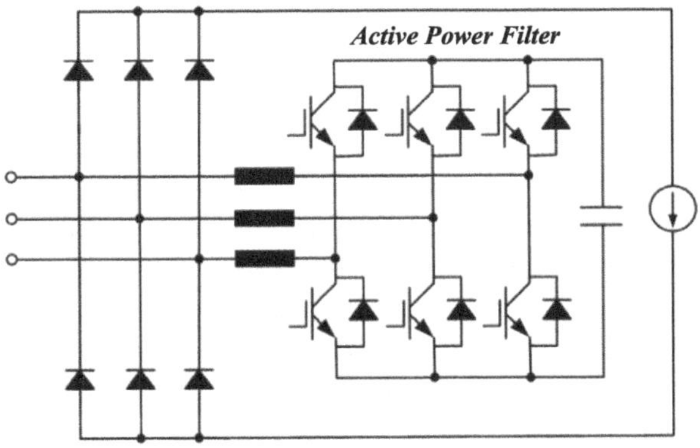

synchronization in modulation between the rectifier and the inverter is necessary for IMCs. That is, the inversion stage must experience a synthesis of the output space vector in every sub-switching cycle of the rectification stage. Thus the input and output performances may be degraded due to the nonlinearity caused by narrow pulse width [16].

Regarding the second solution, two schemes have been proposed and researched. One is to install an extra three-phase active power filter in parallel to the diode rectifier front-end converter [17, 18], as shown in Fig. 2.2. The other is to employ the active third-harmonic injection technique to improve the input current quality [19–27], as presented in Fig. 2.3. The active third-harmonic injection concept is first introduced in [19], where a three-phase grid-connected photovoltaic inverter without electrolytic capacitors for improved lifespan is proposed. In [20] this circuit is applied to the conventional diode rectifier-fed inverter that has a small DC-link capacitor for the purpose of reducing the input current harmonics. Although the quality of the input current is improved greatly, the overvoltage problem may occur when feeding low output power factor loads or in regenerative operation mode due

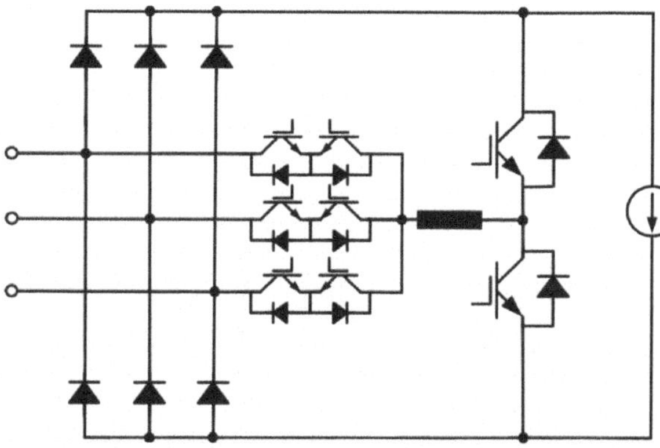

Fig. 2.3 Buck-type active third-harmonic injection converter

to the lack of energy storage components in the DC-link and the inherent characteristic of unidirectional power flow. In [27] the design and implementation of the unidirectional active third-harmonic injection circuit, feeding a buck converter for DC distribution systems, are presented, and the derivation of the bidirectional AC-AC converter, based on active third-harmonic injection circuit, is also investigated. However, as in the situation of [20], a film capacitor is placed in the DC-link to provide the required current path. In theory, it causes inrush transient in the DC-link capacitor and input capacitors and distorts the input currents, especially when the capacitance of the DC-link capacitor and the input capacitor are comparable. Besides, so far the detailed studies of the bidirectional third-harmonic injection-based AC-AC converter have not been reported in the literature, including the input reactive power control range, voltage transfer ratio, and experimental verifications of the full functionality of the topology.

In this chapter, the functionality and performance of the 3HIMC are investigated systematically. It is demonstrated that the DC-link capacitor could be removed through bidirectional implementation and proper control of the rectifier. Besides, the operating principles of 3HIMC are analyzed in detail. An algorithm for achieving sinusoidal input currents and significantly extending the input reactive power control range without lowering the voltage transfer ratio is developed, and mathematical proof of the correctness of the algorithm is also provided. Based on the analysis of the operating principles, a detailed comparison between IMCs and 3HIMC is presented. Moreover, modulation strategies of 3HIMC are discussed, and the narrow pulse issue is analyzed and compared with that of IMCs. In addition, the system design is carried out, and the design criteria of the third-harmonic injection inductor, considering both the current ripple and the third-harmonic current tracking performance, are introduced. Finally the proposed method and all the theoretical analyses are verified by simulation and experimental studies.

This chapter is organized as follows: Sect. 2.2 introduces the topology and operating principles of 3HIMC, presents the solution to cancel the DC-link capacitor, and discusses the algorithm that can achieve sinusoidal input currents and extend the input reactive power significantly; Sect. 2.3 presents the modulation strategies in detail, along with the discussion of narrow pulse width issue; Sect. 2.4 introduces the system design including the design criteria of the third-harmonic injection inductor, followed by the power loss analysis and efficiency evaluation; Sect. 2.5 shows the simulation and experimental results to verify the presented method; Sect. 2.6 draws the final conclusion of this chapter.

2.2 Topology and Operating Principles of 3HIMC

2.2.1 Topology

The topology of 3HIMC is shown in Fig. 2.4. The main circuit consists of an input *LC* filter, a bidirectional rectifier, a clamp circuit consisting of a fast recovery diode D_c and a film capacitor C_c, an active third-harmonic current injection circuit and a conventional VSI. The active third-harmonic current injection circuit is composed of three bidirectional switches, a third-harmonic injection inductor, and a bridge leg. Under normal operation conditions, only one of the three bidirectional switches is turned on to inject the third-harmonic current into the corresponding input phase. According to the requirements of the load, the VSI provides three-phase output voltages with variable frequency and amplitude. The clamp circuit is used to absorb the energy stored in the leakage inductance of the load when the system shutdown. Regarding the topologies presented in [20] and [27], a DC-link capacitor is needed to provide the required current path of the third-harmonic injection inductor. Nevertheless, it causes inrush transient and distorts the input currents. Thus these topologies are not well structured, and the proof of sinusoidal input currents in [27] is not applicable. Actually, through bidirectional implementation and proper control of the rectifier, the DC-link capacitor can be eliminated. In this chapter, commutation of the rectifier is controlled to occur at the moment of zero DC-link current. For example, commutation between the upper switches S_{a+}, S_{b+}, and S_{c+} is arranged when the current i_p is equal to zero. Thus the DC-link capacitor is removed and the switching losses of the rectifier can also be ignored. Besides, the clamp circuit can also provide the required current path for sector transition.

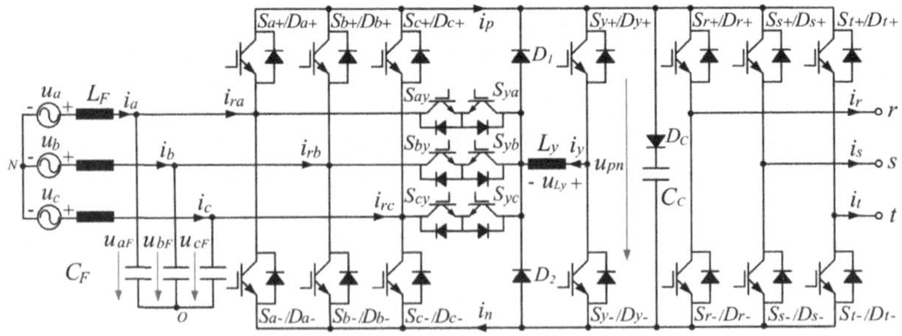

Fig. 2.4 Schematic diagram of 3HIMC

Compared with conventional IMCs, although two more switches are added and an extra third-harmonic injection inductor is introduced, the amplitude of the current imposed on the third-harmonic injection inductor is relatively low and the input current quality of the rectifier is improved due to the existence of the third-harmonic injection inductor. Besides, the switches of the rectifier commutate at the line frequency. Thus low-cost switches with low switching speeds are applicable. This helps to reduce the system cost especially for high power converter applications.

2.2.2 *Operating Principles*

The operating principles of 3HIMC are described as follows: For the switches S_{a+}, S_{b+}, and S_{c+} of the rectifier, the switch connected to the input phase with the highest voltage is kept on; for the switches S_{a-}, S_{b-}, and S_{c-}, the switch connected to the input phase with the lowest voltage is kept on. The switches S_{y+} and S_{y-} in third-harmonic current injection circuit are controlled to form the third-harmonic current i_y flowing through the inductor L_y. The bidirectional switch connected to the input phase with the lowest absolute voltage is kept on to inject the third-harmonic current. For example, when the input voltages satisfy $u_a > u_b > u_c$ (denoted as sector 1), switches S_{a+} and S_{c-} in the rectifier and the bidirectional switches S_{yb} and S_{by} in the third-harmonic current injection circuit are turned on, as shown in Fig. 2.5, and so forth. In this manner, sinusoidal three-phase input currents and controllable input power factor are attainable. The key waveforms of the rectifier and third-harmonic current injection circuit are shown in Fig. 2.6.

2.2.3 *PFC Algorithm and Its Mathematical Proof*

Assume the converter has no power loss and the output power is constant, an algorithm for achieving sinusoidal input currents and controllable input power factor is developed, along with the mathematical proof of the correctness of the algorithm.

Three-phase symmetrical input voltages can be expressed as

$$\begin{cases} u_a &= U\cos(\omega_i t) \\ u_b &= U\cos(\omega_i t - 2\pi/3) \\ u_c &= U\cos(\omega_i t + 2\pi/3) \end{cases} \tag{2.1}$$

where U and ω_i are the amplitude and angular frequency of the input voltages, respectively.

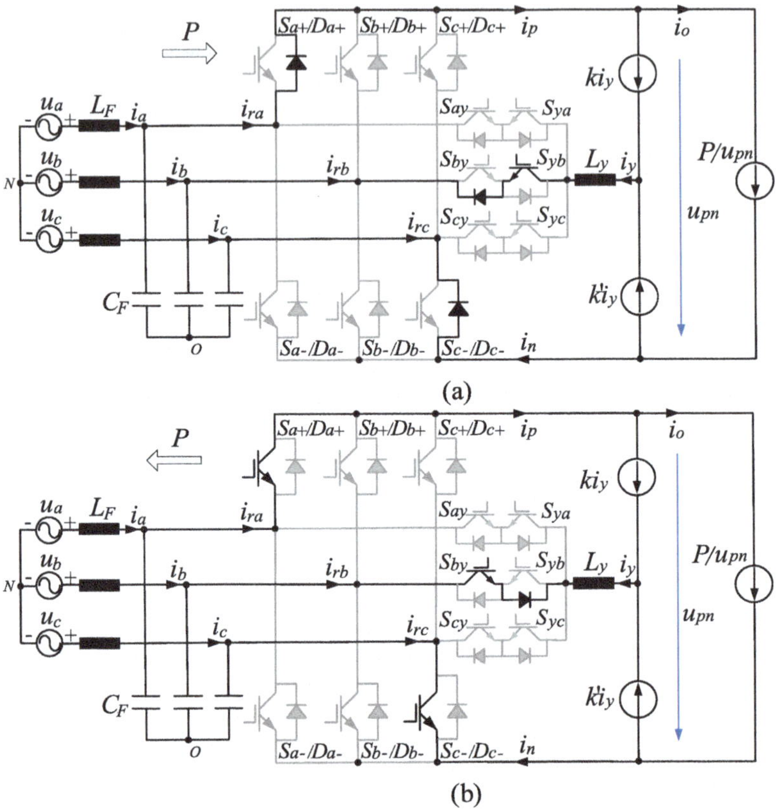

Fig. 2.5 Equivalent circuit diagram of 3HIMC (assuming $u_a > u_b > u_c$ and $u_b < 0$). (**a**) Forward power flow. (**b**) Reverse power flow

Fig. 2.6 Key waveforms of the rectifier and the third-harmonic injection circuit of 3HIMC (assuming forward power flow and unity input power factor). (**a**) Three-phase input voltages and DC-link voltage u_{pn}. (**b**) Three-phase input currents. (**c**) Current at the input of the rectifier in phase a. (**d**) Third-harmonic injection current i_y

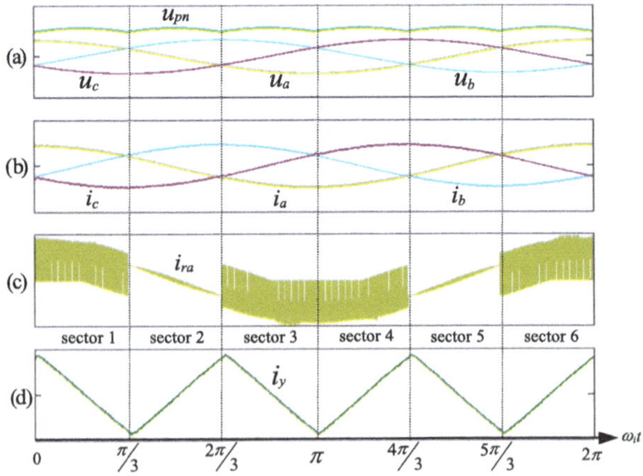

Sort the input voltages according to the relationships of the instantaneous values as follows:

$$\begin{cases} u_{max} = \max(u_a, u_b, u_c) \\ u_{mid} = \text{mid}(u_a, u_b, u_c) \\ u_{min} = \min(u_a, u_b, u_c) \end{cases} \tag{2.2}$$

where u_{max}, u_{mid}, and u_{min} represent the maximum value, medium value, and minimum value among the three-phase input voltages; max, mid, and min are the operators of the maximum value, medium value, and minimum value, respectively.

In each switching period, according to the voltage-second balance of L_y, a duty ratios equation can be obtained as follows:

$$\begin{cases} u_{mid} = k u_{max} + k' u_{min} \\ k + k' = 1 \end{cases} \tag{2.3}$$

where k and k' are steady-state duty ratios of the switches S_{y+} and S_{y-}, respectively. And the duty ratios are solved as

$$\begin{cases} k = \dfrac{u_{mid} - u_{min}}{u_{max} - u_{min}} \\ \\ k' = \dfrac{u_{max} - u_{mid}}{u_{max} - u_{min}} \end{cases} \tag{2.4}$$

In sector 1, from the switching states, it is found that the current i_{rb} is equal to the opposite value of the third-harmonic injection current i_y. Suppose the input currents are sinusoidal and the input power factor is controllable, and then the average value of i_{rb} can be obtained as

$$\begin{cases} \bar{i}_{rb} = -i_y = I_{pm} \cos(\omega_i t - 2\pi/3) + I_{qm} \sin(\omega_i t - 2\pi/3) \\ = GU \cos(\omega_i t - 2\pi/3) + I_{qm} \sin(\omega_i t - 2\pi/3) \\ = \dfrac{GU \cos(\omega_i t - 2\pi/3 + \varphi)}{\cos\varphi} \\ \tan\varphi = \dfrac{I_{qm}}{I_{pm}} \end{cases} \tag{2.5}$$

where I_{pm} and I_{qm} are amplitudes of the active and reactive components of the input current, respectively, $G = 2P/3U^2$ is the equivalent input conductance, P is the active power of the converter, and φ is the desired input displacement angle. According to Fig. 2.5, the average current of phase a, denoted as $\bar{i}_{ra}$, can be expressed as

$$\bar{i}_{ra} = i_o + k i_y \tag{2.6}$$

where i_o is the equivalent output current of the rectifier.

In sector 1, the DC-link voltage u_{pn} is equal to the input line-to-line voltage u_{ac}. According to (2.4), the duty ratio k is equal to u_{bc}/u_{ac}. Substitute (2.4) and (2.5) into (2.6), $\bar{i}_{ra}$ is deduced further as

$$\bar{i}_{ra} = P/u_{ac} - GU\cos(\omega_i t - 2\pi/3 + \varphi)\, u_{bc}/(u_{ac}\cos\varphi) = \frac{GU\cos(\omega_i t + \varphi)}{\cos\varphi} \tag{2.7}$$

As the three-phase input currents satisfy $\bar{i}_{ra} + \bar{i}_{rb} + \bar{i}_{rc} = 0$, the average current of phase c is obtained as

$$\bar{i}_{rc} = \frac{GU\cos(\omega_i t + 2\pi/3 + \varphi)}{\cos\varphi} \tag{2.8}$$

According to Eqs. (2.5), (2.7), and (2.8), the three-phase input currents are symmetry and sinusoidal with a desired displacement angle φ. When the active power P is zero, the active component of the input current is also zero. In that situation, the converter is essentially a STATCOM. Therefore, the constraint of the input displacement angle is

$$\varphi \in [-\pi/2,\ \pi/2] \tag{2.9}$$

For the remaining sectors, similar results can be obtained, and the derivation process will not be repeated here. Thus three-phase sinusoidal input currents and controllable input power factor are achieved via the algorithm above. It should be noted that, in the aforementioned derivation process, the control method is applicable to both directions of the power flow. Positive P indicates forward power flow and the contrary is the case when P is negative.

2.2.4 Performance Comparisons

The differences in topology and operating principles between IMCs and 3HIMC would result in some different external characteristics, such as the input reactive power control range, the voltage transfer characteristic, as well as requirements of the input filter.

It can be seen from the mathematical proof of the PFC algorithm that the input displacement angle range in 3HIMC is quite different from that of conventional IMCs (assume conventional modulation is used in IMCs), since the input displacement angle range in IMCs is only from $-\pi/3$ to $\pi/3$. Consequently, the reactive power control range at the input side of IMCs and 3HIMC can therefore be calculated as shown in (2.10) and (2.11), respectively:

$$Q = P \tan \varphi \in \left[-\frac{P}{\sqrt{3}}, \frac{P}{\sqrt{3}} \right] \tag{2.10}$$

$$Q = P \tan \varphi \in (-\infty, \infty) \tag{2.11}$$

Compared with that of IMCs, the reactive power control range at the input side in 3HIMC is extended significantly and there is no intrinsic limitation. However, for a practical 3HIMC, the reactive power generated is generally limited by current rating of the devices and some other factors such as the value of L_y, as will be discussed in later sections. Besides, as can be seen from Fig. 2.6, the DC-link voltage exhibits a fixed 6-pulse shape waveform with the value varies from $1.5U$ to $\sqrt{3}U$, irrelevant to operation conditions. Therefore for linear PWM modulation the voltage transfer ratio q can be calculated as

$$q = \frac{U_o}{U} = \frac{\sqrt{3}}{2}m,\ m \in [0, 1] \tag{2.12}$$

where U_o is the amplitude of the output phase voltage and m is the modulation index of the inverter. Obviously the maximum voltage transfer ratio of 0.866 can be achieved under all operation conditions. This is another significant advantage of 3HIMC since the maximum voltage transfer ratio of IMCs decreases with the deceasing of the input power factor [11], as illustrated in (2.13):

$$q = \frac{U_o}{U} = \frac{\sqrt{3}}{2}m \cos \varphi,\ m \in [0, 1] \tag{2.13}$$

Moreover, unlike the pure high-frequency PWM current in IMCs, the input current before filtering in 3HIMC shows a quasi-sinusoidal shape due to the prefiltering effect of the inductor L_y, as shown in Fig. 2.6c. In the same input and output circumstances (with the switching frequency being 32 kHz), the spectrum analysis of the input current before filtering in 3HIMC and conventional IMCs is shown in Fig. 2.7. As shown from the spectrums, the dominant harmonic current is around the second-order switching frequency in 3HIMC and about the first-order switching frequency in IMCs. Besides, the magnitudes of the dominant harmonic current in 3HIMC are less than that of IMCs. The THDs of the current i_{ra} in 3HIMC and IMCs are 57% and 153%, respectively. Thus a higher corner frequency of the input filter can be selected and a lower attenuation of the dominant harmonic current is required in 3HIMC [28], which imply a smaller size of the input filter. The detailed comparisons between IMCs and 3HIMC are shown in Table 2.1.

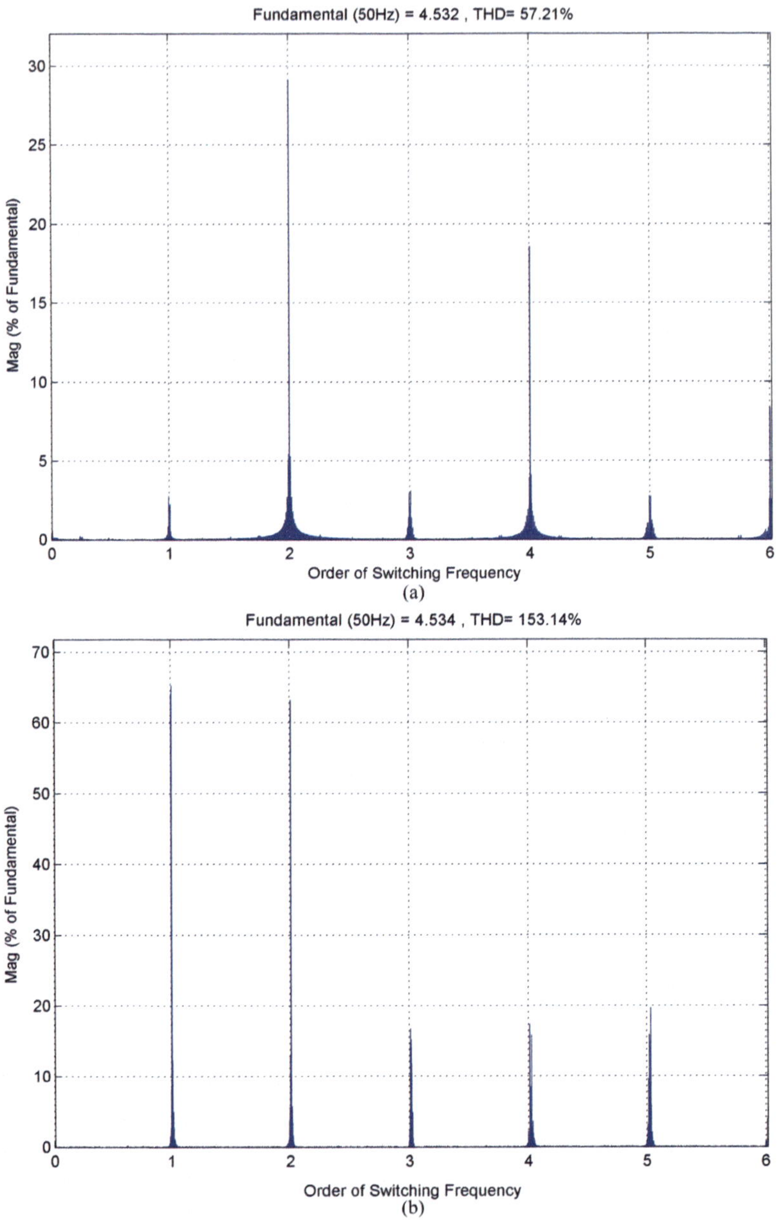

Fig. 2.7 Spectrum of the input current before filtering. (**a**) 3HIMC. (**b**) IMCs

Table 2.1 Detailed comparisons between IMCs and 3HIMC

Items	IMCs	3HIMC
Number of transistors	18	20
Number of diodes	18	20
Number of inductors	3	4
Number of capacitors	3	3
Input displacement angle range	$[-\pi/6, \pi/6]$	$[-\pi/2, \pi/2]$
Maximum voltage transfer ratio	$\frac{\sqrt{3}}{2}\cos\varphi$	$\frac{\sqrt{3}}{2}$
THD of the current before filtering	Relatively high (153%)	Relatively low (57%)
Implementation effort of the input filter	Relatively high	Relatively low

2.3 Modulation Strategies

2.3.1 *Rectifier and Third-Harmonic Current Injection Circuit*

As described before, the switches in the rectifier and the bidirectional switches in the third-harmonic current injection circuit operate at a low frequency, and the states of the switches are determined only by the input voltages. Table 2.2 shows the modulation of the rectifier and the third-harmonic current injection circuit in one line cycle. The switch states of the bridge leg in the third-harmonic current injection circuit are complementary, and the steady-state duty ratio of switch S_{y+} is k.

For the third-harmonic current injection circuit, to prevent short circuit of the input side and open circuit of the inductor L_y, only one of the three bidirectional switches is kept on at any time, and thus a proper commutation strategy should be developed to guarantee the safe operation of the converter. Similar to that of DMC, the current-based four-step commutation strategy is utilized in this project [7].

2.3.2 *Inverter*

As shown in Fig. 2.6, the 6-pulse shape DC-link voltage of 3HIMC is quite different from the high-frequency PWM DC-link voltage in conventional IMCs [11]. Therefore the modulation strategy of the inverter is different from that of conventional IMCs.

For the inversion stage of 3HIMC, the carrier modulation is employed and the concrete analysis is given as follows:

Assume the symmetrical three-phase output reference voltages are

$$\begin{cases} u_r^* = U_o \cos(\omega_o t + \phi) \\ u_s^* = U_o \cos(\omega_o t - 2\pi/3 + \phi) \\ u_t^* = U_o \cos(\omega_o t + 2\pi/3 + \phi) \end{cases} \tag{2.14}$$

Table 2.2 Modulation of the rectifier and the third-harmonic current injection circuit

$\omega_i t$	k	S_{ya} S_{ay}	S_{yb} S_{by}	S_{yc} S_{cy}	S_{a+}	S_{a-}	S_{b+}	S_{b-}	S_{c+}	S_{c-}
$0-\pi/3$	u_{bc}/u_{ac}	0	1	0	1	0	0	0	0	1
$\pi/3-2\pi/3$	u_{ac}/u_{bc}	1	0	0	0	0	1	0	0	1
$2\pi/3-\pi$	u_{ca}/u_{ba}	0	0	1	0	1	1	0	0	0
$\pi-4\pi/3$	u_{ba}/u_{ca}	0	1	0	0	1	0	0	1	0
$4\pi/3-5\pi/3$	u_{ab}/u_{cb}	1	0	0	0	0	0	1	1	0
$5\pi/3-2\pi$	u_{cb}/u_{ab}	0	0	1	1	0	0	1	0	0

where ω_o and ϕ are the angular frequency and initial phase of the expected output voltages, respectively.

To improve the utilization of the DC-link voltage, a zero sequence voltage u_{no} is added to the phase leg reference voltages without impacting the output fundamental line-line voltages. Then the phase leg reference voltages, known as the modulation signals, are modified as follows:

$$u_{io}^* = u_i^* + u_{no}, \ (i = r, s, t) \tag{2.15}$$

where u_{io}^* are the modified modulation signals.

The quasi-third-harmonic zero sequence voltage is selected identically as that used in [29] and can be expressed as

$$u_{no} = -[\max(u_r^*, u_s^*, u_t^*) + \min(u_r^*, u_s^*, u_t^*)]/2 \tag{2.16}$$

According to the fundamental principles of the carrier modulation, the phase leg modulation signals should be normalized referring to the DC-link voltage u_{pn} before calculating the phase leg duty ratios. It should be noted that the maximum output voltage of a phase leg (refers to the virtual middle point of the DC-link) is half of the DC-link voltage, and therefore the normalized modulation signals $\overline{u}_{io}^*$ are obtained as follows:

$$\overline{u}_{io}^* = 2u_{io}^*/u_{pn}, \ (i = r, s, t) \tag{2.17}$$

Based on the carrier modulation schematic diagram illustrated in Fig. 2.8 and the geometry theory, the phase leg duty ratios d_i can be calculated as

$$d_i = \frac{1 + \overline{u}_{io}^*}{2}, \ (i = r, s, t) \tag{2.18}$$

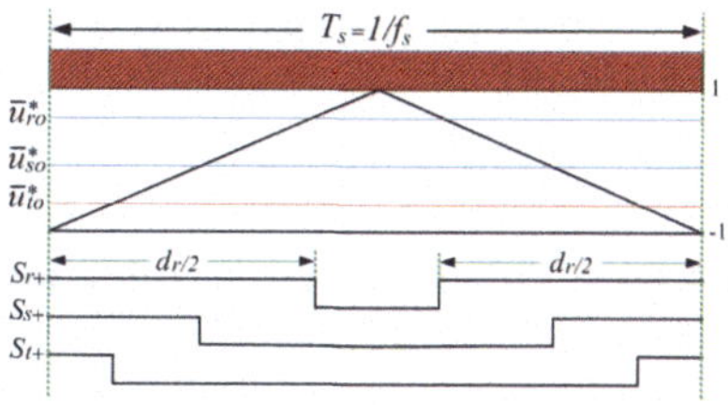

Fig. 2.8 Modulation diagram of the inverter of 3HIMC

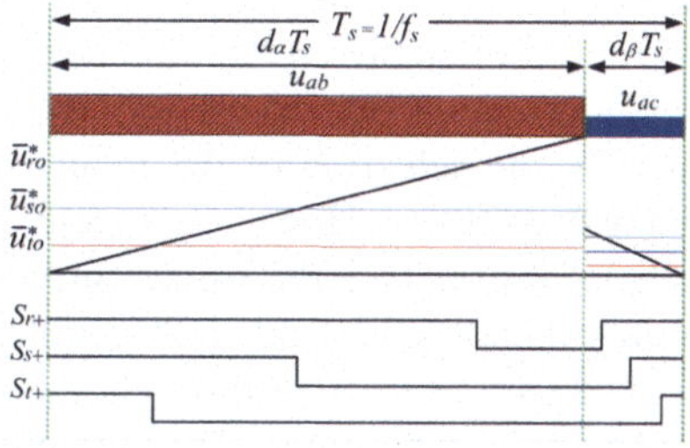

Fig. 2.9 Modulation diagram of the inverter of conventional IMCs

According to the analysis above, the modulations of the rectifier and inverter in 3HIMC are independent. Thus synchronization between the rectifier and inverter is not required. Compared with conventional IMCs, realization of the modulation in 3HIMC is therefore more flexible, which has potential advantages in multiple drive systems [30]. Meanwhile, the nonlinearity caused by narrow pulse width in conventional IMCs is improved. The modulation diagram of the inverter of 3HIMC is illustrated in Fig. 2.8, where T_s is the switching period and f_s is the switching frequency. As a comparison, modulation diagram of the inverter of IMCs is also shown in Fig. 2.9 to illustrate the narrow pulse width issue, where d_α and d_β are the duty cycles of the rectification stage.

The essential reason of conventional IMCs generating narrow pulses is that the modulation of the rectifier and inverter must be synchronized strictly due to the absence of energy storage elements in the DC-link. When the desired input current vector is located in the sector boundaries, one of the duty cycles in the rectification stage would be very short. As a result the modulation of the inversion stage is difficult to realize in such a short time due to the physical limits in a practical system. Also, adopting the round-up method to ignore narrow pulses causes distortions of input and output waveforms [16].

Different from that of IMCs, the 6-pulse waveform DC-link voltage in 3HIMC can be considered as a constant voltage source in one switching period. Therefore, the inversion stage of 3HIMC is modulated only once in each switching period, and then the narrow pulse problem is relieved.

2.4 System Design and Power Loss Analysis

2.4.1 System Design

2.4.1.1 Third-Harmonic Injection Inductor Design

An important issue in system design is the design of the inductor L_y, since the inductor concerns the tracking performance of third-harmonic current, as well as the quality of input current. The design criteria are as follows: (i) The ripple current of the inductor should be suppressed at a reasonable level. (ii) The tracking performance of the third-harmonic injection current should be excellent enough. In [27], the selection of the inductor L_y is analyzed with only the ripple current being considered. However, it should be noted that an inductor of large inductance is helpful to reduce the ripple current, but it makes the current response slow, so a trade-off should be made between these two aspects.

The current flowing through L_y consists of a quasi-third harmonic current I_{3rd} and a ripple current Δi_y. The maximum value of Δi_y is hoped to be limited to γI_{3rd}, where γ is the ripple index and I_{3rd} is the amplitude of i_y. The ripple index γ is defined as

$$\gamma = \Delta i_{ymax} / I_{3rd} \tag{2.19}$$

where Δi_{ymax} is the maximum value of Δi_y.

According to the operating principles mentioned previously, the third-harmonic injection current i_y increases when the switch S_{y+} is turned on and decreases when the switch S_{y+} is turned off. Thus for a given inductor, the ripple current is determined by the voltage across the inductor when the switch S_{y+} is turned on and its duration. Take sector 1 ($0 \leq \omega_i t \leq \pi/3$) for example, the ripple current Δi_y, and its maximum value Δi_{ymax} during the whole sector can be calculated as

$$\begin{cases} \Delta i_y &= \dfrac{u_{Ly}}{L_y} k T_s = \dfrac{u_{ab} u_{bc}}{L_y f_s u_{ac}} \\ \Delta i_{ymax} &= \dfrac{\sqrt{3}U}{4 L_y f_s} \end{cases} \tag{2.20}$$

where u_{Ly} is the voltage across the third-harmonic injection inductor, and u_{ab}, u_{bc}, and u_{ac} are the input line-line voltages.

Therefore the lower limit of the inductor L_y is governed by

$$L_y \geq \frac{\sqrt{3}U}{4\gamma I_{3rd} f_s} \tag{2.21}$$

Equation (2.21) shows that, for the given input voltage and power rating, the lower limit of the inductor L_y depends on the switching frequency f_s and the ripple index γ. In this chapter, $\gamma = 0.5$ is adopted to make a compromise between the ripple performance and volume of the inductor, and $f_s = 32$ kHz is selected to make a trade-off between the switching losses and volume of the passive components.

By considering the unity input power factor and forward power flow, the current slope of i_y in sector 1 is derived as

$$k_i = \omega_i I_{im} \sin(\omega_i t - 2\pi/3) \tag{2.22}$$

where I_{im} is the amplitude of the input current. And for a given L_y, the maximum achievable current changing rate is determined by

$$k_{ia} = \frac{u_{Ly}}{L_y} \tag{2.23}$$

For the case of sector 1, (2.23) can be deduced as

$$k_{ia} = \frac{u_{cb}}{L_y} = \frac{\sqrt{3}U \cos(\omega_i t + \pi/2)}{L_y} \tag{2.24}$$

To achieve a good current tracking performance, (2.25) should be satisfied:

$$|k_{ia}| \geq |k_i| \tag{2.25}$$

Substituting (2.23) and (2.24) into (2.25) leads to the upper limit of the inductor:

$$L_y \leq \frac{\sqrt{3}U \cos(\omega_i t + \pi/2)}{\omega_i I_{im} \sin(\omega_i t - 2\pi/3)}, \quad \omega_i t \in [0, \pi/3] \tag{2.26}$$

The minimum inductance in (2.26) is zero when $\omega_i t = 0$. Thus for a practical inductor, always there exists a zone that the practical third-harmonic injection current cannot track the reference. To minimize distortions of the third-harmonic injection current as well as the input currents, the width of the zone should be limited to a small range of the width of sector 1:

$$\xi = \frac{w}{\pi/3} \tag{2.27}$$

where w is the width of the zone and ξ is the width index.

Thus the upper limit of the inductor is solved as

$$L_y \leq \frac{\sqrt{3}U \cos(\xi\pi/3 + \pi/2)}{\omega_i I_{im} \sin(\xi\pi/3 - 2\pi/3)} \tag{2.28}$$

For the remaining sectors, the same results can be obtained, and the derivation process will not be repeated here. In this chapter, $\xi = 0.01$ is selected to keep the distortion of the third-harmonic injection current at an acceptable level. By considering the input voltage, the rated output power, the switching frequency, the value of the third-harmonic injection inductor is determined as $2m\text{H} \leq L_y \leq 2.4m\text{H}$ based on (2.21) and (2.28). In this chapter a 2 mH inductor is chosen.

2.4.1.2 Control of the Third-Harmonic Current

As discussed before, normal operation of 3HIMC is dependent on injecting the third-harmonic current i_y appropriately. According to Fig. 2.4, the mathematical model of the third-harmonic injection circuit in 3HIMC can be described as follows:

$$u_{Ly} = L_y \frac{di_y}{dt} \tag{2.29}$$

For simplicity reason, influences of the input filter and the output side on the third-harmonic injection circuit are omitted, and the third-harmonic current injection circuit is treated as a first-order system in this chapter. A conventional PI controller is used to control the third-harmonic current, and the detailed design procedure using conventional frequency-domain analysis tools is not given here for the sake of brevity. In order to improve the dynamic tracking speed, a feed-forward term k obtained by the mains voltages, also known as the duty ratio of the switch S_{y+} in steady state, is added to the output of third-harmonic injection current controller. The combination of PI controller and feed-forward term ensures good control performance of the system. The overall control scheme is shown in Fig. 2.10.

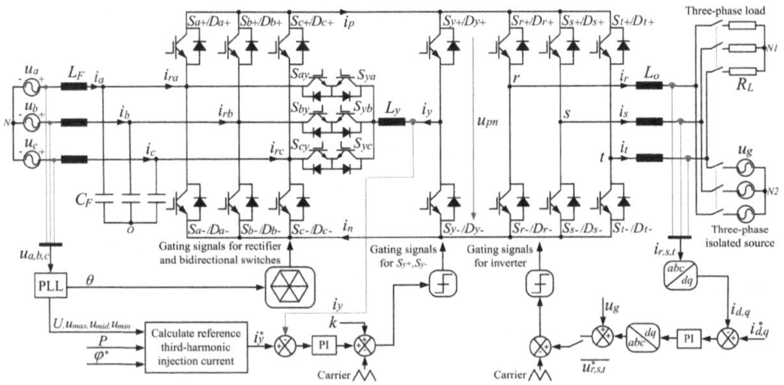

Fig. 2.10 Schematic diagram of 3HIMC

2.4.2 *Power Loss Analysis*

2.4.2.1 Voltage and Current Stresses Analysis

To evaluate the losses of the switching devices of 3HIMC, the voltage and current stresses of the switches are calculated first. The analysis of the voltage and current stresses of the unidirectional third-harmonic injection rectifier is presented in[27]. In this chapter, based on the system parameters shown in Table 2.3, the voltage and current stresses of 3HIMC are analyzed by considering the directions of the power flow and the inversion stage.

According to the operating principles of 3HIMC and parameters listed in Table 2.3, the maximum voltage stresses on bidirectional switches S_{iy}, S_{yi}, and $(i = a, b, c)$ are defined by

$$u_{sy\max} = \frac{3}{2}U \tag{2.30}$$

The maximum voltage stress on the remaining semiconductors is $\sqrt{3}U$. The current stresses of the transistors and diodes are summarized in Table 2.4, where I_{AVG} is the average current, I_{RMS} is the RMS current, and φ_o is the output displacement angle.

Table 2.3 System parameters of 3HIMC

Characteristics	Specifications
Input phase voltage $u_{a,b,c}$	115 Vrms
Phase voltage of the isolated power source u_g	86 Vrms
Mains frequency f_n	50 Hz
Switching frequency f_s	32 kHz
Rated output power	1.1kW
Inductor L_F	150 μH
Capacitor C_F	6.6 μF
Inductor L_y	2 mH
Inductor L_o	3 mH
Resistor R_L	25 ohm
S_{i+}/D_{i+}, $S_{i-}/D_{i-}(i = a, b, c)$	IHW30N120R2
S_{iy}, $S_{yi}(i = a, b, c)$	IHW30N120R2
S_{i+}, $S_{i-}(i = y, r, s, t)$	IGW40N120H3
D_{i+}, $D_{i-}(i = y, r, s, t)$	DSEP29-12A

Table 2.4 Current stresses of the switches in 3HIMC of forward power flow

	Forward power flow($\varphi_o \in [-\pi/2, \pi/2]$)	
Component	I_{AVG}[A]	I_{RMS}[A]
S_{i+}, S_{i-} $(i = a, b, c)$	0	0
D_{i+}, D_{i-} $(i = a, b, c)$	$I_{im}\frac{\sqrt{3}}{2\pi}$	$I_{im}\sqrt{\frac{1}{6}+\frac{\sqrt{3}}{8\pi}}$
S_{iy}, S_{yi} $(i = a, b, c)$	$I_{im}\frac{2-\sqrt{3}}{2\pi}$	$I_{im}\sqrt{\frac{1}{12}-\frac{\sqrt{3}}{8\pi}}$
S_{y+}, S_{y-}	$\frac{3I_{im}}{4\pi}\left(2-\sqrt{3}\ln 3\right)$	$I_{im}\sqrt{\frac{1}{8}+\frac{3\sqrt{3}}{4\pi}\ln\frac{3}{4}}$
D_{y+}, D_{y-}	$\frac{3I_{im}}{4\pi}\left(2-2\sqrt{3}+\sqrt{3}\ln 3\right)$	$I_{im}\sqrt{\frac{1}{8}-\frac{3\sqrt{3}}{4\pi}\ln\frac{3}{4}-\frac{3\sqrt{3}}{8\pi}}$
S_{i+}, S_{i-} $(i = r, s, t)$	$I_{om}\left(\frac{1}{2\pi}+\frac{3m\cos\varphi_o}{8\pi}\ln\left(2+\sqrt{3}\right)\right)$	$I_{om}\sqrt{\frac{1}{8}+\frac{m\cos\varphi_o}{4\pi}\left(\frac{3\sqrt{3}}{8}\ln 2-\frac{9\sqrt{3}}{16}\ln 3-\frac{5\pi}{16}\right)}$
D_{i+}, D_{i-} $(i = r, s, t)$	$I_{om}\left(\frac{1}{2\pi}-\frac{3m\cos\varphi_o}{8\pi}\ln\left(2+\sqrt{3}\right)\right)$	$I_{om}\sqrt{\frac{1}{8}-\frac{m\cos\varphi_o}{4\pi}\left(\frac{3\sqrt{3}}{8}\ln 2-\frac{9\sqrt{3}}{16}\ln 3-\frac{5\pi}{16}\right)}$

Table 2.5 Current stresses of the switches in 3HIMC of reserve power flow

	Reserve power flow($\varphi_o \in [-\pi, -\pi/2)and(\pi/2, \pi]$)	
Component	I_{AVG}[A]	I_{RMS}[A]
S_{i+}, S_{i-} $(i = a, b, c)$	$I_{im}\frac{\sqrt{3}}{2\pi}$	$I_{im}\sqrt{\frac{1}{6}+\frac{\sqrt{3}}{8\pi}}$
D_{i+}, D_{i-} $(i = a, b, c)$	0	0
S_{iy}, S_{yi} $(i = a, b, c)$	$I_{im}\frac{2-\sqrt{3}}{2\pi}$	$I_{im}\sqrt{\frac{1}{12}-\frac{\sqrt{3}}{8\pi}}$
S_{y+}, S_{y-}	$\frac{3I_{im}}{4\pi}\left(2-2\sqrt{3}+\sqrt{3}\ln 3\right)$	$I_{im}\sqrt{\frac{1}{8}-\frac{3\sqrt{3}}{4\pi}\ln\frac{3}{4}-\frac{3\sqrt{3}}{8\pi}}$
D_{y+}, D_{y-}	$\frac{3I_{im}}{4\pi}\left(2-\sqrt{3}\ln 3\right)$	$I_{im}\sqrt{\frac{1}{8}+\frac{3\sqrt{3}}{4\pi}\ln\frac{3}{4}}$
S_{i+}, S_{i-} $(i = r, s, t)$	$I_{om}\left(\frac{1}{2\pi}+\frac{3m\cos\varphi_o}{8\pi}\ln\left(2+\sqrt{3}\right)\right)$	$I_{om}\sqrt{\frac{1}{8}+\frac{m\cos\varphi_o}{4\pi}\left(\frac{3\sqrt{3}}{8}\ln 2-\frac{9\sqrt{3}}{16}\ln 3-\frac{5\pi}{16}\right)}$
D_{i+}, D_{i-} $(i = r, s, t)$	$I_{om}\left(\frac{1}{2\pi}-\frac{3m\cos\varphi_o}{8\pi}\ln\left(2+\sqrt{3}\right)\right)$	$I_{om}\sqrt{\frac{1}{8}-\frac{m\cos\varphi_o}{4\pi}\left(\frac{3\sqrt{3}}{8}\ln 2-\frac{9\sqrt{3}}{16}\ln 3-\frac{5\pi}{16}\right)}$

2.4.2.2 Voltage and Current Stresses Analysis

The losses of switching devices can be calculated according to Tables 2.4 and 2.5. For the switching devices, the blocking losses can be neglected. Thus only the conduction losses and switching losses are considered in this chapter.

The conduction losses of the IGBT are determined by the saturation voltage across the device and the current through the device. Usually the saturation voltage is approximated by the linear dependence on the current [31]. Thus the average conduction losses over time interval T can be estimated using the integral:

$$P_{C,S} = \frac{1}{T}\int_0^T u_{CE}\, i_s\, dt = \frac{1}{T}\int_0^T (U_{CE0} + i_s R_s)\, i_s\, dt$$
$$= I_{S,AVG}\, U_{CE0} + I_{S,RMS}^2 R_s \tag{2.31}$$

where u_{CE} is the saturation voltage across the IGBT for current i_s, R_s is the dynamic on-resistance of the IGBT, U_{CE0} is the forward voltage when the current is zero, and $I_{S,AVG}$ and $I_{S,RMS}$ are the average and RMS currents of the IGBT. The parameters of the device can be obtained from datasheet.

The switching energy is generally assumed proportional to the switching voltage and the conducting current in a switching period:

$$E = E_{on} + E_{off} = E_{on\_R}\frac{u_s i_s}{V_R I_R} + E_{off\_R}\frac{u_s i_s}{V_R I_R} \tag{2.32}$$

where E_{on} and E_{off} are the turn-on and turn-off energy, E_{on_R} and E_{off_R} are the turn-on and turn-off energy at the reference voltage V_R and reference current I_R, and u_s is the actual switching voltage. Then the average switching energy is calculated as

$$\overline{E} = \frac{1}{T}\int_0^T E\, dt = \frac{(E_{on\_R} + E_{off\_R})}{V_R I_R}\frac{1}{T}\int_0^T u_s\, i_s\, dt \tag{2.33}$$

Since u_s and i_s can be expanded to Fourier series, the average switching energy is deduced as

$$\overline{E} = \frac{(E_{on\_R} + E_{off\_R})}{V_R I_R} V_{S,AVG}\, I_{S,AVG} = k_s\, I_{S,AVG} \tag{2.34}$$

where $V_{S,AVG}$ is the average switching voltage of the IGBT and k_s is the switching losses coefficient. Then the average switching losses of the IGBT are developed as

$$P_{S,S} = \overline{E}\, f_s = k_s\, I_{S,AVG}\, f_s \tag{2.35}$$

Similarly, the conduction losses of the diodes are calculated by

$$P_{C,D} = I_{D,AVG}\, U_{D0} + I_{D,RMS}^2 R_D \tag{2.36}$$

where $I_{D,AVG}$ and $I_{D,RMS}$ are the average and RMS currents of the diode, respectively, U_{D0} is the forward voltage when the current is zero, and R_D is the dynamic on-resistance of the diode.

The loss of an inductor can be divided into two parts: core loss and winding resistance loss. Since the inductors applied in 3HIMC operate at low frequency, only

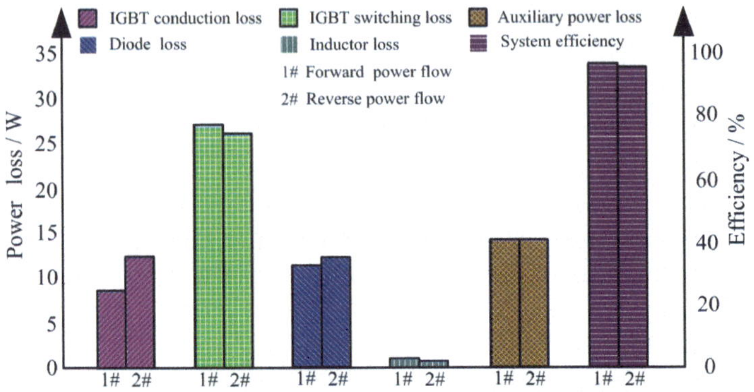

Fig. 2.11 Calculated power loss distribution and efficiency diagram of 3HIMC

low-frequency loss in winding resistance is considered in this chapter. The power loss of the inductor is calculated as

$$P_{ind} = I_{L,RMS}^2 R_{ind} \tag{2.37}$$

where $I_{L,RMS}$ is the RMS current of the inductor and R_{ind} is the dc resistance of the inductor.

According to the above formulas, the system loss distribution is derived as shown in Fig. 2.11. The nominal system efficiency calculated reaches 95.3% in forward power flow operation condition, while in reverse power flow operation condition, an efficiency of 94.8% is achieved. It should be noted that in the efficiency diagram, the auxiliary power consumed by the control electronics and IGBT drivers is also counted in.

2.5 Simulation and Experimental Results

In this section, the functionality and performance of 3HIMC were preliminarily evaluated by simulation using Matlab/Simulink software and then were validated experimentally. Simulation study was planned such that it used the same parameters as the final implemented prototype, whose values are listed in Table 2.3. The only difference between simulation and experiments is that ideal switches and power source were used in simulation. The performance of 3HIMC is evaluated adequately by two different loads: passive resistive load and three-phase isolated power source. For convenience, the resistive load is used to verify the function of third-harmonic current injection circuit, the output capability, and input power factor regulation of 3HIMC. And the bidirectional power flow capability and dynamic tests were conducted on the isolated power source load, as shown in Fig. 2.10. When the output connects to the isolated power source, the output currents are controlled in

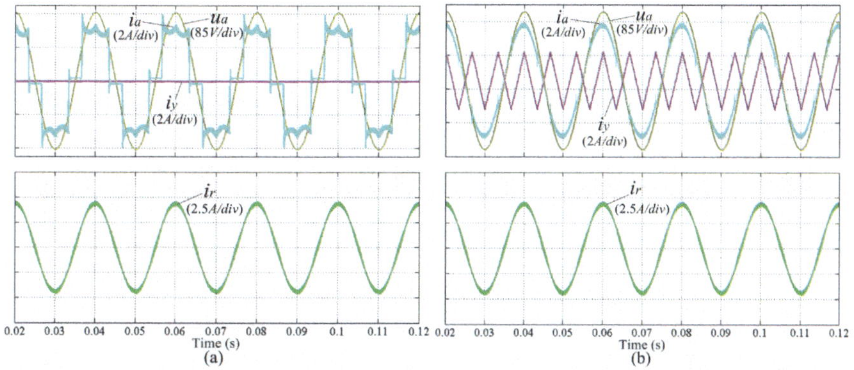

Fig. 2.12 Simulated waveforms of 3HIMC with the third-harmonic injection circuit: (**a**) disabled and (**b**) enabled

synchronous dq reference frame, and the output power capacity and direction of power flow are controlled by changing the active current reference.

2.5.1 Simulation

Figure 2.12 shows the waveforms without and with the third-harmonic current injection circuit enabled. In the simulation, the desired displacement angle φ at input side is set as 0, and the modulation index m and output frequency f_o were set as 0.9 and 40 Hz, respectively. As can be seen from Fig. 2.12a, the input current exhibits a quasi-square shape and contains a lot of low-frequency harmonics when the third-harmonic current injection circuit is disabled. In Fig. 2.12b, the third-harmonic current injection circuit is enabled and a quasi-third-harmonic current i_y is injected. As a result, the input current is sinusoidal and in phase with the input voltage, and thus the effectiveness of the third-harmonic current injection circuit is verified.

Figure 2.13 illustrates the capability of producing three-phase output with controllable amplitude and frequency of 3HIMC. The output parameters were set as $m = 0.7$ and $f_o = 40$ Hz in Fig. 2.13a and $m = 1$ and $f_o = 100$ Hz in Fig. 2.13b, respectively, and the desired input displacement angle φ is set as 0. The results again show that the input current is sinusoidal and in phase with the input voltage in different output conditions. When $m = 1$, which means a maximum voltage transfer ratio of 0.866, the output current is sinusoidal with an amplitude of 5.6A. This is correct since an inductor $L_o = 1\,m$H and a resistor $R_L = 25$ ohm are used as the load.

Figure 2.14 shows the simulation results with different input power factors. The output parameters were set as $m = 1$ and $f_o = 40$ Hz. In Fig. 2.14a, the desired input displacement angle φ is set as $\pi/6$, which means a lag power factor of 0.866

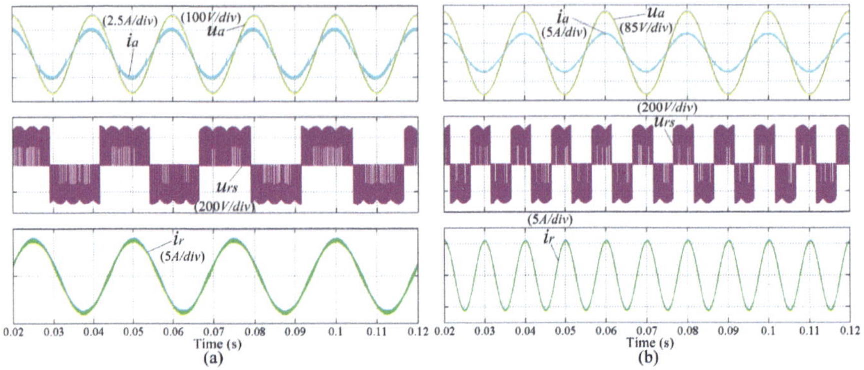

Fig. 2.13 Simulated waveforms of 3HIMC with different m and f_o. (**a**) $m = 0.7$ and $f_o = 40$ Hz. (**b**) $m = 1$ and $f_o = 100$ Hz

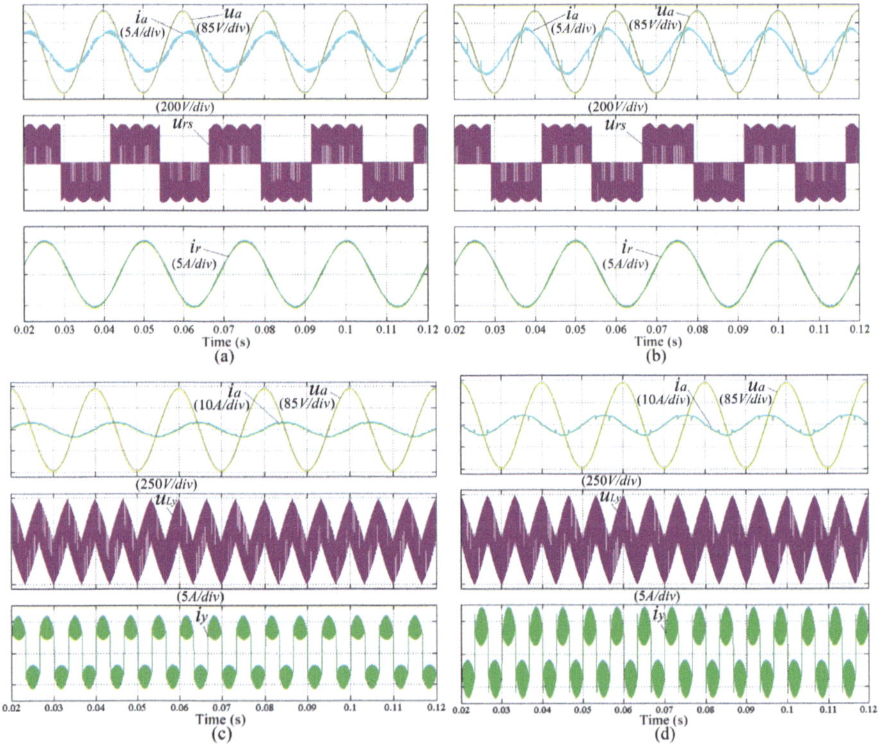

Fig. 2.14 Simulated waveforms of 3HIMC with different input power factors. (**a**) The displacement angle is $\pi/6$. (**b**) The displacement angle is $-\pi/6$. (**c**) The displacement angle is $\pi/2$. (**d**) The displacement angle is $-\pi/2$

at the input side. And in Fig. 2.14b, the desired input displacement angle φ is set as $-\pi/6$, which means a lead power factor of 0.866 at the input side. It is clear that the input current is still sinusoidal with the desired displacement angle in both conditions. Besides, the amplitude of the output current remains unchanged with a value of 5.6A, which confirms the controllable input power factor and that a maximum voltage transfer ratio of 0.866 can be achieved in various input power factor operation conditions. In Fig. 2.14c and d, with the amplitude of the input reactive current reference being 4A, the inverter is disabled, and the desired input displacement angles are set as $\pi/2$ and $-\pi/2$, respectively. Thus the converter runs as a STATCOM under these conditions. It can be seen that pure reactive currents with the desired amplitude are appeared at the input side. When the input displacement angle is $\pi/2$, the input current lags the input voltage by $\pi/2$ and the converter generates pure inductive reactive power at the input side; while the input displacement angle is $-\pi/2$, the input current leads the input voltage by $\pi/2$ and the converter generates pure capacitive reactive power. The results in Fig. 2.14 verified the wide input reactive power control range of 3HIMC.

In Figs. 2.15 and 2.16, a three-phase isolated power source with a frequency of 50 Hz is used as the load of 3HIMC. In these simulations, φ and the reactive current reference i_{q_ref} at output side are both set as 0. Figure 2.15 demonstrates the capability of bidirectional power of 3HIMC. In this simulation, the active current references i_{d_ref} at output side are set as +6A in Fig. 2.15a and −6A in Fig. 2.15b, respectively. It can be seen that the output current tracks the reference very well. The input current is sinusoidal and in phase with the input voltage, while i_{d_ref} is positive (denoted as motoring operation mode). When under negative i_{d_ref} (denoted as regenerating operation mode), the input current is in antiphase with the input voltage. The simulation results verify the bidirectional power flow characteristic of 3HIMC.

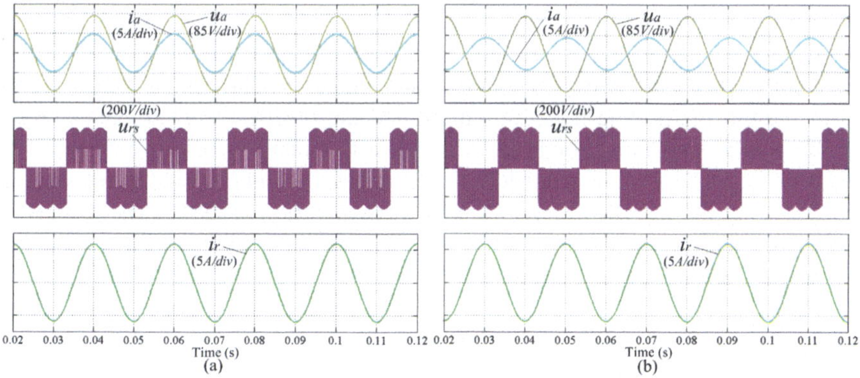

Fig. 2.15 Simulated waveforms of 3HIMC when in (**a**) forward and (**b**) reverse power flow operation conditions

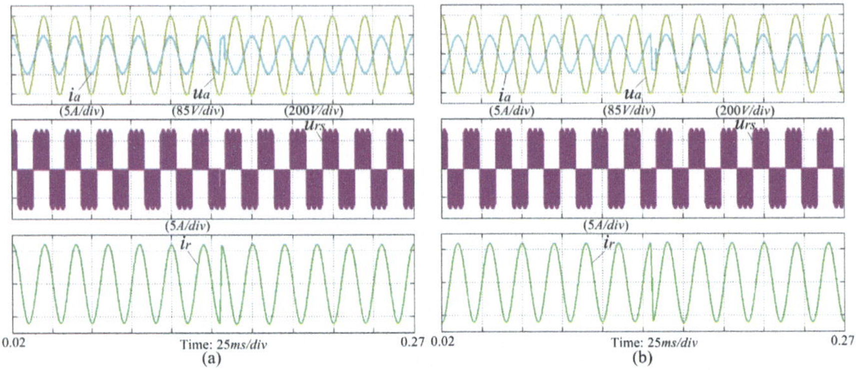

Fig. 2.16 Simulated waveforms of 3HIMC with the output active current reference stepping from (**a**) +6A to −6A and (**b**) −6A to +6A

Figure 2.16 shows the dynamic test of the system. In Fig. 2.16a, i_{d_ref} is +6A at first. At 0.15s, i_{d_ref} is step from +6A to −6A. The setup in Fig. 2.16b is the same as that in Fig. 2.16a except for an initial i_{d_ref} of −6A and a final i_{d_ref} of +6A. As can be seen from Fig. 2.16, the system can change over between the two operation modes fast and smoothly, demonstrating the good dynamic performance of the system.

2.5.2 Experimental Results

To validate the theoretical analyses and simulated results, a laboratory prototype of 3HIMC with the specifications given in Table 2.3 is built, as shown in Fig. 2.17. For the rectifier switches and bidirectional switches in the third-harmonic injection circuit, the conduction losses are the major part, and thus IGBT IHW30N120R2 (1200V/30A, Infineon) with low saturation voltage is selected. Switches on the legs of the third-harmonic injection circuit and the inverter stage are switching at high frequency, and thus, combined with ultra-fast recovery diode DSEP29-12A (1200V/30A, IXYS), high speed IGBT IGW40N120H3 (1200V/40A, Infineon) is used. The control of 3HIMC is realized by a combination of DSP TMS320F28335 and FPGA EP2C8T144C8N.

The experimental results shown in Figs. 2.18, 2.19, 2.20, 2.21, and 2.22 correspond to the simulated results shown in Figs. 2.12, 2.13, 2.14, 2.15, and 2.16, and the experimental conditions and commands are exactly the same as those in the simulations. As can be seen from Figs. 2.18, 2.19, 2.20, 2.21, and 2.22, the experimental results match the simulated results very well, except for a slight reduction in the amplitudes of the input and output current when fed resistive load. The subtle difference is primarily caused by the power losses in the prototype. In Fig. 2.21, THDs of the input current and input power factor measured by

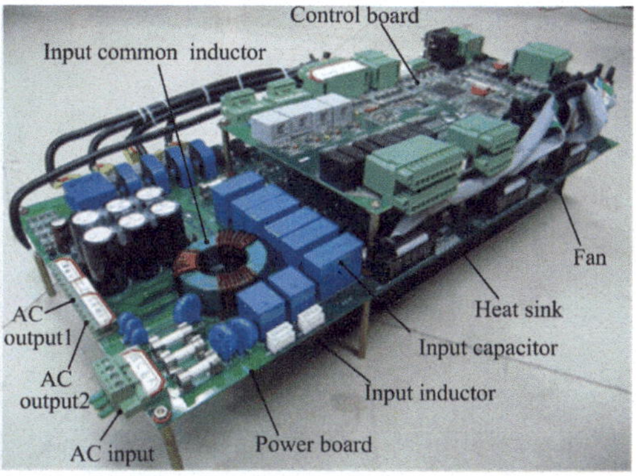

Fig. 2.17 Laboratory prototype of 3HIMC

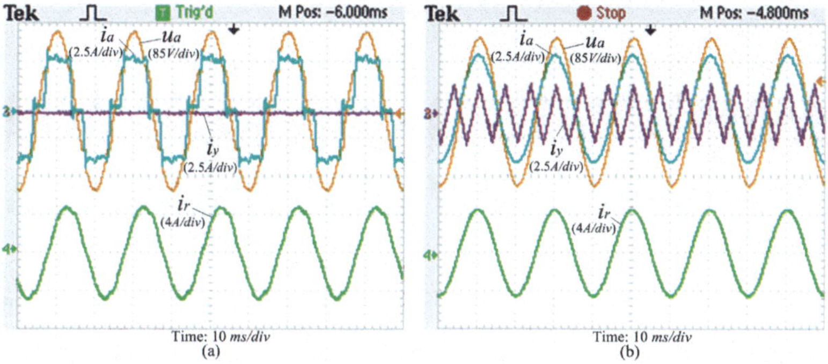

Fig. 2.18 Experimental waveforms of 3HIMC with the third-harmonic injection circuit (**a**) disabled and (**b**) enabled. CH1 is the input voltage u_a, CH2 is the input current i_a, CH3 is the third-harmonic injection current i_y, and CH4 is the output current i_r

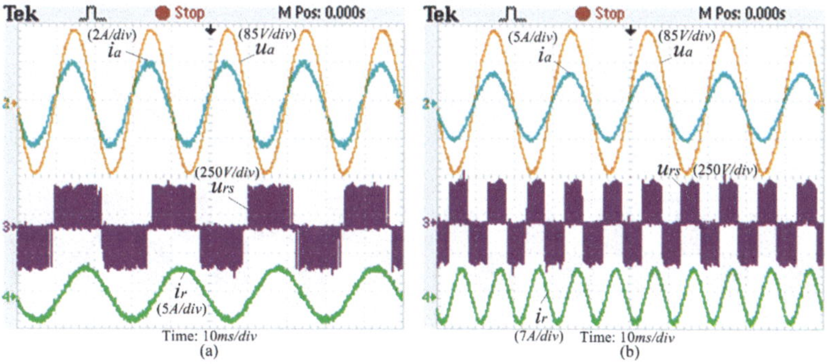

Fig. 2.19 Experimental waveforms of 3HIMC with different m and f_o. (**a**) $m = 0.7$ and $f_o = 40$ Hz. (**b**) $m = 1$ and $f_o = 100$ Hz. CH1 is the input voltage u_a, CH2 is the input current i_a, CH3 is the output line-line voltage u_{rs}, and CH4 is the output current i_r

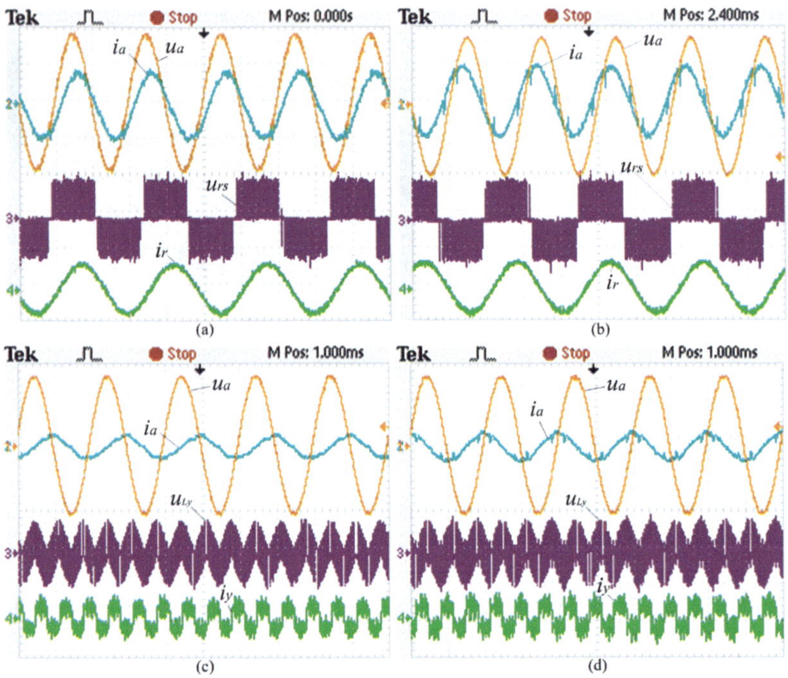

Fig. 2.20 Experimental waveforms of 3HIMC with different input power factors. (**a**) The displacement angle is $\pi/6$. (**b**) The displacement angle is $-\pi/6$. (**c**) The displacement angle is $\pi/2$. (**d**) The displacement angle is $-\pi/2$. CH1 is the input voltage u_a (85 V/div), CH2 is the input current i_a (5 A/div), CH3 is the output line-line voltage u_{rs} (250 V/div), CH4 is the output current i_r (7 A/div), and time scale is 10 ms/div in (**a**) and (**b**). CH1 is the input voltage u_a (85 V/div), CH2 is the input current i_a (10 A/div), CH3 is the voltage across the inductor L_y (250 V/div), CH4 is the third-harmonic injection current (10 A/div), and time scale is 10 ms/div in (**c**) and (**d**)

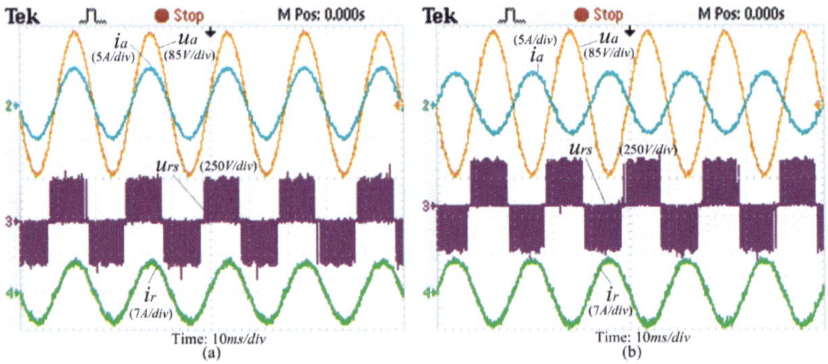

Fig. 2.21 Experimental waveforms of 3HIMC when in (**a**) forward and (**b**) reverse power flow operation conditions. CH1 is the input voltage u_a, CH2 is the input current i_a, CH3 is the output line-line voltage u_{rs}, and CH4 is the output current i_r

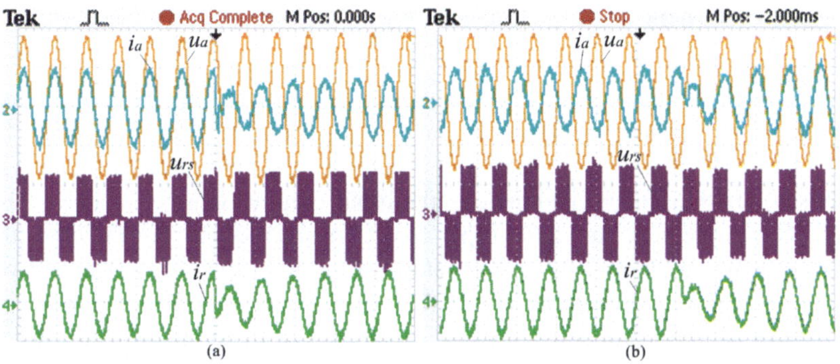

Fig. 2.22 Experimental waveforms of 3HIMC with the output active current reference stepping from (**a**) +6A to −6A and (**b**) −6A to +6A. CH1 is the input voltage u_a (85 V/div), CH2 is the input current i_a (5 A/div), CH3 is the output line-line voltage u_{rs} (250 V/div), CH4 is the output current i_r (7 A/div), and time scale is 25 ms/div

Fig. 2.23 Measured efficiency of 3HIMC prototype

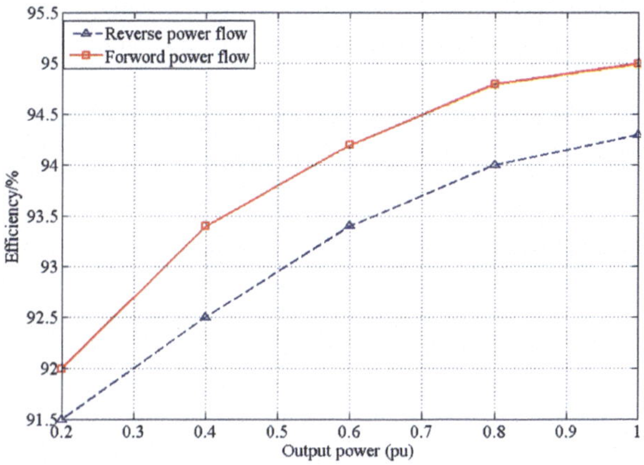

power quality analyzer FLUKE430 are 2.74% and 0.997 for motoring operation mode. When in regenerating operation mode, a THD of 3.69% and a power factor of 0.995 are achieved. Thus the excellent characteristics of 3HIMC are verified experimentally.

To verify the accuracy of the losses modeling, the system efficiency is measured by a three-phase wattmeter in the experiments as shown in Fig. 2.23. And the nominal efficiency measured is 95.0% in motoring operation mode and 94.3% in regenerating operation mode, which are a bit lower than the calculated ones. This can be explained by the fact that the losses dissipated on input capacitors, PCB traces, and current sensors are not considered in theoretical calculation.

2.6 Conclusions

In this chapter, the major problems of the buck-type active third-harmonic injection converter are solved through systematical investigation of 3HIMC topology based on third-harmonic injection technique. It is demonstrated that the DC-link capacitor that causes inrush transient and distorts the input currents can be removed through bidirectional implementation and proper control of the rectifier. A PFC algorithm is developed, which addresses the disadvantage of limited capability of generating the input reactive power in conventional IMCs. Thus the input reactive power control range is extended significantly, and the input power factor angle can vary from $-\pi/2$ to $\pi/2$, irrelevant to the output conditions. By considering both the current ripple and current tracking performance, the design criteria of the third-harmonic injection inductor are discussed, and a general guideline for designing and realizing of those converters is provided. The 3HIMC possesses the same essential features as conventional IMCs such as sinusoidal input and output currents, and bidirectional power flow. Besides, it has some superior performance such as enhanced input reactive power capability, a maximum voltage transfer ratio of 0.866 under all operation conditions, improved input currents, as well as flexible implementation of the modulation strategies. All the advantages make the 3HIMC an attractive candidate for AC drives, WECS, and other applications.

References

1. IEEE Recommended Practices and Requirements for Harmonic Control in Electric Power Systems. IEEE Standard 519, 1992.
2. A. Kuperman, U. Levy, J. Goren, A. Zafranski, A. Savernin, High power Li-ion battery charger for electric vehicle, in *Proceedings of the 7th International Compatibility and Power Electronics Conf.-Workshop CPE'11* (2011), pp.342–347
3. P.D. Ziogas, Y.G. Kang, V.R. Stefanavic, Rectifier-inverter frequency changer with suppressed DC-link components. IEEE Trans. Ind. Appl. **IA-22**(6), 1027–1036 (1986)
4. S. Kim, S.K. Sul, T.A. Lipo, AC to AC power conversion based on matrix converter topology with unidirectional switches, in *Proc. 1998 IEEE Appl. Power Electron. Conf. (APEC'98)*, Anaheim. (1998), pp. 301–307
5. J. Holtz, U. Boelkens, Direct frequency convertor with sinusoidal line currents for speed-variable AC motors. IEEE Trans. Ind. Electron. **36**(4), 475–479 (1989)
6. L. Huber, D. Borojevic, Space vector modulated three-phase to three-phase matrix converter with input power factor correction. IEEE Trans. Ind. Appl. **31**(6), 1234–1246 (1995)
7. P. Wheeler, J. Rodriguez, J. Clare, L. Empringham, A. Weinstein, Matrix converters: a technology review. IEEE Trans. Ind. Electron. **49**(2), 276–288 (2002)
8. P. Zwimpfer, H. Stemmler, Modulation and realization of a novel two-stage matrix converter, in *Proceedings of the Brazilian Power Electronics Conference* (2001), pp. 485–490
9. L. Wei, T.A. Lipo, A novel matrix converter topology with simple commutation, in *Proceedings of IEEE Industry Applications Society: Annual Meeting*, vol. 3 (2001), pp. 1749–1754
10. L. Wei, T.A. Lipo, Matrix converter with reduced number of switches, in *ConJ Record of the 2th W'EMPEC Anniversary Meeting*, Madison (2001)

11. J.W. Kolar, F. Schafmeister, S.D. Round, H. Ertl, Novel three-phase ac-ac sparse matrix converters. IEEE Trans. Power Electron. **22**(5), 1649–1661 (2007)
12. R. Cardenas, R. Pena, P. Wheeler, J. Clare, G. Asher, Control of the reactive power supplied by a WECS based on an induction generator fed by a matrix converter. IEEE Trans. Ind. Electron. **56**(2), 429–438 (2009)
13. J. Monteiro, J. Silva, S. Pinto, J. Palma, Matrix converter-based unified power-flow controllers: advanced direct power control method. IEEE Trans. Power Del. **26**(1), 420–430 (2011)
14. F. Schafmeister, J.W. Kolar, Novel hybrid modulation schemes significantly extending the reactive power control range of all matrix converter topologies with low computational effort. IEEE Trans. Ind. Electron. **52**(1), 194–210 (2012)
15. M. Rivera, J. Rodriguez, J. Espinoza, T. Friedli, J. Kolar, A. Wilson, C. Rojas, Imposed sinusoidal source and load currents for an indirect matrix converter. IEEE Trans. Ind. Electron. **59**(9), 3427–3435 (2012)
16. S. Yao, S. Mei, W. Hui, et al., Nonlinearity analysis and compensation strategies for two-stage matrix converter. Proc. CSEE **30**(12), 20–27 (2010) (in Chinese)
17. H. Akagi, A. Nabae, S. Atoh, Control strategy of active power filters using multiple voltage source PWM converters. IEEE Trans. Ind. Appl. **IA-22**(3), 460–465 (1986)
18. D. Pedder, A. Brown, J. Ross, A. Williams, A parallel-connected active filter for the reduction of supply current distortion. IEEE Trans. Ind. Electron. **47**(5), 1108–1117 (2000)
19. M. Jantsch, C.W.G. Verhoeve, Inverters with three-phase output and without electrolyte capacitor for improved lifetime, efficiency and costs of grid connected systems, in *Presented at the 14th European Photovoltaic Solar Energy Conference* (1997)
20. H. Yoo, S. Sul, A new circuit design and control to reduce input harmonic current for a three-phase ac machine drive system having a very small dc-link capacitor, in *Proceedings of the 25th Annual IEEE Applied Power Electronics Conference and Exposition* (2010), pp. 611–618
21. T. Soeiro, T. Friedli, J.W. Kolar, Three-phase high power factor mains interface concepts for electric vehicle battery charging systems, in *Proceedings of the 27th Annual IEEE Applied Power Electronics Conference and Exposition* (2012), pp. 2603–2610
22. S. Kim, P.N. Enjeti, D. Rendusara, L.J. Pitel, A new method to improve THD and reduce harmonics generated by a three phase diode rectifier type utility interface, in Proceedings of the IEEE IAS Annual Meeting (1995), pp. 1071–1077
23. P. Pejovic, Low harmonic three phase rectifier, in *Power Electronics Specialists Conference*, Galway (2000), pp. 1029–1034
24. N. Vazquez, H. Rodriguez, C. Hernandez, E. Rodriguez, J. Arau, Three-phase rectifier with active current injection and high efficiency. IEEE Trans. Ind. Electron. **56**(1), 110–119 (2009)
25. Jankovic, M., et al., Hybrid three-phase rectifier with switched current injection device, in *Proceedings of the 15th International Power Electronics and Motion Control Conference (ECCE Europe 2012)*, Novi Sad (2012)
26. J. Kolar, T. Friedli, The essence of three-phase PFC rectifier systems, in *Proceedings of the 3rd IEEE INTELEC* (2011), pp. 1–27
27. T. Soeiro, F. Vancu, J.W. Kolar, Hybrid active 3rd harmonic current injection mains interface concept for DC distribution systems. IEEE Trans. Power Electron. **28**(1), 7–13 (2013)
28. T. Nussbaumer, M. Heldwein, J. Kolar, Differential mode input filter design for a three-phase buck-type PWM rectifier based on modeling of the EMC test receiver. IEEE Trans. Ind. Electron. **53**(5), 1649–1661 (2006)
29. D.G. Holmes, T.A. Lipo, *Pulse Width Modulation for Power Converters: Principles and Practice* (IEEE Press, Piscataway, 2003)
30. C. Klumpner, F. Blaabjerg, Modulation method for a multiple drive system based on a two-stage direct power conversion topology with reduced input current ripple. IEEE Trans. Power Electron. **20**(4), 922–929 (2005)
31. B. Wang, G. Venkataramanan, Analytical modeling of semiconductor losses in matrix converters, in *IEEE Power Electronics and Motion Control Conference (IPEMC)* (2006), pp. 1–8

Chapter 3
Third-Harmonic Injection Matrix Converter with Multiple-Outputs

3.1 Introduction

Advances in the field of power electronic semiconductor devices and microprocessors have greatly promoted the wide use of ASD in industrial drive applications. A standard ASD system includes a rectifier, an intermediate DC-link unit, a VSI, a microprocessor control unit, and an AC motor. However, the requirements of driving multiple motors or multiphase motor (such as dual three-phase winding motor) [1] are existing in such applications as diesel-electric traction [2], MEA [3], marine ship propulsion [4, 5], and steel processing [6]. A direct but not economic solution to this application is to use the standard ASD. One widely accepted alternative for reducing the volume and cost is to share a common rectifier and DC-link [7]. Further reduction in the cost can be achieved by employing converter topologies with a reduced number of power switching devices, such as the topologies sharing a common leg or the neutral point of the DC capacitor [7–10]. Unfortunately, the main drawbacks of such topologies include low utilization of DC voltage and relatively large current stress on the switches of the common leg or the capacitors. Besides, the bulky and failure premature electrolytic capacitors in such VSI-based topologies hinder its applications in some occasions that have requirements of high reliability and compactness, such as MEA, marine ship propulsion, and so on.

As a direct AC-AC converter, the conventional IMC has received considerable attentions in recent years due to its merits such as controllable input power factor, sinusoidal input and output waveforms, four-quadrant operation, no need for large energy storage components, simplicity in commutation and clamp circuit, and so on [11–15]. In addition, due to the presence of physical DC-link, multiple VSIs can share a common rectifier easily, which makes it suitable for multiple drive applications [16–19]. To reduce the cost, IMC topologies with nine switches inverter and five legs inverter are reported in [17] and [18], respectively. However, the shortcoming of low utilization of DC voltage remains. Moreover, it is difficult to implement such IMC-based multiple drive system when the number of inverters is

H. Wang, *Advanced Matrix Converters*, Power Systems,
https://doi.org/10.1007/978-3-031-69902-3_3

large since the modulation of the rectifier and inverters need strict synchronization in conventional IMC.

In some multiple drive applications such as MEA, strict requirements on the line current quality and volume/weight of the converter must be met. To reduce the current ripple and THD, as well as the weight/volume of the passive components, usually the interleaved carrier modulation is adopted, which is done by phase shifting the carrier signals of the multiple inverters [19–22]. In [19] to reduce the input current ripple of an IMC-based multiple drive system, the multiple inverters are divided into two groups according to the power flow direction and the gate signals of the inverters are interleaved. Though the reduction of input current ripple is achieved, a complex commutation is needed.

To overcome the disadvantages of VSI/IMC-based multiple drive system mentioned above, an active 3HIMC topology with multiple-outputs for two different applications, named as independent operation mode and cooperative operation mode, is proposed in this chapter. The proposed 3HIMC has almost all the advantages as conventional IMC. Most importantly, the modulation of the rectifier and multiple inverters in 3HIMC are independent, and the modulation and control task can be implemented flexibly in a separate controller for each inverter, which is impossible in IMC-based multiple drive system. Besides, the narrow pulse width problem existing in conventional IMC is avoided, and the input current ripple, as well as the size of passive components, is reduced dramatically when in cooperative operation mode. All of the above features make the proposed 3HIMC topology very suitable for multiple drive applications such as diesel-electric traction, MEA, marine ship propulsion, and steel processing. The topology structure, operation principles, modulation strategy, and control algorithm of the proposed 3HIMC are studied in detail, and the experimental results based on a 3HIMC system containing two VSIs demonstrate the validity and effectiveness of the proposed topology and modulation strategy.

3.2 Circuit Topology and Operation

3.2.1 Independent Operation Mode

The application of the multi-drive system is different, and the working characteristics of each inverter are not the same, so the operation mode of the multi-drive system needs to be selected according to different application occasions to optimize the performance of the multi-drive system in some aspects. For MEA and some manufacturing industries including winding processes (such as paper making, printing, and dyeing, etc.), generally speaking, the working conditions of each inverter are not the same, and the operation mode of the multi-drive system should be designed to be independently controlled, that is, the modulation and control of each inverter are controlled by a separate controller, which is unrelated to each other.

In order to give full play to the advantages of 3HIMC, it is especially necessary when there are many inverters in multi-drive system.

3.2.1.1 Topology

The topology of 3HIMC with independent inverters is shown in Fig. 3.1a. The main circuit consists of a line-commutated current-source-type rectifier, an active third-harmonic current injection circuit, and two inverters denoted as inverter 1 and inverter 2. As the same as that of conventional IMCs, an *LC* input filter is required for normal operation. The active third-harmonic current injection circuit is composed of three bidirectional switches, a third-harmonic injection inductor, and a bridge leg. Compared with conventional IMC with dual VSIs in [21], although the total number of the switches is increased by two and an extra inductor is required, the power density of 3HIMC can still be kept at a relatively high level. As will be discussed in the operating principles, the current stress of the third-harmonic injection circuit is relatively small. Thus the third-harmonic injection inductor with low peak current can be selected for low power rating applications and its volume is small, while for the case of high power ratings, the third-harmonic injection inductor with low value should be selected, and the volume of the inductor can still be kept at a reasonable level. Besides, bidirectional switches with low current rating can be selected. Even though a third-harmonic injection circuit is needed in 3HIMC, it can bring some benefits. First, the currents at the input of the rectifier (before filtering) are improved significantly. As a result, an input filter with higher corner frequency can be selected, which implies a smaller size of the input filter. Moreover, the switches of the rectifier and the bidirectional switches commutate at line frequency. Thus the switching losses of the rectifier can be ignored, and the size of the heat sink can be reduced. The features above are helpful for increasing the power density, making 3HIMC attractive in aspect of power density compared to conventional IMCs.

3.2.1.2 Operating Principles

To explain the operating principles, first the input voltage interval is divided into six sectors, and the principles of the sector division are as follows: The interval with the input voltages satisfying $u_a > u_b > u_c$ is denoted as sector 1; the interval with the input voltages satisfying $u_a > u_b > u_c$ is denoted as sector 2, and so forth. Due to the symmetry of the three-phase input voltages, for simplicity, only one special example is used to explain the operating principles of the 3HIMC. For the case of sector 1, the input voltage of phase *a* has the maximal instantaneous value, the input voltage of phase *b* has the medium instantaneous value, and the input voltage of phase *c* has the minimal instantaneous value. The positive output rail of the rectifier is connected to phase *a*, the negative output rail of the rectifier is connected to phase *c*, and the DC-link voltage is therefore the input line-line voltage u_{ac}. To form the

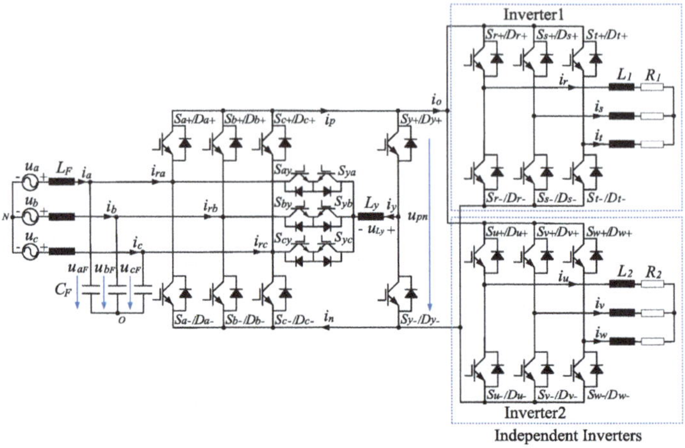

(a)

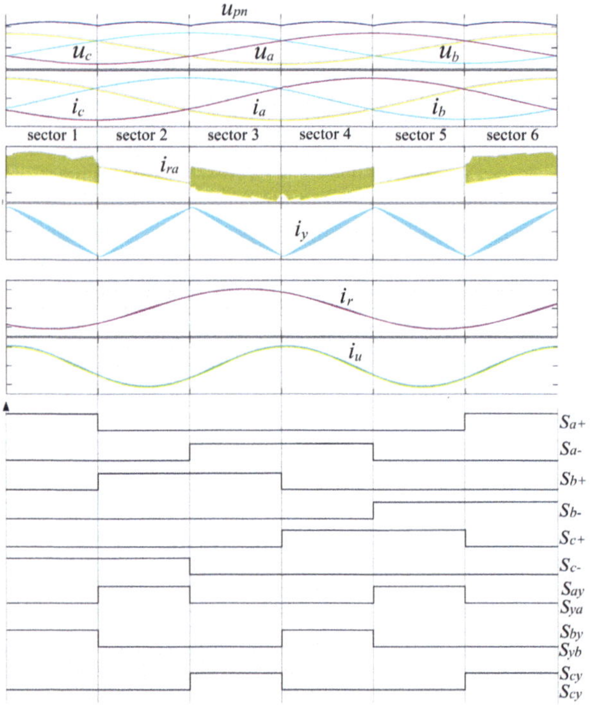

(b)

Fig. 3.1 (**a**) 3HIMC with independent operation mode. (**b**) Key waveforms of the 3HIMC (assuming forward power flow and unity input power factor) conducting paths and avoid short circuit of the input side, the switch S_{a+} in the upper bridge legs and the switch S_{c-} in the lower bridge legs are turned on, and the switches S_{b+} and S_{b-} in the bridge leg of phase b are turned off during this interval. The switches S_{y+} and S_{y-} in third-harmonic current injection circuit are controlled to form the third-harmonic current i_y flowing through the third-harmonic injection inductor L_y. Because there is no current flowing through the bridge leg of phase b during this interval, so the bidirectional switches S_{yb} and S_{by} connected to phase b are kept being on-state to inject the third-harmonic current. The analysis in other sectors is similar to the aforementioned descriptions in sector 1 and will not be repeated here. The switching state of the rectifier stage and the bidirectional switches during an input period are shown in Fig. 3.1b. The two VSIs provide output voltages with variable frequency and amplitude according to the requirements of the loads. By operating in this manner, sinusoidal three-phase input currents and controllable input power factor are attainable. The key waveforms of 3HIMC are shown in Fig. 3.1b.

From Fig. 3.1b, we can find a significant difference between 3HIMC and IMC. The DC-link voltage of 3HIMC exhibits a fixed 6-pulse shape waveform with its value varies from $1.5U$ to $\sqrt{3}U$, irrelevant to operation conditions, where U is the amplitude of the input phase voltage. Therefore the operations of the rectifier and the inverters are independent, and for linear modulation, the voltage transfer ratio q can be determined easily as

$$q = \frac{U_o}{U} = \frac{\sqrt{3}}{2}m, \ m \in [0, 1] \tag{3.1}$$

where U_o is the amplitude of the output phase voltage and m is the modulation index of the inverter. In contrast, the maximal voltage transfer ratio in conventional IMCs decreases with the deceasing of the input power factor [23].

3.2.2 Cooperative Operation Mode

For electrified traction and battleship electrical propulsion applications, since the load of each inverter is the same, the working conditions of each inverter such as output frequency and output amplitude are also the same, at this time, the operation mode of the multi-drive system should be designed as centralized control, that is, one controller controls all inverters. And the working parameters of each inverter such as switching frequency and output frequency are completely consistent. In this way, it is possible to reduce the input current ripple and improve the input performance by using the carrier shift method.

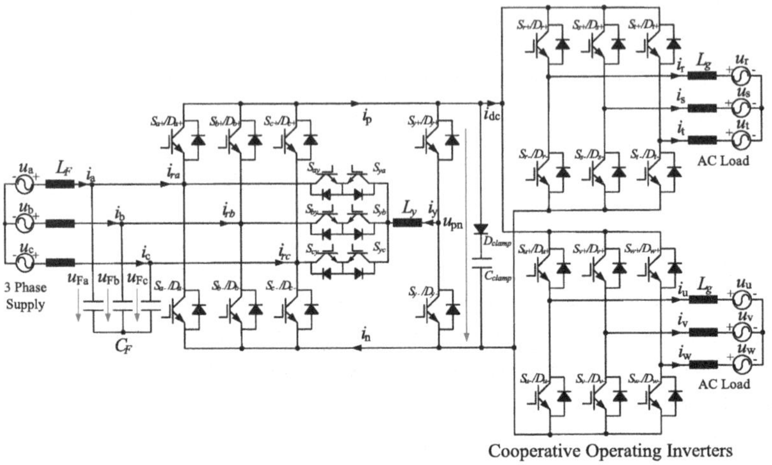

Fig. 3.2 3HIMC with cooperative operation mode

3.2.2.1 Topology

The topology of 3HIMC in cooperative operation mode is shown in Fig. 3.2. The main circuit consists of input LC filter, front-end rectifier, active third-harmonic current injection circuit, clamp circuit, and multiple VSIs. For the purpose of simplicity, a 3HIMC system containing two VSIs is analyzed in this chapter, and the analysis and conclusions can be generalized to other 3HIMC topologies containing a plurality of inverters (more than 2). From the point view of structure, the proposed topology is similar to the VBBC, but the difference lies in that the front-end rectifier of the 3HIMC is essentially a CSR. Active third-harmonic current injection circuit is composed of three bidirectional switches, an inductor, and a current shaping bridge leg. Similar to the conventional IMC, the clamp circuit of the proposed topology is comprised by a diode and capacitor to absorb the energy stored in leakage inductance of the load when the system shutdown. Multiple VSIs are connected to the common DC-link, and each inverter can drive a three-phase load independently.

3.2.2.2 Operating Principle

As discussed above, the main circuit of the proposed topology consists of front-end rectifier, active third-harmonic current injection circuit, and multiple inverters. So the operating principle of the topology can be analyzed by the three sub-circuits, respectively. For the front-end rectifier, only two switches, one of the three switches S_{a+}, S_{b+}, S_{c+} on the upper bridge arm corresponding to the maximum input voltage, and one of the three switches S_{a-}, S_{b-}, S_{c-} on the lower bridge arm corresponding to the minimum input voltage, are turned on at any time. When the power flows forwardly, the current flows through the diodes, while in condition

Table 3.1 Modulation of the rectifier and the third-harmonic current injection circuit

$\omega_i t$	k	S_{ya} S_{ay}	S_{yb} S_{by}	S_{yc} S_{cy}	S_{a+}	S_{a-}	S_{b+}	S_{b-}	S_{c+}	S_{c-}
$0-\pi/3$	u_{bc}/u_{ac}	0	1	0	1	0	0	0	0	1
$\pi/3-2\pi/3$	u_{ac}/u_{bc}	1	0	0	0	0	1	0	0	1
$2\pi/3-\pi$	u_{ca}/u_{ba}	0	0	1	0	1	1	0	0	0
$\pi-4\pi/3$	u_{ba}/u_{ca}	0	1	0	0	1	0	0	1	0
$4\pi/3-5\pi/3$	u_{ab}/u_{cb}	1	0	0	0	0	0	1	1	0
$5\pi/3-2\pi$	u_{cb}/u_{ab}	0	0	1	1	0	0	1	0	0

of reverse power flow, the IGBTs are conducted. The DC voltage u_{pn} is always the maximum input line-line voltage. The active third-harmonic current injection circuit is essentially an active power filter, and the third-harmonic injection current i_y flowing through inductor L_y is formed by controlling the switches S_{y+} and S_{y-} in the current shaping leg. At any time, the bidirectional switch corresponding to the input phase with the smallest absolute voltage is kept being on-state to conduct the third-harmonic current i_y. For example, when the input voltages satisfy $u_{Fa} > u_{Fb} > u_{Fc}$, switches S_{a+} and S_{c-} are turned on, and the bidirectional switches S_{yb} and S_{by} in the third-harmonic current injection circuit are always being on-state, and so forth. Switching state of the rectifier and third-harmonic injection circuit switches is shown in Table 3.1 for all operating conditions. The VSIs provide variable frequency and amplitude output voltages according to the requirements of the loads. By operating in this manner, sinusoidal three-phase input currents and unity input power factor are attainable. As can be seen from the above analysis, compared with the conventional IMC, only the switches of the inverter and the current shaping leg operate at high frequency in the proposed topology, while the others operate at line frequency.

3.3 DPWM Strategy for Reducing Input Ripple

The intermediate DC voltage u_{pn} of 3HIMC is a 6-pulse shape waveform with its value ranging from $1.5U$ to $\sqrt{3}U$, which is quite different from the high-frequency DC voltage in conventional IMC. Obviously, the maximum voltage transfer ratio of 3HIMC is 0.866, the same as that in conventional IMC. However, the 6-pulse shape characteristic of the DC voltage makes the modulation strategy for the inverter and the output waveforms different from that in conventional IMC.

As analyzed in previous section, the input currents of 3HIMC are synthesized by the DC current i_{dc} directly. Therefore, the reduction of peak-to-peak value and RMS value of i_{dc} will help to reduce the input currents ripple and THDs. For the VSI, i_{dc}

can be expressed as follows:

$$i_{dc} = \sum S_i i_i, \ (i = r, s, t, u, v, w) \tag{3.2}$$

where i_i is the inverter output current, and S_i is the switching function and defined as

$$S_i = \begin{cases} 1, & S_{i+} \text{ is on} \\ 0, & S_{i+} \text{ is off} \end{cases}, \ (i = r, s, t, u, v, w) \tag{3.3}$$

In the case of multiple VSIs sharing a common DC-link, suppose that each inverter operates under the same conditions and the gate signals are synchronized, and the peak-to-peak value of the total i_{dc} is the sum of the peak value of DC current generated by each inverter. However, by shifting the gate signals of the inverters with a predetermined angle, the peak-to-peak and RMS value of the total i_{dc} are expected to be reduced, as well as the input current ripple and THD. Taking 3HIMC with two VSIs as an example, and assuming that the two VSIs operate under the same conditions, which is the actual situation in applications such as marine ship propulsion and steel processing, a detailed analysis of DPWM for multiple inverters in 3HIMC is presented below.

The expected output voltage can be expressed as

$$\begin{cases} u_r = u_u = U_o \cos \omega_o t \\ u_s = u_v = U_o \cos (\omega_o t - 2\pi/3) \\ u_t = u_w = U_o \cos (\omega_o t + 2\pi/3) \end{cases} \tag{3.4}$$

The modulation signal is given by

$$u_{io}^* = u_i + u_{no}, \ (i = r, s, t, u, v, w) \tag{3.5}$$

where u_{no} is the zero sequence voltage and can be set as

$$u_{no} = -[max(u_r, u_s, u_t) + min(u_r, u_s, u_t)]/2 \tag{3.6}$$

Then the normalized modulation signal can be expressed as

$$\overline{u}_{io}^* = 2u_{io}^*/u_{pn}, \ (i = r, s, t, u, v, w) \tag{3.7}$$

And the modified normalized modulation signals are derived as

$$\overline{u}_{io}^{*'} = \overline{u}_{io}^* + 1 - max(\overline{u}_{ro}^*, \overline{u}_{so}^*, \overline{u}_{to}^*), (i = r, s, t) \tag{3.8}$$

$$\overline{u}_{io}^{*'} = \overline{u}_{io}^* - 1 - min(\overline{u}_{uo}^*, \overline{u}_{vo}^*, \overline{u}_{wo}^*), \ (i = u, v, w) \tag{3.9}$$

According to the carrier modulation principle, the duty ratio of each phase can be obtained as

$$d_i = \frac{1 + \overline{u}_{io}^{*'}}{2}, \ (i = r, s, t, u, v, w) \tag{3.10}$$

From the analysis above, the modulation signal of each phase leg is clamped at the carrier peak or carrier trough as long as 1/3 output period, as shown in Fig. 3.3. In other words, it implies that one of the switches in the corresponding phase leg is always being on-state during that time. Consequently, the switching losses of the inverter are reduced dramatically, especially in high output power factor region, despite a slight increase in the THD of the output currents due to the reduction of equivalent switching frequency. On the other hands, the short duration pulse in high modulation index region is avoided due to the clamping characteristic of the DPWM. The DPWM strategy of the multiple inverters is presented in Fig. 3.4.

Compared to the modulation strategy of the conventional IMC, the implementation of DPWM is simpler since the rectifier stage and inverter stage are controlled independently, and the advantage is especially obvious in multiple inverters system. Meanwhile, the narrow pulse and nonlinear problem are prevented. As illustrated in Fig. 3.5, the essential reason of the conventional IMC generating narrow pulses is that the duration time of one of the input line-line voltage used to synthesize the DC-link voltage is very short when the input current vector locates in sector boundary. Adopting approximate methods of ignoring the narrow pulse will cause distortions of the input and output waveforms [24].

In contrast, the DC-link voltage of the 3HIMC is 6-pulse waveform and can be considered as a constant voltage source in one switching period. Therefore the inverter stage of the 3HIMC is modulated only once in each switching period, and

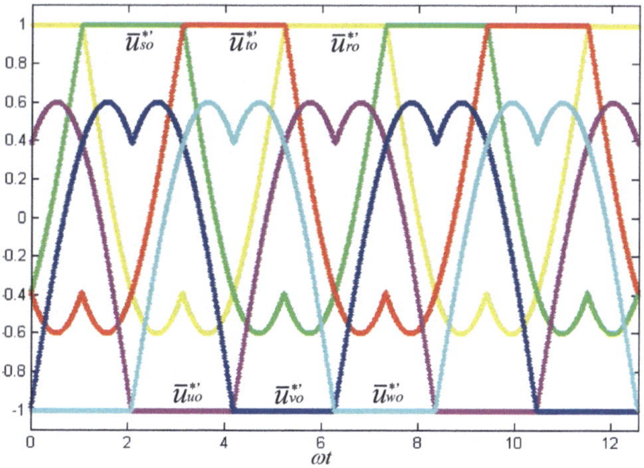

Fig. 3.3 Modulation signals for the inverters of the proposed 3HIMC

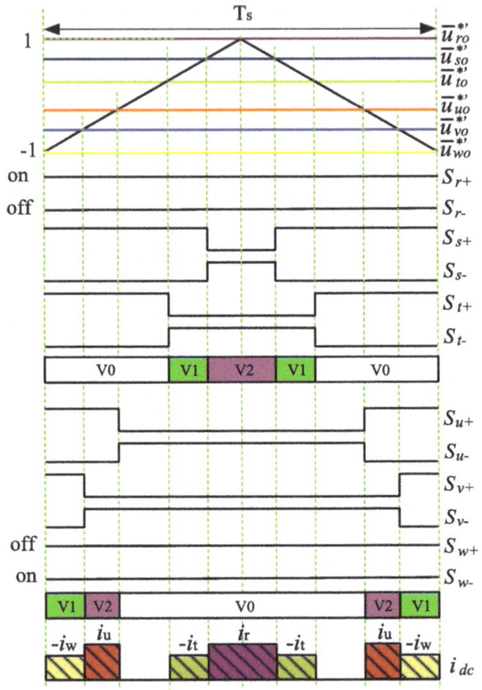

Fig. 3.4 DPWM for the inverters of the proposed 3HIMC

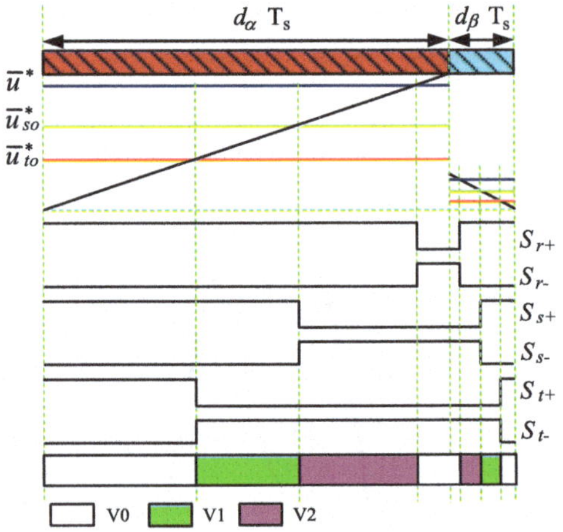

Fig. 3.5 Modulation scheme for the inverter of the conventional IMC

then the narrow pulse problem is avoided with the combined effect of clamping action in DPWM.

3.4 Control of the Third-Harmonic Injection Current

3.4.1 *Extraction of the Third-Harmonic Injection Current Reference Based on Voltage Harmonic Analysis*

As discussed before, normal operation of 3HIMC is dependent on injecting appropriate third-harmonic current i_y, and thus the controlling of the third-harmonic current i_y is the key to implement 3HIMC. For the third-harmonic current injection circuit, the object is to control the actual third-harmonic current i_y to track the third-harmonic injection reference current i_y^*. Thus P^* and i_y^* should be obtained first.

According to the operating principles, the calculation of the third-harmonic injection reference current is dependent on the output power P. So P must be obtained first before solving i_y^*.

It can be seen from the operating principles above that sinusoidal input currents can be obtained only under the premise that the reference output power P^* is exactly the same as the actual output power P, and any deviation from the actual value will cause distortions of the input currents, and thus the input currents harmonic spectrum can reflect the deviation information.

Look back to the proof for sinusoidal input currents in previous section, and assume that the reference power P^* is not equal to the actual power P and the deviation is defined as

$$\Delta P = P^* - P \tag{3.11}$$

Then the average input current of phase a can be rewritten as

$$\bar{i}_{ra} = i_{ra}^* + \Delta i_{ra} \tag{3.12}$$

$$\Delta i_{ra} = \begin{cases} (\frac{2\Delta P}{3U})\{\frac{[\cos(\omega_i t-2\pi/3)+\tan\varphi\sin(\omega_i t-2\pi/3)]\sin(\omega_i t)}{\cos(\omega_i t+5\pi/6)}\}, & 0\le\omega_i t<\frac{\pi}{3} \\ (\frac{2\Delta P}{3U})\{[\cos(\omega_i t)+\tan\varphi\sin(\omega_i t)]\}, & \frac{\pi}{3}\le\omega_i t<\frac{2\pi}{3} \\ (\frac{2\Delta P}{3U})\{\frac{[\cos(\omega_i t+2\pi/3)+\tan\varphi\sin(\omega_i t+2\pi/3)]\sin(\omega_i t)}{\cos(\omega_i t+\pi/6)}\}, & \frac{2\pi}{3}\le\omega_i t<\pi \\ (\frac{2\Delta P}{3U})\{\frac{[\cos(\omega_i t-2\pi/3)+\tan\varphi\sin(\omega_i t-2\pi/3)]\sin(\omega_i t)}{\cos(\omega_i t+5\pi/6)}\}, & \pi\le\omega_i t<\frac{4\pi}{3} \\ (\frac{2\Delta P}{3U})\{[\cos(\omega_i t)+\tan\varphi\sin(\omega_i t)]\}, & \frac{4\pi}{3}\le\omega_i t<\frac{5\pi}{3} \\ (\frac{2\Delta P}{3U})\{\frac{[\cos(\omega_i t+2\pi/3)+\tan\varphi\sin(\omega_i t+2\pi/3)]\sin(\omega_i t)}{\cos(\omega_i t+\pi/6)}\}, & \frac{5\pi}{3}\le\omega_i t<2\pi \end{cases} \tag{3.13}$$

where i_{ra}^* is the desired fundamental current of phase a and Δi_{ra} is the harmonic current.

It can be seen from (3.13) that the current harmonic spectrum is fixed, and the amplitudes of the harmonic currents are proportional to ΔP and the initial phases of the harmonics are dependent on the polarity of ΔP. So a feedback closed-loop control algorithm can be developed based on the harmonic currents. By using FFT theory $\bar{i}_{ra}$ can be expressed in Fourier series as

$$\begin{aligned}\bar{i}_{ra} &= i_{ra}^{*} + I_5 \cos(5\omega_i t + \phi_{i5}) + I_7 \cos(7\omega_i t + \phi_{i7}) \\ &+ I_{11} \cos(11\omega_i t + \phi_{i11}) + I_{13} \cos(13\omega_i t + \phi_{i13}) + \ldots \\ &+ I_n \cos(n\omega_i t + \phi_{in}) + \ldots, n = 5, 7, 11, 13, \ldots\end{aligned} \tag{3.14}$$

$$i_{hn} = I_n \cos(n\omega_i t + \phi_{in}), n = 5, 7, 11, 13, \ldots \tag{3.15}$$

where i_{hn} is the nth harmonic current, and I_n and ϕ_{in} are the amplitude and initial phase of the nth harmonic current, respectively. From (3.14), the amplitude and initial phase of the $5th$ harmonic current can be solved as

$$\begin{aligned}I_5 &= \left(\frac{2\Delta P}{3U}\right)\left(1 - \frac{3\sqrt{3}}{2\pi}\right), \\ \phi_{i5} &= \begin{cases} 0, & \Delta P \geq 0 \\ \pi, & \Delta P < 0 \end{cases}\end{aligned} \tag{3.16}$$

So the item $I_5 \cos(\phi_{i5})$ can be used to evaluate ΔP and form the closed loop to obtain P^*.

Since i_{ra} is discontinuous, extracting $I_5 \cos(\phi_{i5})$ is not an easy task. As can be seen from the single-phase harmonic equivalent circuit shown in Fig. 3.6a, if the current i_{ra} is distorted, the voltage across the capacitor will also contain the relevant harmonic spectrum, so an alternative is to analyze the voltage harmonic spectrum. From Fig. 3.6a, the nth harmonic voltage across the capacitor can be expressed as

$$u_{hn} = -\frac{L_F s}{L_F C_F s^2 + 1} i_{hn} \tag{3.17}$$

It can be seen from (3.17) that u_{hn} lags behind i_{hn} by $\pi/2$ and the amplitude of u_{hn} is proportional to I_n as long as (3.18) is satisfied:

$$n\omega_i < \omega_c = \frac{1}{\sqrt{L_F C_F}} \tag{3.18}$$

For the input filter parameters used in this chapter, (3.18) is satisfied, so the item $U_5 \cos(\phi_{U5})$is used instead of $I_5 \cos(\phi_{i5})$, where U_5 and ϕ_{U5} are the amplitude and initial phase of the $5th$ harmonic voltage, respectively.

The control algorithm based on voltage harmonic analysis is shown in Fig. 3.6b in detail. First the capacitor voltage u_{Fa} is passed a BPF to obtain the $5th$ harmonic

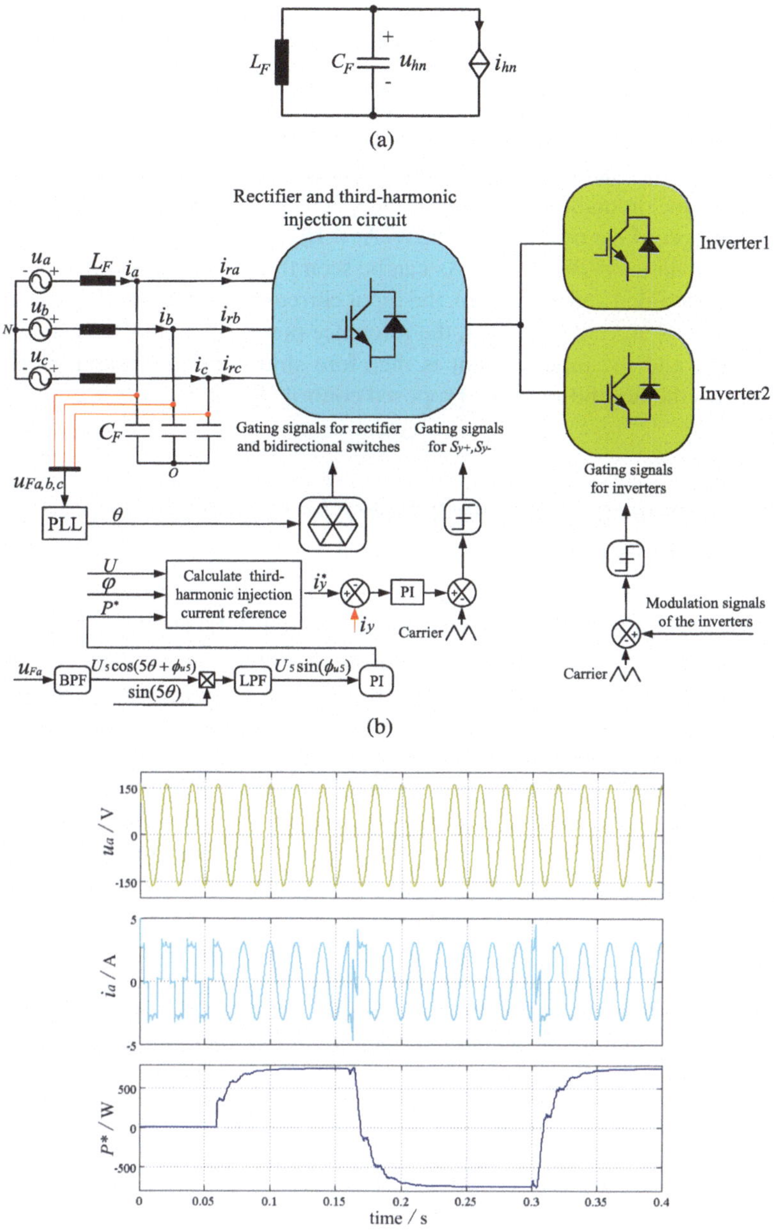

(c)

Fig. 3.6 (**a**) Single-phase harmonic equivalent circuit. (**b**) System control schematic diagram of 3HIMC with dual outputs. (**c**) Simulated result of the control algorithm for extracting the third-harmonic reference current voltage u_{h5}, and then u_{h5} is multiplied by $\sin(5\theta)$ and passed an LPF to get U_5 and ϕ_{U5}, which is the input of the PI controller. Here $\theta = \omega_i t$ is the phase of the input voltages. The output of the PI controller is the reference power P^*, and finally the third-harmonic injection reference current is obtained.

To verify the correctness of the proposed control algorithm, numerical simulation is carried out in Matlab/Simulink environment and the result is illustrated in Fig. 3.6c. In the simulation the control algorithm is disabled before 0.06s. At first the actual power of the converter is 750W. At 0.16s it is changed to -750W and then is changed again to 750W at 0.3s. As can be seen from the result, when the control algorithm is disabled and P^* is zero, the input current exhibits a quasi-square shape. After the control algorithm is used, the reference power P^* tracks the actual power P very well, and the input current is therefore sinusoidal. The simulation result demonstrates the feasibility of the proposed control algorithm.

3.4.2 Estimation of the Input Power by Reconstructing the Input Currents

According to the above operating principles and control block diagram, the extraction of the reference third-harmonic injection current i_y^* is dependent on the fundamental component of input currents $\bar{i}_{ra,rb,rc}$. However, it is not easy to obtain $\bar{i}_{ra,rb,rc}$ directly by measuring $i_{ra,rb,rc}$ since it is discontinuous. Though the input currents $i_{a,b,c}$ can be measured by current sensors easily at the expense of extra volume/cost of the hardware, the effect of the input filter may affect the estimation accuracy of $\bar{i}_{ra,rb,rc}$. In [24] the input power of the converter was estimated through the measured output power and the premeasured rated conversion efficiency η. However, the conversion efficiency η may vary with operating conditions. Consequently the performance of the input currents also degrades to a certain extent. Similar to the conventional IMC, the input currents of the rectifier in 3HIMC are coupled with the output currents of the inverter, so it is possible to reconstruct the input currents by using the output currents, and then the input power can be estimated. According to the operating principle of VSI, one can get

$$i_p = \sum S_i \cdot i_i, \ (i = r, s, t, u, v, w, y) \tag{3.19}$$

After averaging over one switching period, the average input current be given as

$$\overline{i_p} = \sum d_i \cdot i_i \tag{3.20}$$

where d_i is the duty ratio of switch S_{i+}.

Table 3.2 Current reconstruction of the rectifier in 3HIMC

	$u_{Fa} > u_{Fb} > u_{Fc}$	$u_{Fb} > u_{Fa} > u_{Fc}$	$u_{Fb} > u_{Fc} > u_{Fa}$	$u_{Fc} > u_{Fb} > u_{Fa}$	$u_{Fc} > u_{Fa} > u_{Fb}$	$u_{Fa} > u_{Fc} > u_{Fb}$
$\overline{i}_{ra}^{*}$	$\overline{i}_p$	$-\overline{i}_y$	$\overline{i}_y - \overline{i}_p$	$\overline{i}_y - \overline{i}_p$	$-\overline{i}_y$	$\overline{i}_p$
$\overline{i}_{rb}^{*}$	$-\overline{i}_y$	$\overline{i}_p$	$\overline{i}_p$	$-\overline{i}_y$	$\overline{i}_y - \overline{i}_p$	$\overline{i}_y - \overline{i}_p$
$\overline{i}_{rc}^{*}$	$\overline{i}_y - \overline{i}_p$	$\overline{i}_y - \overline{i}_p$	$-\overline{i}_y$	$\overline{i}_p$	$\overline{i}_p$	$-\overline{i}_y$

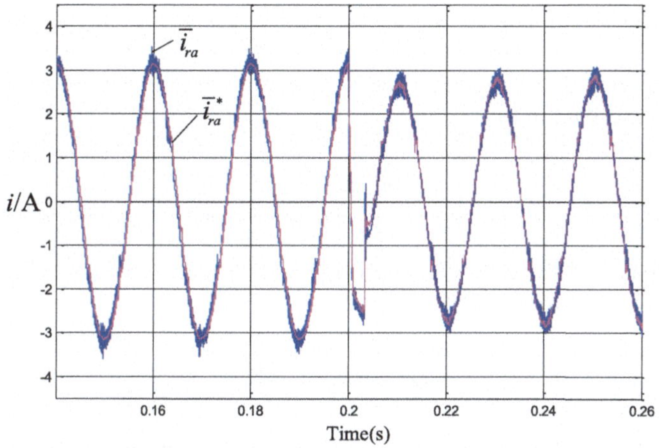

Fig. 3.7 Simulation result of input current reconstruction of the rectifier in 3HIMC

According to the modulation strategy mentioned before, when input voltages satisfy $u_{Fa} > u_{Fb} > u_{Fc}$, then

$$i_{ra} = i_p, i_{rb} = -i_y, i_{rc} = -i_{ra} - i_{rb} \tag{3.21}$$

The derivation is similar in other conditions. The reconstructed fundamental components of input currents $\overline{i}_{ra,rb,rc}^{*}$ are shown in Table 3.2, and the input power of the rectifier can be estimated as

$$P_r = u_{Fa}\overline{i}_{ra}^{*} + u_{Fb}\overline{i}_{rb}^{*} + u_{Fc}\overline{i}_{rc}^{*} \tag{3.22}$$

Then the reference third-harmonic injection current i_y^* can be calculated according to the control block diagram. The numerical simulation result for the reconstruction method is illustrated in Fig. 3.7. Before $t = 0.2$s, the direction of power flow is forward, and after $t = 0.2$s the direction is reversed. It can be seen that the reconstructed current of phase a is in good agreement with the actual value $\overline{i}_{ra}^{*}$, which demonstrates the validity of the current reconstruction strategy.

3.4.3 Controller Design

After extracting the third-harmonic injection reference current, the next step is to control the actual third-harmonic injection current to track the reference exactly. Although the third-harmonic injection circuit is a time-varying nonlinear system, it is simplified as a time-invariant linear system by treating it to be in "quasi-steady state" around each instant of the third-harmonic injection current cycle. This is justified due to the relatively slower variation of the third-harmonic injection current waveform (150 Hz) in comparison to the control frequency f_{con} (32 kHz) of the third-harmonic injection current. The third-harmonic injection current circuit can be modeled as a control voltage input to the third-harmonic injection current transfer function, as described in (3.23):

$$G_o = \frac{1}{L_y s} \tag{3.23}$$

Thus a conventional PI controller given by (3.24) is used for controlling the third-harmonic injection current in this chapter

$$G_c = K_p + \frac{K_i}{s} \tag{3.24}$$

where K_p is the proportional gain and K_i is the integration gain. So the design work becomes to selecting proper K_p and K_i.

The choice of K_p and K_i is mainly based on the following considerations: (1) The gain at the third-harmonic frequency should be high enough and the BW of the control loop should be wide enough; (2) The crossover frequency f_c should be selected properly to maintain sufficient phase margin and keep the loop BW to be less than $f_{con}/2$. In this chapter, the crossover frequency f_c of the control loop is selected as $f_c = f_{con}/15 = 2.13\,\text{kHz}$, and the phase margin is chosen as 45°. Thus the designed PI controller is given by

$$G_c = 9.48 + \frac{127065}{s} \tag{3.25}$$

and the open-loop transfer function of the control loop is

$$G = G_o G_c = \frac{9480s + 127065000}{s^2} \tag{3.26}$$

The Bode diagram of the control loop is shown in Fig. 3.8. It can be seen from the diagram that a BW of 4.6 kHz is obtained. Besides, the gain of the closed-loop transfer function at the third-harmonic frequency (150 Hz) is almost 0 dB, and thus the tracking error of the third-harmonic injection current in steady state is close to 0.

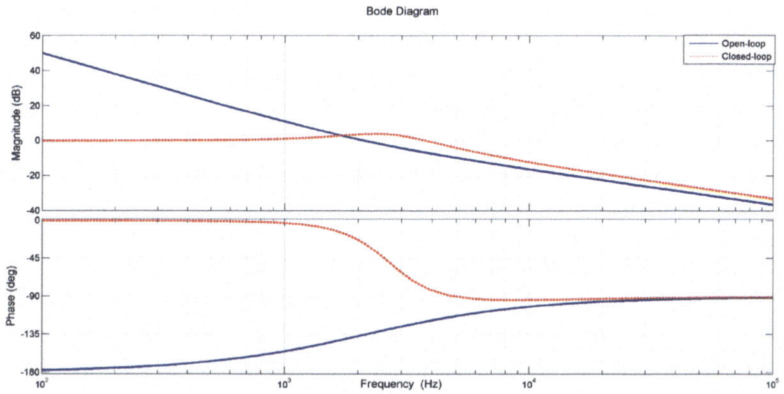

Fig. 3.8 Bode diagram of the third-harmonic injection current control loop

In this chapter, a high control/switching frequency is selected for the purpose of reducing the volume of the third-harmonic injection inductor, and the switching frequency of the inverters is also set as 32 kHz for convenience consideration. However, it should be noted that a lower control/switching frequency of several kHz would be sufficient to get a reasonable steady-state error of the third-harmonic current. Besides, since the operations of the third-harmonic injection circuit and the inverters in 3HIMC are independent, the inverters can operate at a different and relatively low switching frequency. Therefore 3HIMC topology is also suitable for high power applications.

3.5 Simulation and Experimental Results

3.5.1 Simulation Results

Figure 3.9 shows the simulated results under CF mode. In this simulation the desired displacement angle at input side is set as 0, and the desired output frequencies and modulation indices are set as follows: $f_1 = 40\,\text{Hz}$, $m_1 = 0.8$ for inverter 1 and $f_2 = 40\,\text{Hz}$, and $m_2 = 0.8$ for inverter 2. From the top to the bottom, input phase voltage u_a, input current i_a, output current i_r of inverter 1, and output current i_u of inverter 2 are sequentially displayed. Figure 3.9a presents the waveforms under CF mode when the desired output voltages of the two inverters are in phase. It can be seen that both the input and output currents are sinusoidal. Because the desired input displacement angle is set as 0, the input current is in phase with the input voltage. Since the loads of the two inverters are identical, the output currents of the two inverters are also in phase and with the same amplitude of approximately 4.5A. Figure 3.9b shows the simulated waveforms under CF mode with different initial phases of the desired output voltages, during which the initial phase of the desired output voltage in inverter 2 lags behind that of the inverter 1 by $\pi/2$. The other setups remain the same as that in Fig. 3.9a. As can be seen from the result in Fig. 3.9b, the amplitude of the output currents are still the same for both inverters except for a $\pi/2$ phase shift. Since the total output power and other conditions are kept unchanged, the input current also remains the same as that in Fig. 3.9a.

Figure 3.10 presents the simulated results under DF mode. In this simulation, both steady and dynamic conditions are tested. In Fig. 3.10a the parameters of the inverters are set as follows: $f_1 = 40\,\text{Hz}$, $m_1 = 1$ for inverter 1 and $f_2 = 100\,\text{Hz}$, $m_2 = 1$ for inverter 2. In this simulation the modulation indices are both set as 1, which means a voltage transfer ratio of 0.866 for both inverters. Similar to that in CF mode, it can be seen again that the actual outputs follow the desired outputs very well. Besides, the input and output currents also have good qualities and unity input power factor is achieved. Figure 3.10b shows the waveforms under DF mode

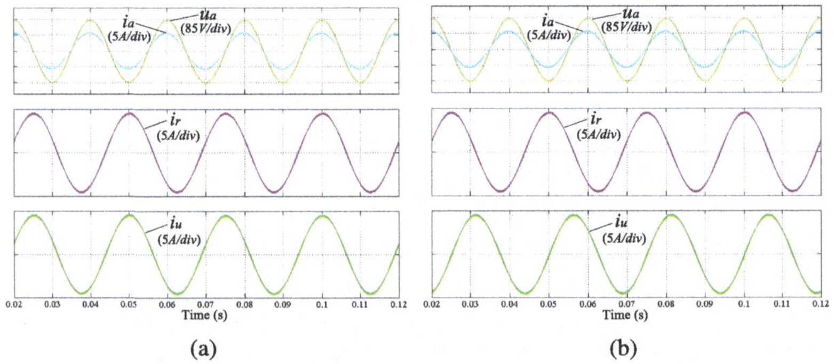

Fig. 3.9 Simulated waveforms of 3HIMC under CF mode when the output voltages of the two inverters (**a**) are in phase and (**b**) have a $\pi/2$ phase shift

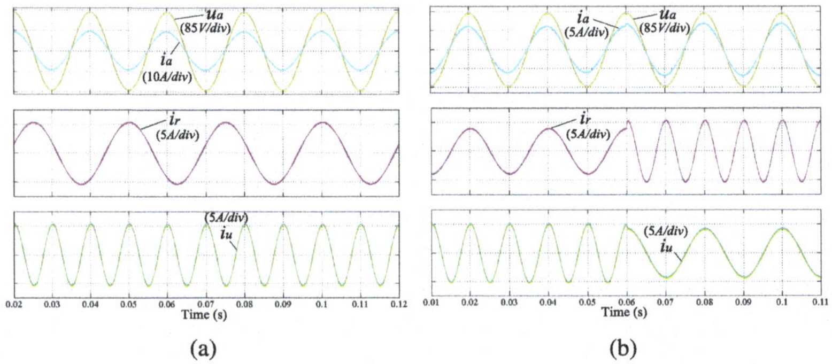

Fig. 3.10 Simulated waveforms of 3HIMC under DF mode (**a**) in steady condition with $f_1 = 40$ Hz, $m_1 = 1$ for inverter 1 and $f_2 = 100$ Hz, $m_2 = 1$ for inverter 2 and (**b**) at load step: Output parameters of inverter 1 change from $f_1 = 50$ Hz, $m_1 = 0.7$ to $f_1 = 100$ Hz, $m_1 = 0.95$ and the output parameters of inverter 2 change from $f_2 = 100$ Hz, $m_2 = 0.9$ to $f_2 = 50$ Hz, $m_2 = 0.75$

when the output parameters of inverter 1 change from $f_1 = 50\,\text{Hz}$, $m_1 = 0.7$ to $f_1 = 100\,\text{Hz}$, $m_1 = 0.95$ and the output parameters of inverter 2 change from $f_2 = 100\,\text{Hz}$, $m_2 = 0.9$ to $f_2 = 50\,\text{Hz}$, $m_2 = 0.75$. It is clear that the input and output currents are sinusoidal and the dynamic performance is good in the presented 3HIMC. From the results shown in Figs. 3.9 and 3.10, we can see that the two sets of three-phase outputs in 3HIMC can be controlled independently with arbitrary frequency and amplitude within a maximal voltage transfer ratio of 0.866. Besides, sinusoidal input and output currents are achievable under various operation conditions.

To verify the input reactive power control feature of 3HIMC, 3HIMC with different input power factors are tested as shown in Fig. 3.11. In Fig. 3.11a and b, the output parameters were set as $f_1 = f_2 = 40\,\text{Hz}$ and $m_1 = m_2 = 0.8$, and the desired input displacement angles are set as $\pi/6$ and $-\pi/6$, respectively. As can be seen from the results, the output currents remain sinusoidal with amplitude of approximately 4.5A, which means a voltage transfer ratio of 0.693 and thus the irrelevance between the voltage transfer ratio and the input power factor. This is very different from that of IMC, since only a voltage transfer ratio of 0.6 is available

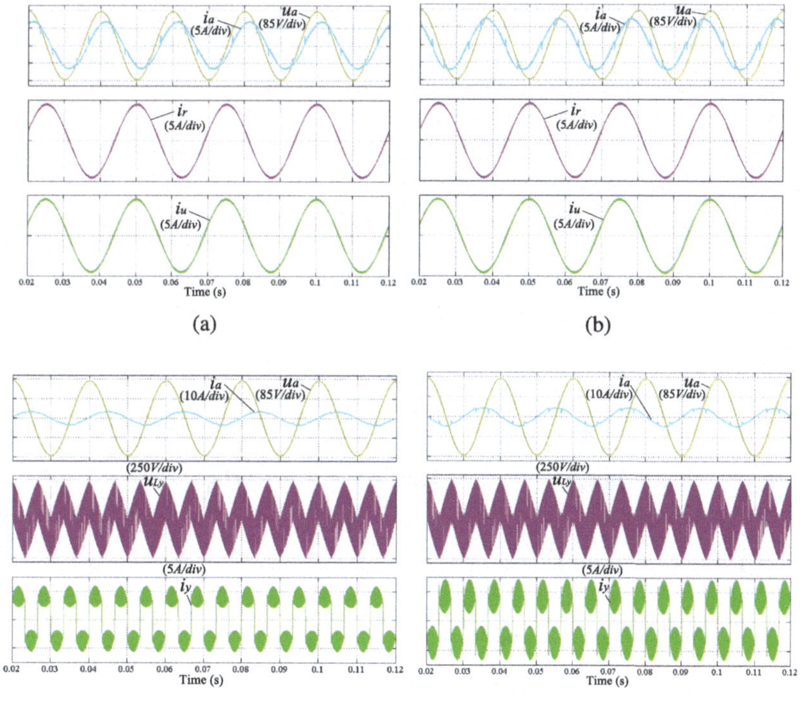

Fig. 3.11 Simulated waveforms of 3HIMC with different input power factors. (**a**) The displacement angle is $\pi/6$. (**b**) The displacement angle is $-\pi/6$. (**c**) The displacement angle is $\pi/2$. (**d**) The displacement angle is $-\pi/2$

under such situation in IMC. Besides, the input current is still sinusoidal with the desired displacement angle. Figure 3.11c and d demonstrates the STATCOM operation mode of 3HIMC. In Fig. 3.11c and d, both the two VSIs are disabled, and the desired input displacement angles are set as $\pi/2$ and $-\pi/2$ with the amplitude of the input reactive reference current being 4A, respectively. It can be seen that pure reactive current with the desired amplitude and displacement angle is appeared at the input side. When the input displacement angle is $\pi/2$, the input current lags the input voltage by $\pi/2$, and the converter generates pure inductive reactive power at the input side. While the input displacement angle is $-\pi/2$, the input current leads the input voltage by $\pi/2$ and the converter generates pure capacitive reactive power. The results in Fig. 3.11 verified the excellent voltage transfer ratio feature and wide input reactive power control range of 3HIMC.

Figure 3.12 illustrates the input current waveforms without and with the third-harmonic current injection circuit enabled. As can be seen, the input current exhibits a quasi-square wave shape and includes a lot of low-frequency harmonics. At 0.04s, the third-harmonic current injection circuit is enabled, and the injected current i_y exhibits a quasi-third-harmonic wave shape, and the input current is sinusoidal and in phase with the input voltage. So PFC at input side is achieved with the third-harmonic current injection circuit enabled.

Figure 3.13 presents the simulation results in steady state under forward and reverse power flows. From the top, input voltage u_a, input current i_a, third-harmonic injection current i_y, output line-line voltage u_{rs}, and output current i_r are sequentially displayed. In this simulation, the RMS phase voltage of the isolated gird connected to the output of 3HIMC is set to 48V, and the reference reactive current i_{q_ref} is set to 0. At the beginning the reference active current i_{d_ref} is 5A and the power flows forwardly. After 0.1s, the reference active current i_{d_ref} is changed to -5A and the power flows reversely. It is clear that the quality of input and output current is good with the proposed modulation strategies and control algorithm. Under forward power flow, the input current is in phase with the input

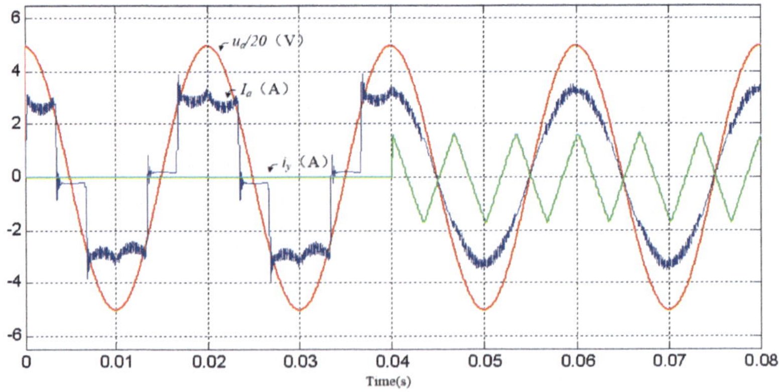

Fig. 3.12 Input current comparison without and with the third-harmonic injection circuit enabled

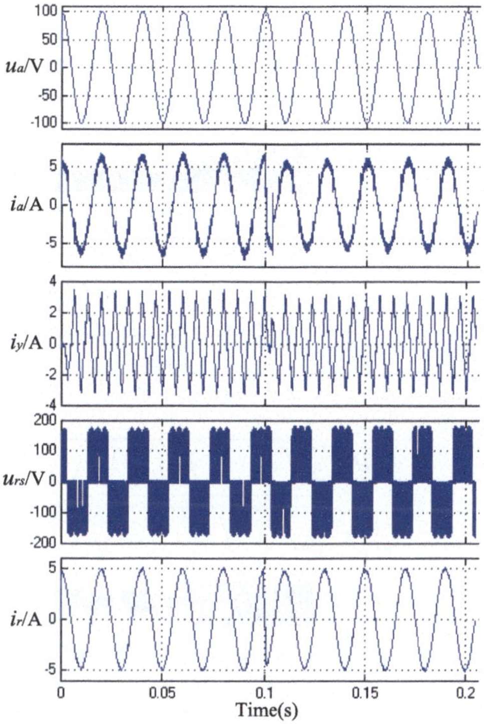

Fig. 3.13 Simulation results of bidirectional power flow in 3HIMC

voltage, while under reverse power flow, the input current and input voltage are opposite in phase.

Figures 3.14 and 3.15 illustrate the comparisons of input current performance between the proposed DPWM and the third-harmonic added SPWM [25], referred to here as TSPWM. Under TSPWM, the carrier signals of the two inverters are in phase with each other. In each simulation, the parameters of the two inverters are identical.

Figure 3.14 shows the simulation results with different modulation indices m_v. The reference active current and reactive current are set as $i_{d\_ref} = 5\text{A}$, $i_{q\_ref} = 0$ and the output power factor angle is nearly 0. From Fig. 3.14a–c, the RMS phase voltages of the isolated gird are 12V, 24V, 36V, and 48V, respectively, and the corresponding modulation indices m_v are 0.2, 0.4, 0.6, and 0.8, respectively. At 0.1s, the modulation strategy is changed from TSPWM to DPWM. As shown in Fig. 3.14, compared with TSPWM, the proposed DPWM can achieve lower peak DC current and input current ripple under different modulation indices, and therefore better quality of the input waveforms is obtained. For TSPWM, the peak DC current is about 10A, and the THDs of the input current are 7.28%, 4.7%, 3.46%, and 2.71% when the modulation indices m_v are 0.2, 0.4, 0.6, and 0.8, respectively. When DPWM is applied, the peak DC current is reduced to about 5A, and the THDs of input current are reduced to 5.23%, 3.18%, 2.08%, and 2.16% when the modulation

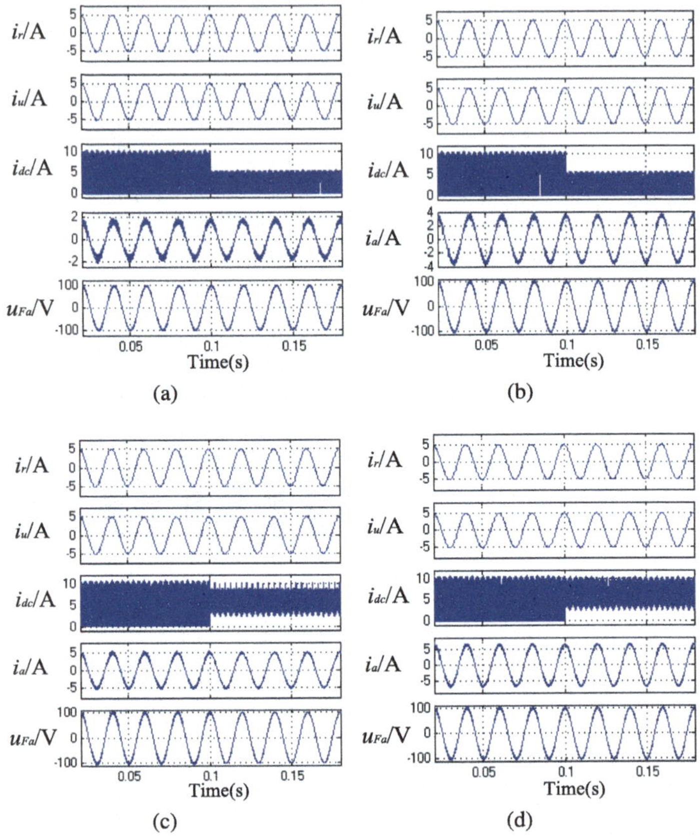

Fig. 3.14 Comparisons of waveforms in 3HIMC between TSPWM and the proposed DPWM under different modulation indices (under unity output power factor). (**a**) Modulation index m_v is 0.2. (**b**) Modulation index m_v is 0.4. (**c**) Modulation index m_v is 0.6. (**d**) Modulation index m_v is 0.8. From top to bottom is the output current i_r, output current i_u, DC current i_{dc}, input current i_a, and voltage u_{Fa} across the capacitor of the input filter, respectively

indices m_v are 0.2, 0.4, 0.6, and 0.8, respectively. It can also be seen that the voltage ripple of the capacitor in the input filter is reduced due to the reduction of the peak DC current, which is beneficial for selecting smaller size passive components such as capacitor and inductor of the input filter.

Figure 3.15 shows the simulation results under different output power factor angles φ. The amplitude of reference output current is set to 5A, and the RMS phase voltage of the isolated gird is 48V, corresponding to a 0.8 modulation index m_v. From Fig. 3.15a–e, the corresponding output power factor angles φ are $-3\pi/4$, $-\pi/4$,$\pi/4$, $3\pi/4$, and π, respectively. At 0.1s, the modulation strategy is changed from TSPWM to DPWM. Under TSPWM, THDs of input current are 4.66%, 3.42%, 3.64%, 3.92%, and 3.35%, respectively, while under DPWM they are reduced to 4.16%, 2.69%, 2.64%, 3.69%, and 2.72%, respectively. Similar to the results in

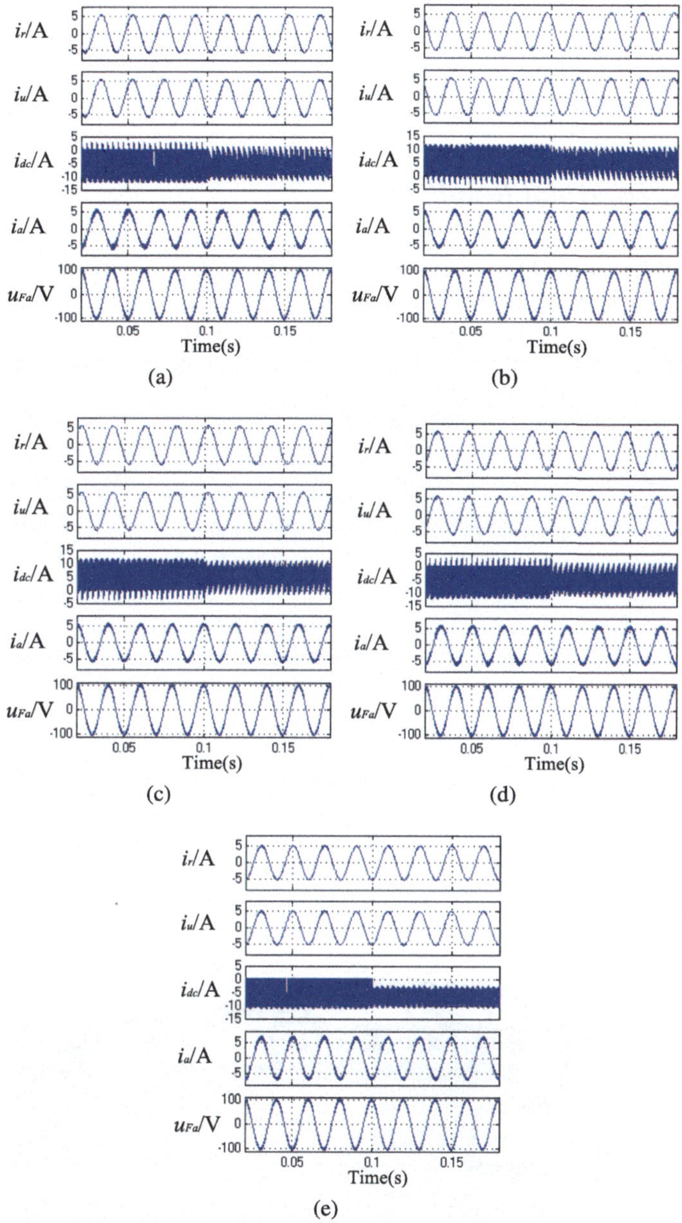

Fig. 3.15 Comparisons of waveforms in 3HIMC between TSPWM and the proposed DPWM under different output power factor angles (under a 0.8 modulation index). (**a**) Output power factor angle φ is $-3\pi/4$. (**b**) Output power factor angle φ is $-\pi/4$. (**c**) Output power factor angle φ is $\pi/4$. (**d**) Output power factor angle φ is $3\pi/4$. (**e**) Output power factor angle φ is π. From top to bottom is the output current i_r, output current i_u, DC current i_{dc}, input current i_a, and voltage u_{Fa} across the capacitor of the input filter, respectively

Fig. 3.14, it can be seen that under different φ, the proposed DPWM still has a lower peak DC current, THD of input current, and voltage ripple of the capacitor in the input filter compared to TSPWM.

3.5.2 Experimental Results

To validate the proposed topology and simulated results, a laboratory prototype of 3HIMC with the specifications given in Table 3.3 is built, as shown in Fig. 3.16. The control platform of 3HIMC is realized by a combination of DSP TMS320F28335 and FPGA EP2C8T144C8N. DSP is tasked to handle with ADC sampling, control

Table 3.3 System parameters of 3HIMC

Characteristics	Specifications
Input line-line voltage u_i	200 V_{RMS}
Mains frequency f_n	50 Hz
Switching frequency f_s	32 kHz
Input inductor L_F	0.4 mH
Input capacitor C_F	6.6 μF
Inductor L_y	2 mH
Inductor L_1	3 mH
Inductor L_2	3 mH
Resistor R_1	25 ohm
Resistor R_2	25 ohm
S_{i+}/D_{i+}, $S_{i-}/D_{i-}(i = a, b, c)$	IHW30N120R2
S_{iy}, $S_{yi}(i = a, b, c)$	IHW30N120R2
S_{i+}, $S_{i-}(i = y, r, s, t)$	IGW40N120H3
D_{i+}, $D_{i-}(i = y, r, s, t)$	DSEP29-12A

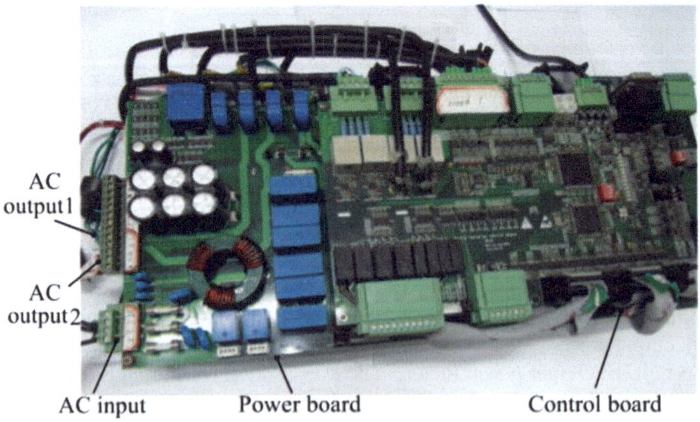

Fig. 3.16 Prototype of 3HIMC

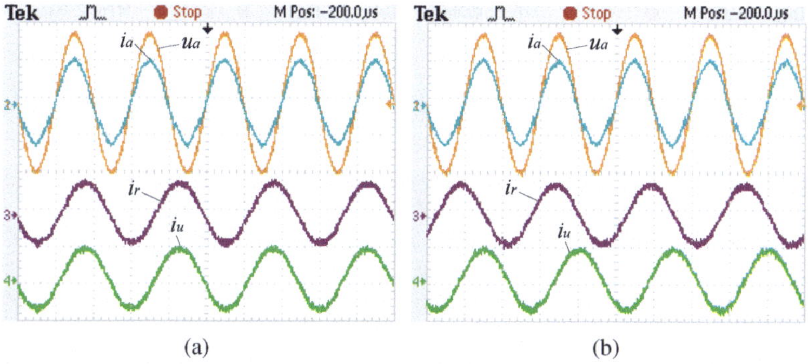

Fig. 3.17 Experimental waveforms of 3HIMC under CF mode when the output voltages of the two inverters (**a**) are in phase and (**b**) have a $\pi/2$ phase shift. CH1 is the input voltage u_a (85 V/div), CH2 is the input current i_a (5 A/div), CH3 is the output current i_r of inverter 1 (5 A/div), CH4 is the output current i_u of inverter 2 (5A/div), and time scale is 10 ms/div

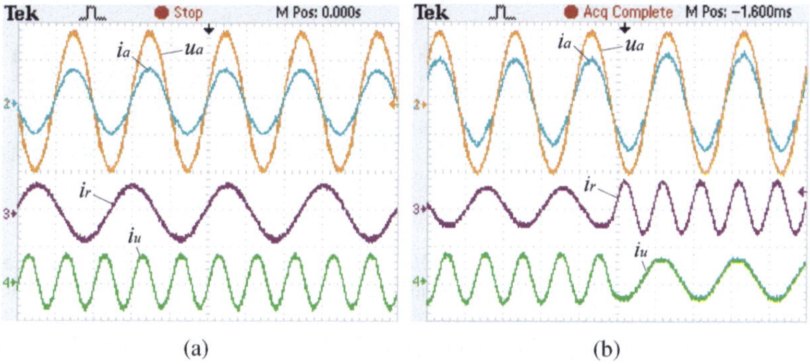

Fig. 3.18 Experimental waveforms of 3HIMC under DF mode (**a**) in steady condition and (**b**) at load step: Output parameters of inverter 1 change from $f_1 = 50$ Hz, $m_1 = 0.7$ to $f_1 = 100$ Hz, $m_1 = 0.95$ and the output parameters of inverter 2 change from $f_2 = 100$ Hz, $m_2 = 0.9$ to $f_2 = 50$ Hz, $m_2 = 0.75$. CH1 is the input voltage u_a (85 V/div), CH2 is the input current i_a (10 A/div in (**a**) and 5 A/div in (**b**)), CH3 is the output current i_r of inverter 1 (7 A/div), CH4 is the output current i_u of inverter 2 (7A/div), and time scale is 10 ms/div

algorithm, and so on, while the modulation strategy and IO control are realized by FPGA.

The experimental results shown in Figs. 3.17, 3.18, and 3.19 correspond to the simulated results shown in Figs. 3.9, 3.10, and 3.11, and the experimental conditions and commands are exactly the same as those in the simulations.

Figure 3.17 shows the corresponding experimental results under CF mode, during which the input and output parameters are set as the same as those in Fig. 3.9. As can be seen from the results, the actual outputs track the references exactly, and the input and output currents are sinusoidal with almost unity input power factor. The slight phase difference between the input voltage and the input current is contributed

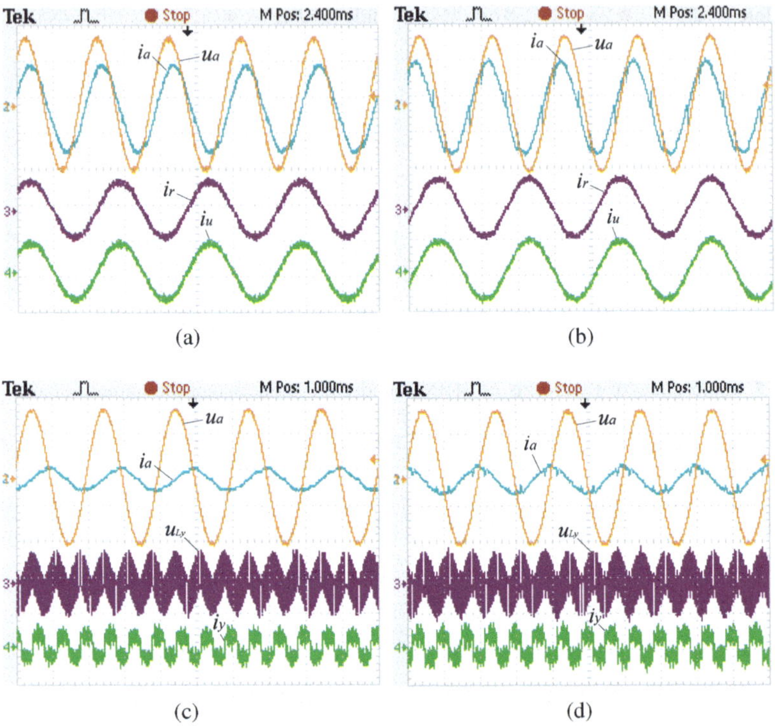

Fig. 3.19 Experimental waveforms of 3HIMC with different input power factors. (**a**) The displacement angle is $\pi/6$. (**b**) The displacement angle is $-\pi/6$. (**c**) The displacement angle is $\pi/2$. (**d**) The displacement angle is $-\pi/2$. CH1 is the input voltage u_a (85 V/div), CH2 is the input current i_a (5 A/div), CH3 is the output current i_r of inverter 1 (5 A/div), CH4 is the output current i_u of inverter 2 (5A/div), and time scale is 10 ms/div in (**a**) and (**b**). CH1 is the input voltage u_a (85 V/div), CH2 is the input current i_a (10 A/div), CH3 is the voltage across the inductor L_y (250 V/div), CH4 is the third-harmonic injection current (10 A/div), and time scale is 10 ms/div in (**c**) and (**d**)

by the input *LC* filter. Under CF mode, the actual rated efficiency is measured as 93.8% by a wattmeter.

Figure 3.18 presents the experimental results under DF mode. As the same as the simulation in Fig. 3.10, it can be seen from the experimental results that the proposed 3HIMC has the capability of providing two sets of three-phase outputs independently. Similar to the results in Fig. 3.17, the amplitude of the input and output currents are slightly lower than that of the expected due to the power loss of the converter.

Figure 3.19 shows the corresponding experimental results under different input power factors. From the top to the bottom, waveforms under input displacement angle of $\pi/6$, $-\pi/6$, $\pi/2$, and $-\pi/2$ are sequentially displayed. Similar to the simulation in Fig. 3.11, slight spike exists in the input current. This can be explained as the follows. Since the third-harmonic injection reference current is not continuous derivative at the points of input voltage sector transits, it is a challenge to achieve high tracking performance of the reference current. Theoretically, zero tracking error of the third-harmonic injection current in steady state can be achieved at very high switching frequency and very low inductance of the third-harmonic injection inductor. However, the switching frequency cannot be selected too high due to the limits of the switching speed and power loss, and the value of the third-harmonic injection inductor also cannot be chosen too small due to the requirement of the current ripple. In practice a trade-off should be made between the third-harmonic current tracking performance and the system efficiency as well as the third-harmonic current ripple.

As can be seen from Figs. 3.17, 3.18, and 3.19, the experimental results match the simulated results very well, and sinusoidal input/output currents and controllable input power factor are achieved in the experiments. The tiny differences between the simulation and the experiments can be attributed to the nonlinearities in the practical converter such as power losses, dead time effect, and so on. Thus the effectiveness of the proposed topology and control algorithm are verified experimentally.

Figure 3.20 shows the comparative waveforms of 3HIMC without and with the third-harmonic injection circuit being enabled. It can be seen from Fig. 3.20a that the waveform of input current is quasi-square when the third-harmonic injection circuit is disabled, which exhibits the same characteristics as the diode rectifier in series with a large inductor in DC-link, and the THD of input current measured by power quality analyzer FLUKE430 is 25.2%. Figure 3.20b shows the experimental waveforms when the third-harmonic injection circuit is enabled. Due to the injection of third-harmonic current, the input current is sinusoidal and in phase with the input voltage without affecting the output performance of the inverter, as can be seen from the output current in channel 4. The measured THD of input current is 2.53%, and the input power factor is 0.997. The above comparative experiments validate

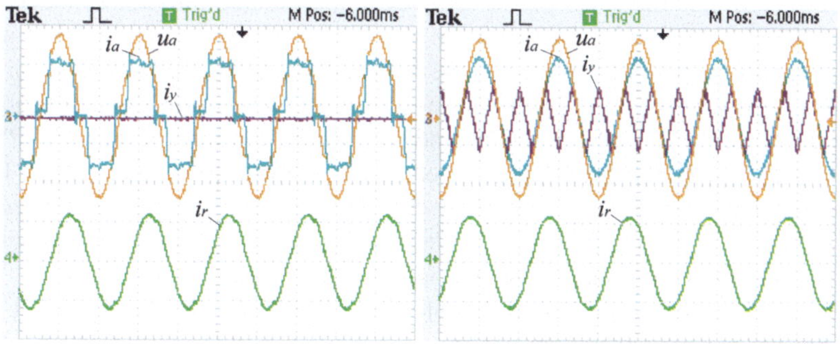

(a) (b)

Fig. 3.20 Experimental results of3HIMC. (**a**) Without and (**b**) with the active third-harmonic circuit enabled. CH1 is the input voltage u_a (50 V/div), CH2 is the input current i_a (3.36 A/div), CH3 is the third-harmonic injection current i_y (3.36 A/div), CH4 is the output current i_r (5 A/div), and time scale is 10 ms/div

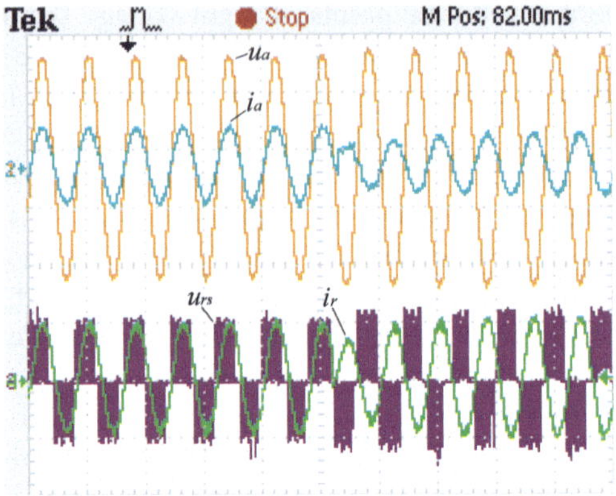

Fig. 3.21 Experimental waveforms of positive and negative direction power flows in3HIMC. CH1 is the input voltage u_a (50 V/div), CH2 is the input current i_a (5 A/div), CH3 is the output line-line voltage u_{rs} (150 V/div), CH4 is the output current i_r (5 A/div), and time scale is 25 ms/div.

the correctness and feasibility of the proposed 3HIMC topology and modulation strategy.

Figure 3.21 presents the bidirectional power flow capability of the proposed 3HIMC. At the beginning, the reference active current i_{d_ref} is 5 A, and then it is changed to -5A at 0.13s. In both conditions the reference reactive current i_{q_ref} is 0. It can be seen that the actual current tracks the reference well, and the THD of the output current measured by power quality analyzer is 3.69%. The input current is in phase with the input voltage when the reference active current is positive. Under negative reference active current, the input current is in antiphase with the input voltage and the power flows reversely. The experimental results verify the bidirectional energy flow characteristics of 3HIMC.

Figure 3.22 is the experimental waveforms with different modulation indices m_v under TSPWM and DPWM. The reference active current and reactive current are set as $i_{d_ref} = 5$A and $i_{q\_ref} = 0$. Figure 3.22a and c are the waveforms under TSPWM, and the RMS phase voltages of the isolated gird are 21V and 42V, corresponding to 0.4 and 0.8 modulation indices, respectively. Figure 3.22b and d are the waveforms under DPWM, and the corresponding modulation indices m_v are 0.4 and 0.8, respectively. The THDs of the output current measured by power quality analyzer are 4.27% and 3.51% for 0.4 and 0.8 modulation indices under TSPWM, while under DPWM, the THDs are reduced to 3.57% and 2.86%, respectively. It can also be seen that the voltage ripple of the capacitor in the input filter is reduced due to the reduction of the input current ripple, which is beneficial for selecting smaller size passive components such as capacitor and inductor of the input filter.

Figure 3.23 is the experimental result about THDs of the input current with different output power factor angles φ and modulation index under TSPWM and DPWM. As can be seen from the chart, the THD of the input current under DPWM is always lower than that under TSPWM in the whole output power factor angle range. Therefore, it is safe to say that the proposed DPWM shows a better performance from the point view of input waveforms.

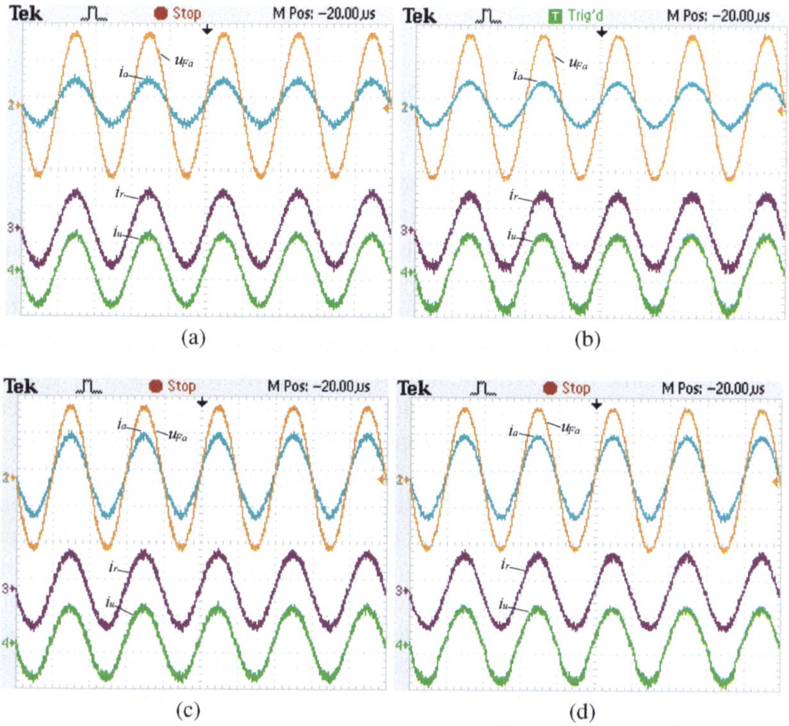

Fig. 3.22 Experimental comparisons of waveforms in 3HIMC between TSPWM and DPWM under different modulation indices (under unity output power factor). (**a**) TSPWM and modulation index m_v is 0.4. (**b**) DPWM and modulation index m_v is 0.4. (**c**) TSPWM and modulation index m_v is 0.8. (**d**) DPWM and modulation index m_v is 0.8. CH1 is the voltage across the capacitor of the input filter u_{Fa} (50 V/div), CH2 is the input current i_a (5 A/div), CH3 is the output current i_r (5 A/div), CH4 is the output current i_u (5A/div), and time scale is 10 ms/div

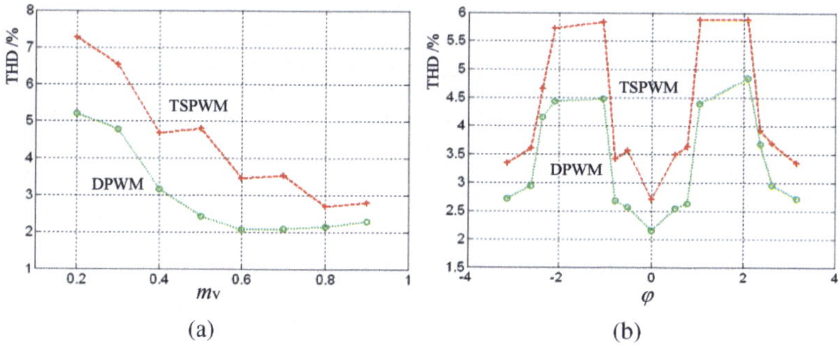

Fig. 3.23 Comparisons of THDs of the input current in 3HIMC between TSPWM and DPWM. (**a**) Different modulation index and a fixed output power factor angle $\varphi = 0$. (**b**) Different output power factor angle and a fixed modulation index $m_v = 0.8$

3.6 Conclusions

This chapter presents a 3HIMC topology for multiple drive applications and a DPWM strategy with reduced input current ripple. The proposed 3HIMC topology has such merits as sinusoidal input and output currents, bidirectional power flow, simplicity in commutation, and all-silicon solution, the same as the conventional IMC. Also, the multiple inverters in the proposed topology can be controlled independently, and each inverter can achieve a 0.866 voltage transfer ratio simultaneously. The proposed DPWM strategy is easily implemented by using only one triangular carrier signal to generate the gate signals for all the switches in 3HIMC. At the same time, it is helpful to reduce the input current ripple and avoid narrow pulse width problems. What is more, it presents a 3HIMC topology with dual three-phase outputs. Besides, a control algorithm for solving the third-harmonic injection reference current is developed. The proposed 3HIMC topology has the merits of sinusoidal input and output currents, bidirectional power flow, controllable input power factor, flexible implementation, and can provide two sets of independent three-phase outputs. Compared with the conventional IMC, the input reactive power control range is extended significantly, and a maximal voltage transfer ratio of 0.866 is achievable under all operation conditions for both outputs. Additionally, the scalability of the system is enhanced as the synchronization between the rectifier stage and the inverter stage is not required. All of the features above make 3HIMC very suitable for the applications that require multiple-outputs such as diesel-electric traction drives, MEA, marine ship propulsion, steel processing, and so on. Simulation and experiments verified the correctness of the proposed topology.

References

1. R. Bojoi, F. Farina, G. Griva, F. Profumo, A. Tenconi, Direct torque control for dual three-phase induction motor drives. IEEE Trans. Ind. Appl. **41**(6), 1627–1636 (2005)
2. T.M. Jahns, V. Blasko, Recent advances in power electronics technology for industrial and traction machine drives. Proc. IEEE **89**(6), 963–975 (2001)
3. A. Griffo, J. Wang, Large signal stability analysis of 'more electric' aircraft power systems with constant power loads. IEEE Trans. Aerosp. Electron. Syst. **48**(1), 477–489 (2012)
4. J.M. Apsley, A. Gonzalez-Villasenor, M. Barnes, A.C. Smith, S. Williamson, J.D. Schuddebeurs, P.J. Norman, C.D. Booth, G.M. Burt, J.R. McDonald, Propulsion drive models for full electric marine propulsion systems. IEEE Trans. Ind. Appl. **45**(2), 676–684 (2009)
5. S. Kuznetsov, Machine design and configuration of a 7000 HP hybrid electric drive for naval ship propulsion, in *Proceedings of the 2011 Electric Machines & Drives Conference (IEMDC)*, 2011 IEEE International (2011), pp. 1625–1628
6. S.H. Jeon, J.M. Kim, K.C. Jung, S.K. Sul, J.Y. Choi, Decoupling control of bridle rolls for steel mill drive system. IEEE Trans. Ind. Appl. **35**, 119–125 (1999)
7. C.B. Jacobina, E.C. dos Santos, Jr., E.R.C. da Silva, M.B. de Correa, A.M. N. Lima, T.M. Oliveira, Reduced switch count multiple three-phase ac machine drive systems. IEEE Trans. Power Electron. **23**(2), 966–976 (2008)

8. E. Ledezma, B. McGrath, A. Munoz, T.A. Lipo, Dual AC-drive system with a reduced switch count. IEEE Trans. Ind. Appl. **37**(5), 1325–1333 (2001)
9. M. Jones, D. Dujic, E. Levi, A. Batako, O. Mgaloblishvili, A novel five-leg inverter PWM technique for two-motor centre-driven winders, in *Proceedings of the IEEE Electric Machines and Drives Conference, 2007*, vol. 1 (2007), pp. 254–259
10. D. Dujic, M. Jones, S.N. Vukosavic, E. Levi, A general PWM method for a $(2n+1)$-leg inverter supplying n three-phase machines. IEEE Trans. Ind. Electron. **56**(10), 4107–4118 (2009)
11. J. Holtz, U. Boelkens, Direct frequency convertor with sinusoidal line currents for speed-variable AC motors. IEEE Trans. Ind. Electron. **36**(4), 475–479 (1989)
12. K. Shinohara, Y. Minari, T. Irisa, Analysis and fundamental characteristics of induction motor driven by voltage source inverter without dc link components. IEEJ Trans. Ind. Appl. **109**(6), 637–644 (1989)
13. L. Wei, T.A. Lipo, A novel matrix converter topology with simple commutation, in *IEEE Transactions on Industry Applications Society Annual Meeting*, vol. 3 (2001), pp.1749–1754
14. J.W. Kolar, M. Baumann, F. Schafmeister, H. Ertl, Novel three-phase AC-DC-AC sparse matrix converter. Proc. APEC **2**, 777–791 (2002)
15. J.W. Kolar, T. Friedli, J. Rodriguez, P.W. Wheeler, Review of three-phase PWM AC–AC converter topologies. IEEE Trans. Ind. Electron. **58**(11), 4988–5006 (2011)
16. R. Pena, R. Cardenas, E. Reyes, J. Clare, P. Wheeler, A topology for multiple generation system with doubly fed induction machines and indirect matrix converter. IEEE Trans. Ind. Electron. **56**(10), 4181–4193 (2009)
17. X. Liu, P. Wang, P.C. Loh, F. Blaabjerg, A compact three-phase single-input/dual-output matrix converter. IEEE Trans. Ind. Electron. **59**(1), 6–16 (2012)
18. T.D. Nguyen, H.-H. Lee, Dual three-phase indirect matrix converter with carrier-based PWM method. IEEE Trans. Power Electron. **29**(2), 569–581 (2014)
19. C. Klumpner, F. Blaabjerg, Modulation method for a multiple drive system based on a two-stage direct power conversion topology with reduced input current ripple. IEEE Trans. Power Electron. **20**(4), 922–929 (2005)
20. J.S. Siva Prasad, G. Narayanan, Minimization of grid current distortion in parallel-connected converters through carrier interleaving. IEEE Trans. Ind. Electron. **61**(1), 76–91 (2014)
21. D. Zhang, F. Wang, R. Burgos, R. Lai, D. Boroyevich, DC-link ripple current reduction for paralleled three-phase voltage-source converters with interleaving. IEEE Trans. Power Electron. **26**(6), 1741–1753 (2011)
22. D. Zhang, F. Wang, R. Burgos, R. Lai, D. Boroyevich, Interleaving impact on ac passive components of paralleled three-phase voltage-source converters. IEEE Trans. Ind. Appl. **46**(3), 1042–1054 (2010)
23. Kolar, J., Schafmeister, F., Round, S., et al., Novel three-phase ac-ac sparse matrix converters. IEEE Trans. Power Electron. **22**(5), 1649–1661 (2007)
24. M. Jantsch, C.W.G. Verhoeve, Inverters with three-phase output and without electrolyte capacitor for improved lifetime, efficiency and costs of grid connected systems, in *Proceedings of the 14th European Photovoltaic Solar Energy Conference Exhibition* (1997), pp.1–3
25. D.G. Holmes, T.A. Lipo, *Pulse Width Modulation for Power Converters—Principle and Practice* (Wiley-IEEE Press, New York, 2003), p. 219

Chapter 4
Three-Level T-Type Third-Harmonic Injection Matrix Converter

4.1 Introduction

A matrix converter (MC) is an "all-semiconductor" direct AC-AC power converter, which is featured by the advantages such as bidirectional power flow, sinusoidal input and output currents, controllable input power factor as well as high power density. Due to the merits described above, MCs have attracted an increasing attention in recent years [1–7].

As an improvement of the output waveform quality, the multilevel converter concept has been applied to MCs [6, 8–18]. With the ability to synthesize multilevel output voltages, the multilevel MCs could achieve better output power quality, reduced voltage stress on the power switches, and lower du/dt stresses when compared with the conventional two-level MCs. Generally, the multilevel MCs could be classified as three categories: the multimodular MCs [8–10], the capacitor clamped MCs [11–13], and the NPC MCs (also known as the multilevel IMCs) [6, 14–18]. The multimodular MC is formed by cascaded three-phase to single-phase MC modules, and it has the advantages of good output power quality and flexible expansibility, but a bulky multi-winding transformer is necessary. The capacitor clamped MC utilizes flying capacitors to provide the middle voltage levels so as to produce multilevel output voltages. A drawback of the capacitor clamped MC is the need of excessive numbers of capacitors and complicated control methods to balance the flying capacitor voltages.

As for the NPC MC, it is derived from the IMC. Thus, except the ability to generate multilevel output voltages and the possibility of achieving higher conversion efficiency, the NPC MCs also inherit the advantages of IMCs such as simple commutation mechanism and clamp circuit. These advantages make NPC MCs an attractive choice in some applications. In [6], a 3MC topology has been proposed, and the operating principles and SVM scheme of 3MC have been investigated in [14]. However, the developed nearest three virtual SVM aiming at maintaining zero averaged neutral-point current has the drawbacks of relatively high

H. Wang, *Advanced Matrix Converters*, Power Systems,
https://doi.org/10.1007/978-3-031-69902-3_4

computational burden and nonzero-current commutation of the rectifier. In [15], a carrier-based modulation scheme for 3MC has been presented, and the performances of 3MC have been verified experimentally. Another three-level IMC topology, referred to as the I3SMC, has been proposed and studied in [16, 17]. Compared with 3MC, the I3SMC topology has the advantage of a reduced number of switches, but at the cost of a slight degradation of the output performance due to the absence of the medium voltage vectors when synthesizing the output voltages. In [18], a three-level diode-clamped matrix converter has been suggested, which consists of a cascaded-rectifier and a three-level diode-clamped inverter. Although the three-level diode-clamped matrix converter can be easily extended to a generalized multilevel IMC topology, the bulky multi-winding isolated transformer reduces the power density of the converter inevitably.

Besides advantages, multilevel IMCs derived from IMCs also inherit the drawbacks of the conventional IMCs: One is the limited ability to generate input reactive power. The other is the need for synchronization in modulation between the rectification stage and the inversion stage. The first one is more obvious when multilevel IMCs are applied in a WECS or a FACTS [19, 20]. The input reactive power of multilevel IMCs can be regulated to some extent by changing the input displacement angle, but at the cost of a reduced maximum voltage transfer ratio, which is similar to that of IMCs [21, 22]. For the second, the input and output power quality of multilevel IMCs may degrade [23], since the process of synthesizing the output voltage vector occurs in two time-varying subperiods of the rectification stage, and modulation of the inversion stage is difficult to realize when the desired input current vector is located in the sector boundaries.

On the other hand, similar to the conventional NPC converters, neutral-point voltage balancing is a key challenge to the normal operation of multilevel IMCs. The neutral-point voltage imbalance problem in multilevel IMC may become even worse than that of conventional NPC converters with bulky buffer capacitors, since the capacitance of the input capacitors providing the major energy storage is relatively small. For the 3MC and I3SMC topologies studied in [19–22], although natural-balancing of the neutral-point voltage under ideal conditions can be achieved through proper modulation strategies, the nonlinearities in the practical converter such as the nonidealities of the switches, the dead times effect and the transients, may cause the neutral-point potential drift and distort the output waveforms. For the three-level diode-clamped matrix converter presented in [23], the leakage inductance of the isolated transformer may disrupt the self-balance of the neutral-point voltage, since zero averaged neutral-point current cannot be guaranteed by the modulation scheme. As far as we know, the control algorithm for balancing the neutral-point voltage of multilevel IMCs has not been discussed.

In this chapter, a T^3HIMC topology is presented and investigated systematically, which is derived from the 3HIMC [24–26] shown in Fig. 4.1. In addition to possessing the inherent advantages of 3HIMC such as bidirectional power flow capability, sinusoidal inputs and outputs, extended input reactive power control range without lowing the maximum voltage transfer ratio and no need for synchronization between the rectification stage and inversion stage, T^3HIMC could provide an improved

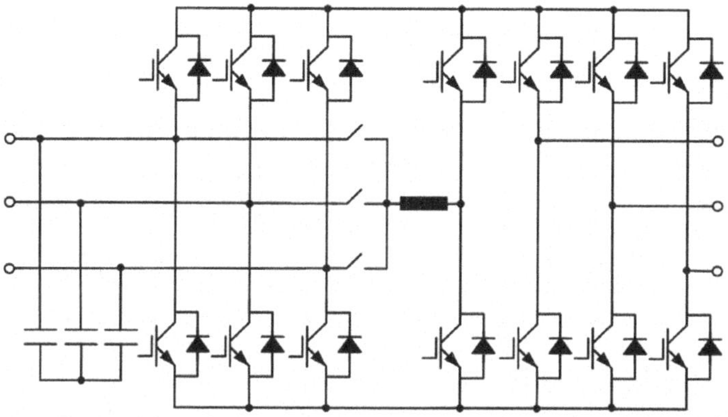

Fig. 4.1 3HIMC topology

output power quality in terms of output harmonic contents. The topology and operating principles of T^{3}HIMC are analyzed in detail. Based on the analysis of the operating principles, a simple carrier-based modulation method is introduced. Moreover, a closed-loop control algorithm for balancing the neutral-point voltage is developed, which overcomes the neutral-point potential drift issue caused by the nonidealities of the practical converter. Finally, the proposed topology and method are verified by simulation and experimental studies, and a comprehensive performance comparison with 3HIMC is carried out. This chapter is organized as follows: Sect. 4.2 introduces the topology and operating principles of T^{3}HIMC; Sect. 4.3 presents the carrier-based modulation scheme in detail; Sect. 4.4 analyzes the factors that influence the balance of the neutral-point voltage in practice, followed by the presentation of the control algorithm for balancing the neutral-point voltage; Sect. 4.5 shows the simulation and experimental results to verify the presented methods and gives a comprehensive performance comparison with 3HIMC; Sect. 4.6 draws the final conclusion of this chapter.

4.2 Topology and Operating Principles of T^{3}HIMC

4.2.1 Topology

The topology of T^{3}HIMC is shown in Fig. 4.2, and the main circuit consists of a bidirectional CSR, a three-level VSI, an active third-harmonic current injection circuit, and an input *LC* filter. Compared with 3HIMC shown in Fig. 4.1, the CSR of T^{3}HIMC is cascaded to a three-level T-type VSI instead of the conventional two-level VSI. The active third-harmonic current injection circuit is composed of three bidirectional switches, a third-harmonic injection inductor, and a bridge leg. The input filter consisting of inductor L_F and film capacitor C_F mainly has three

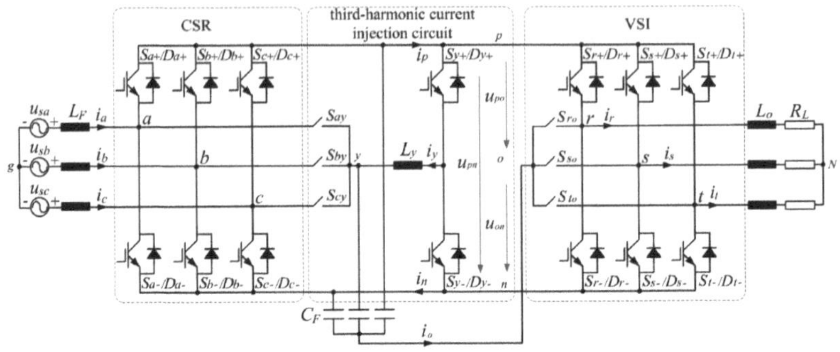

Fig. 4.2 Schematic diagram of T^{3}HIMC topology

functions. First, it is used for filtering the pulse currents generated by the converter so as to produce three-phase sinusoidal input currents. Second, the star point of the capacitors in the filter provides the neutral point of the rear-end three-level T-type inverter. Third, as with the clamp circuit of the conventional IMCs, the capacitors of the filter are used for absorbing the energy stored in the leakage inductance of the load and the third-harmonic injection inductor when the converter shutdown. It should be noted that although the three-level T-type inverter is used in the proposed topology shown in Fig. 4.2, other three-level NPC inverters such as the three-level diode-clamped inverter can also be adopted here for implementing T^{3}HIMC. This can be realized by reallocating the gating signals of the inversion stage, and therefore it is not elaborated here for clarity.

4.2.2 Operating Principles

Assume the three-phase input voltages are symmetrical and sinusoidal, and the operating principles of T^{3}HIMC are described as follows: For the CSR, only the switch in the upper bridge connected to the input phase with the maximum instantaneous voltage and the switch in the lower bridge connected to the input phase with the minimum instantaneous voltage are turned on so as to impose two of the three phase-to-neutral voltages (labeled as u_{aO}, u_{bO}, and u_{cO}) across the intermediate DC-link. For the third-harmonic current injection circuit, switches S_{y+} and S_{y-} are controlled to generate the desired quasi-third-harmonic current i_y flowing through the inductor L_y; and the bidirectional switch that connected to the input phase with the minimum absolute voltage is turned on to inject the third-harmonic current i_y into the corresponding input phase. Table 4.1 shows the switching states of the CSR and the bidirectional switches of the third-harmonic current injection circuit, where θ_{sa} is the phase of the input phase voltage u_{sa}. For example, when the input voltages satisfy $u_{sa} > u_{sb} > u_{sc}$ (denoted as sector 1), switches S_{a+} and S_{c-} of the rectifier and the bidirectional switch S_{by}

Table 4.1 Switching states of the rectifier and the third-harmonic current injection circuit

θ_{sa}	Sector	S_{ay}	S_{by}	S_{cy}	S_{a+}	S_{a-}	S_{b+}	S_{b-}	S_{c+}	S_{c-}
0-π/3	1	0	1	0	1	0	0	0	0	1
π/3-2π/3	2	1	0	0	0	0	1	0	0	1
2π/3-π	3	0	0	1	0	1	1	0	0	0
π-4π/3	4	0	1	0	0	1	0	0	1	0
4π/3-5π/3	5	1	0	0	0	0	0	1	1	0
5π/3-2π	6	0	0	1	1	0	0	1	0	0

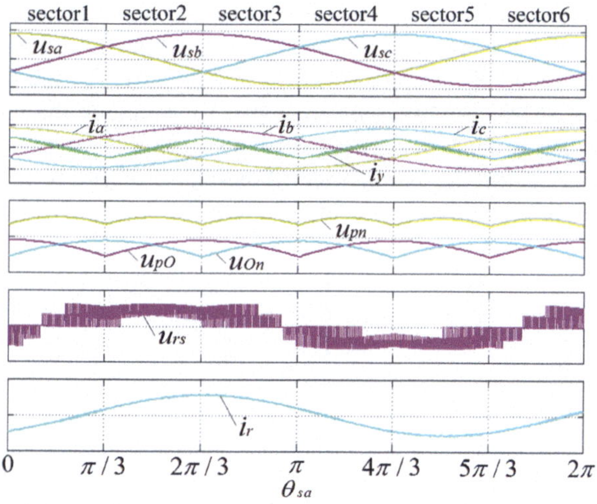

Fig. 4.3 Key waveforms of T^{3}HIMC

of the third-harmonic current injection circuit are turned on, node a connects to the positive terminal p, and node c connects to the negative terminal n of the DC-link. Thus the upper DC source voltage u_{pO} and the lower DC source voltage u_{On} are represented by the phase-to-neutral voltage u_{aO} and -u_{cO}, respectively, and so forth. Consequently, as with the output voltage of a diode rectifier, the DC-link voltage u_{pn} of T^{3}HIMC exhibits a piecewise six-pulse shape waveform. According to the requirements of the load, the three-level T-type inverter provides three-phase three-level output voltages with variable frequency and amplitude. In this manner, sinusoidal three-phase input/output currents and controllable input power factor are attainable. Figure 4.3 shows the key waveforms of T^{3}HIMC, where u_{rs} is the output line-line voltage and i_r is the output current. The validity of sinusoidal currents and controllable power factor at the input side has been presented in detail in [25] and therefore is not elaborated here.

As can be seen from Fig. 4.3, different from the conventional T-type VSI, the split DC source of the T-type VSI in T^{3}HIMC is no longer to have balanced upper and lower DC voltages. Instead, the upper and lower DC voltages in each sector are now represented by two different input phase-to-neutral voltages. Consequently, as will be explained in the next section, the modulation strategies of the traditional T-type VSI cannot be directly used here, and proper modifications should be made for achieving symmetrical and sinusoidal input and output waveforms.

4.3 Carrier-Based PWM Modulation

Similar to 3HIMC, the modulation strategies of the developed T^{3}HIMC can also be divided into two independent parts. Since the switching states of the rectifier and the bidirectional switches of the third-harmonic current injection circuit are determined only by the input voltages, they commutate at line frequency, and the modulation strategies of the rectifier and the third-harmonic current injection circuit are relatively simple, as shown in Table 4.1. For the rear-end T-type VSI, a carrier-based DSPWM [27] is developed for reducing the computational burden, and the concrete analysis is given as follows.

Assume the symmetrical three-phase output reference voltages, also known as the modulation signals, are given by

$$\begin{cases} u_r^* = U_{om}\cos(\omega_o t + \phi) \\ u_s^* = U_{om}\cos(\omega_o t - 2\pi/3 + \phi) \\ u_t^* = U_{om}\cos(\omega_o t + 2\pi/3 + \phi) \end{cases} \tag{4.1}$$

where u_r^*, u_s^*, and u_t^* are the three modulation signals, and U_{om}, ω_o, and ϕ are the magnitude, angular frequency, and initial phase of the expected output voltages, respectively.

First, sort the original modulation signals according to the instantaneous values of the expected output voltages as follows:

$$\begin{cases} u_{max}^* = \max(u_r^*, u_s^*, u_t^*) \\ u_{mid}^* = \text{mid}(u_r^*, u_s^*, u_t^*) \\ u_{min}^* = \min(u_r^*, u_s^*, u_t^*), \end{cases} \tag{4.2}$$

where u_{max}^*, u_{mid}^*, and u_{min}^* represent the maximum value, medium value, and minimum value among the modulation signals; max(), mid(), and min() are the operators of the maximum value, medium value, and minimum value, respectively.

To achieve the maximum range of linear voltage transfer ratio for T^{3}HIMC, each original modulation signal is modified by adding a common zero-sequence voltage:

$$\begin{cases} u_{max}^{*'} = u_{max}^* + u_0 \\ u_{mid}^{*'} = u_{mid}^* + u_0 \\ u_{min}^{*'} = u_{min}^* + u_0 \end{cases} \tag{4.3}$$

where u_0 is the zero-sequence voltage, and $u_{max}^{*'}$, $u_{mid}^{*'}$, and $u_{min}^{*'}$ are the modified modulation signals.

According to DSPWM, the modified modulation signals are then decomposed into two modulation signals for achieving zero averaged neutral-point current. Denote u^*_{maxp}, u^*_{midp}, and u^*_{minp} as the positive modulation signals corresponding to the maximum, medium, and minimum original modulation signals, respectively; u^*_{maxn}, u^*_{midn}, and u^*_{minn} are the negative modulation signals corresponding to the maximum, medium, and minimum original modulation signals, respectively, and then the modified modulation signals are

$$\begin{cases} u^{*'}_{max} = u^*_{maxp} + u^*_{maxn} \\ u^{*'}_{mid} = u^*_{midp} + u^*_{midn} \\ u^{*'}_{minn} = u^*_{minp} + u^*_{minn} \end{cases} \tag{4.4}$$

$$\begin{cases} u^*_{maxp} = u^*_{max} + u_0, \\ u^*_{maxn} = 0 \\ u^*_{midp} = 0.5u^*_{mid} + u_{01}, \\ u^*_{midn} = 0.5u^*_{mid} + u_{02} \\ u^*_{minp} = 0, \\ u^*_{minn} = u^*_{min} + u_0 \end{cases} \tag{4.5}$$

where u_{01} and u_{02} are the zero-sequence voltages of the positive and negative modulation signals corresponding to the medium original modulation signal, respectively.

For the convenience of digital implementation, the positive and negative modulation signals are normalized according to the upper and lower DC voltages, respectively. The normalized modulation signals and the duty ratios are determined by

$$\begin{cases} \overline{u}^*_{ip} = u^*_{ip}/u_{pO} \\ \overline{u}^*_{in} = u^*_{in}/u_{On} \end{cases}, \quad i \in \{max, mid, min\} \tag{4.6}$$

$$\begin{cases} d_{ip} = \overline{u}^*_{ip} \\ d_{in} = -\overline{u}^*_{in} \end{cases}, \quad i \in \{max, mid, min\} \tag{4.7}$$

where $\overline{u}^*_{ip}$ are the normalized positive modulation signals, $\overline{u}^*_{in}$ are the normalized negative modulation signals, d_{ip} represent the duty ratios connected to the positive terminal of the DC-link, and d_{in} denote the duty ratios connected to the negative terminal of the DC-link.

To eliminate the low-frequency oscillations of the neutral-point voltage, the averaged neutral-point current in each switching period (denoted as $\overline{i}_O$) must be kept at zero. Thus, the zero-sequence voltages to ensure zero averaged neutral-point current could be solved from the following equations:

$$\begin{cases} \bar{i}_O = (1 - d_{maxp})\, i_{max} + (1 - d_{midp} - d_{midn})\, i_{mid} + (1 - d_{minn})\, i_{min} = 0 \\ i_{max} + i_{mid} + i_{min} \;\; = 0 \\ u^*_{max} + u^*_{mid} + u^*_{min} \;\; = 0 \\ u_0 = u_{01} + u_{02} \end{cases} \tag{4.8}$$

with the constraints

$$\begin{cases} 0 \le u^*_{maxp} \le u_{pO} \\ 0 \le u^*_{midp} \le u_{pO} \\ -u_{On} \le u^*_{midn} \le 0 \\ -u_{On} \le u^*_{minn} \le 0 \end{cases} \tag{4.9}$$

where i_{max}, i_{mid}, and i_{min} represent the output currents corresponding to the phase having the maximum, medium, and minimum original modulation signals, respectively.

One set of feasible choice of the zero-sequence voltages is given by

$$\begin{cases} u_{01} = \dfrac{-0.5u^*_{mid}\,(u_{On} - u_{pO}) - u_{pO}\, u^*_{min}}{u_{pn}} \\ u_{02} = \dfrac{0.5u^*_{mid}\,(u_{On} - u_{pO}) - u_{On}\, u^*_{max}}{u_{pn}} \\ u_0 = \dfrac{-u_{pO}\, u^*_{min} - u_{On}\, u^*_{max}}{u_{pn}} \end{cases} \tag{4.10}$$

Substitute (4.10) into (4.5), and the unified expression of the normalized modulation signals is given by

$$\begin{cases} \overline{u}^*_{ip} = (u^*_i - u^*_{min})/u_{pn} \\ \overline{u}^*_{in} = (u^*_i - u^*_{max})/u_{pn} \end{cases}, \; i \in \{max, mid, min\} \tag{4.11}$$

Under the condition of $u^*_r = u^*_{max}$, $u^*_s = u^*_{mid}$, $u^*_t = u^*_{min}$, Fig. 4.4 shows the schematic diagram and switching pattern of the modulation scheme for the inverter of T[3]HIMC, where T_s is the switching period, f_s is the switching frequency, $\overline{u}^*_{rp}$, $\overline{u}^*_{sp}$, and $\overline{u}^*_{tp}$ are the normalized positive modulation signals corresponding to the output phase r, s, and t, respectively, and $\overline{u}^*_{rn}$, $\overline{u}^*_{sn}$, and $\overline{u}^*_{tn}$ are the normalized negative modulation signals corresponding to the output phase r, s, and t, respectively.

The voltage transfer characteristic is always the main concerns for IMCs. As a member of the IMCs family, the voltage transfer characteristic of T[3]HIMC is deduced as follows.

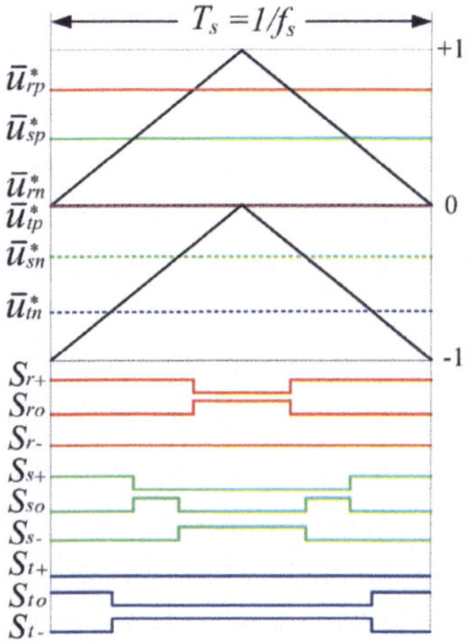

Fig. 4.4 Schematic diagram and switching pattern of the modulation scheme for the inverter of T³HIMC

Assume that the input voltages are given by

$$\begin{cases} u_{sa} &= U_{im}\cos(\theta_{sa}) \\ u_{sb} &= U_{im}\cos(\theta_{sa} - 2\pi/3) \\ u_{sc} &= U_{im}\cos(\theta_{sa} + 2\pi/3) \end{cases} \tag{4.12}$$

where U_{im} is the magnitude of the input voltages. Ignoring the effects of the input filter, the DC-link voltage u_{pn} can be written as

$$u_{pn} = \sqrt{3}U_{im}\cos[\text{rem}(\theta_{sa}, \frac{\pi}{3}) - \frac{\pi}{6}] \tag{4.13}$$

where rem() is the operator of the remainder value.

In linear modulation regions, the following conditions must be satisfied

$$\begin{cases} 0 \le \overline{u}^*_{ip} \le 1 \\ -1 \le \overline{u}^*_{in} \le 0 \end{cases}, \quad i \in \{max, mid, min\} \tag{4.14}$$

Combine (4.1), (4.2), (4.4), (4.11), (4.13), and (4.14), and the linear voltage transfer ratio of T³HIMC could be derived as

$$q = \frac{U_{om}}{U_{im}} = \frac{\sqrt{3}}{2}m_i \le \frac{\sqrt{3}}{2} \tag{4.15}$$

where q is the linear voltage transfer ratio and m_i is the modulation index with the value range from 0 to 1. It can be found from (4.15) that the maximum linear voltage transfer ratio of T³HIMC under all operation conditions is limited to 0.866, which is the same as that of 3HIMC.

4.4 Control Algorithm for Balancing Neutral-Point Potential

From the analysis in Sect. 4.3, it can be found that the developed modulation scheme for the T-type inverter possesses the feature of natural-balancing, since the average value of the neutral-point current in a switching period is zero under ideal conditions, as shown in (4.8). However, the nonlinearities of the practical converter such as the nonidealities of the devices, dead times effect, transients, etc., may cause the neutral-point potential drift and distort the output waveforms. Compared with the conventional NPC VSI with bulky buffer capacitors, the capacitance of the filtering capacitor of T^{3}HIMC is relatively small (usually in the range of several microfarads to tens of microfarads), and thus T^{3}HIMC is more susceptible to the disturbance of the neutral-point current. Consequently, as with the three-level VSIs with small DC-link capacitors [28], a closed-loop control algorithm for balancing the neutral-point voltage is essential for the normal operation of T^{3}HIMC.

4.4.1 Dynamic Modeling

Before designing a closed-loop controller for balancing the neutral-point voltage, the dynamic model of the neutral-point voltage should be obtained first. Using the circuit shown in Fig. 4.2, the mathematical model of the input filter is given by

$$\begin{cases} u_{sa} &= L_F \dfrac{di_a}{dt} + u_{aO} + u_{Og} \\ u_{sb} &= L_F \dfrac{di_b}{dt} + u_{bO} + u_{Og} \\ u_{sc} &= L_F \dfrac{di_c}{dt} + u_{cO} + u_{Og} \\ i_O &= C_F \dfrac{du_{aO}}{dt} + C_F \dfrac{du_{bO}}{dt} + C_F \dfrac{du_{cO}}{dt} \end{cases} \tag{4.16}$$

where u_{aO}, u_{bO}, and u_{cO} are the voltages imposed across the filtering capacitors, and u_{Og} is the neutral-point voltage. For the symmetrical and sinusoidal three-phase system, the average state space model can be derived as

$$\frac{d\overline{u}_{Og}}{dt} = -\frac{\overline{i}_O}{3C_F} \tag{4.17}$$

where $\overline{u}_{Og}$ is the average value of the neutral-point voltage in a switching period.

It can be found from (4.17) that the neutral-point voltage variation is a function of the averaged neutral-point current. Thus the neutral-point voltage can be controlled by changing the averaged neutral-point current. It shows in the first formula of Eq. (4.8) that the averaged neutral-point current can be regulated by changing the modulation signals, which in turn controls the neutral-point voltage. However, there are two criteria when modifying the modulation signals: (1) do not increase the switching frequency of the converter and (2) keep the expected output line-line voltages unchanged, which can be achieved by modifying the zero-sequence voltages of the modulation signals according to the discussion in Sect. 4.3.

By introducing a new zero-sequence voltage u_{off}, the positive and negative modulation signals in (4.5) are modified as

$$\begin{cases} u_{maxp}^{*'} = u_{maxp}^{*} + u_{off} \\ u_{maxn}^{*'} = 0 \\ u_{midp}^{*'} = u_{midp}^{*} + 0.5u_{off} \\ u_{midn}^{*'} = u_{midn}^{*} + 0.5u_{off} \\ u_{minp}^{*'} = 0 \\ u_{minn}^{*'} = u_{minn}^{*} + u_{off} \end{cases} \tag{4.18}$$

where the symbol "′" represents the modified value. It can be verified easily from (4.18) that the expected output line-line voltages remain unchanged after modifying the modulation signals. Besides, (4.14) must be guaranteed to avoid distortions in the output voltages, which leads to the constraint condition of the new zero-sequence voltage u_{off} as follows:

$$\begin{cases} u_{off\_min} \le u_{off} \le u_{off\_max} \\ u_{off\_min} = \max(L_1, L_2, L_3, L_4) \\ u_{off\_max} = \min(U_1, U_2, U_3, U_4) \\ L_1 = -u_{pO}(u_{max}^{*} - u_{min}^{*})/u_{pn} \\ L_2 = -2u_{pO}(u_{mid}^{*} - u_{min}^{*})/u_{pn} \\ L_3 = -2u_{On}(u_{pn} - u_{max}^{*} + u_{mid}^{*})/u_{pn} \\ L_4 = -u_{On}(u_{pn} - u_{max}^{*} + u_{min}^{*})/u_{pn} \\ U_1 = u_{pO}(u_{pn} - u_{max}^{*} + u_{min}^{*})/u_{pn} \\ U_2 = 2u_{pO}(u_{pn} - u_{mid}^{*} + u_{min}^{*})/u_{pn} \\ U_3 = 2u_{On}(u_{max}^{*} - u_{mid}^{*})/u_{pn} \\ U_4 = u_{On}(u_{max}^{*} - u_{min}^{*})/u_{pn} \end{cases} \tag{4.19}$$

where u_{off_min} and u_{off_max} denote the lower and upper limits of the new zero-sequence voltage, respectively.

On the basis of (4.6), (4.7), (4.8), (4.17), and (4.18), the model of the neutral-point voltage is deduced as

$$\frac{d\bar{u}_{Og}}{dt} = \frac{u_{pn}(i_{max} - i_{min})u_{off}}{3C_F u_{pO} u_{On}} \approx \frac{0.84\,(i_{max} - i_{min})u_{off}}{C_F U_{im}} \tag{4.20}$$

Obviously, the model described by (4.20) is a nonlinear system since a time-varying variable (i_{max}-i_{min}) is included. Thus a linearized model should be derived before the linear control theory is applied for designing the controller to balance neutral-point voltage conveniently. Considering the term (i_{max}-i_{min}) can be obtained by measurement, a new linearized model is constructed as

$$\begin{cases} \dfrac{d\bar{u}_{Og}}{dt} = \dfrac{0.84}{C_F U_{im}} u'_{off} \\ \\ u'_{off} = (i_{max} - i_{min}) u_{off} \end{cases} \tag{4.21}$$

where u'_{off} is the modified zero-sequence voltage. Equation (4.21) is treated as the plant model of the neutral-point voltage with the transfer function of the plant given by

$$G_p(s) = \frac{\bar{u}_{Og}(s)}{u'_{off}(s)} = \left(\frac{0.84}{C_F U_{im}}\right) \frac{1}{s} \tag{4.22}$$

4.4.2 Controller Design

It can be seen that a well-designed neutral-point voltage controller is essential for T^3HIMC since the capacitance of the filtering capacitor is small. Based on the plant model described in (4.22), a PI controller given by (4.23) is used in this chapter

$$G_c(s) = K_p + \frac{K_i}{s} \tag{4.23}$$

where K_p is the proportional gain and K_i is the integration gain. And the open-loop transfer function of the control loop is derived as

$$G_o(s) = G_c(s)\, G_p(s) = \frac{0.84}{C_F U_{im}} \left(\frac{K_p s + K_i}{s^2}\right) \tag{4.24}$$

The block diagram of the controller is given in Fig. 4.5, where $D(s)$ is the disturbance in the control system and $G_d(s)$ is the transfer function of the disturbance to the neutral-point voltage.

The selection of K_p and K_i is mainly based on the following considerations: (1) zero steady-state error; and (2) the gain crossover frequency f_c should be selected properly considering both the requirements of stability and dynamic response. Taking the factors such as the switching frequency into consideration, the crossover frequency of the control loop is selected as $f_c = f_s/20 = 1\,k$Hz, and the phase margin is chosen as 60°. Therefore, the parameters of the PI controller are calculated as $K_p = 7.68$ and $K_i = 27851$. The Bode diagrams of the control loop are shown in Fig. 4.6.

It is worth noting that the presented modulation strategy and neutral-point voltage control algorithm are generalized and can be extended to other three-level IMCs

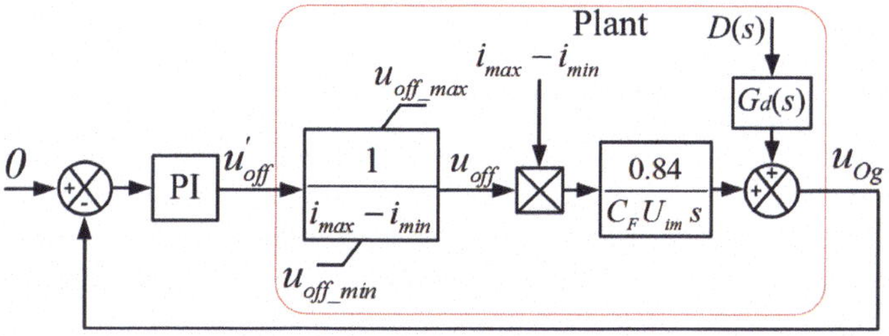

Fig. 4.5 Block diagram of the neutral-point voltage controller

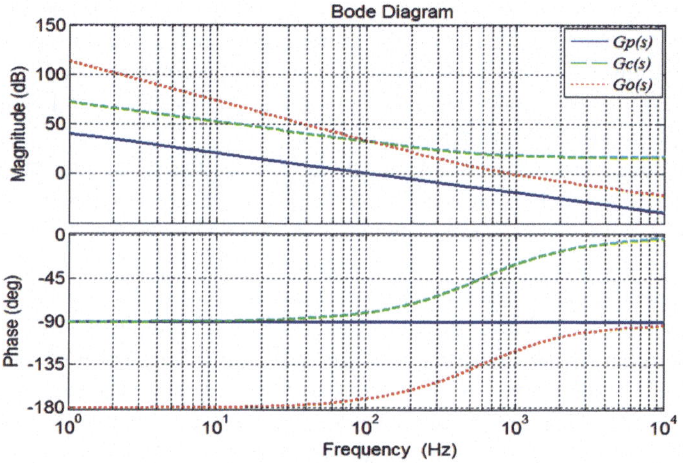

Fig. 4.6 Bode diagrams of the neutral-point voltage control loop

such as 3MC and three-level diode-clamped matrix converter, since the methods are independent of specific DC source voltages. These can be performed according to their respective DC source voltages and therefore are not elaborated here, which are beyond the scope covered by this chapter.

4.5 Simulation and Experimental Results

In this section, the system design of the T^3HIMC was presented first. Then to verify the correctness and feasibility of the proposed methods, the functionality and performance of T^3HIMC were first evaluated by simulation using Matlab/Simulink software and then were validated experimentally.

4.5.1 System Design

1. *Passive Components*
   Similar to the 3HIMC, proper design of the third-harmonic injection inductor is the key challenges to implement the converter [25]. The selection of the third-harmonic injection inductor should consider both the ripple current and current tracking performance. An inductor of large inductance is helpful to reduce the ripple current, but it makes the current response slow and vice versa. So a trade-off should be made between these two aspects. According to the design criteria in [25], the range of the inductance of the third-harmonic injection inductor is determined by many system parameters, such as the amplitude of the input voltage, the switching frequency, the current ripple index, the power rating of the converter, and so on. By considering the system parameters listed in Table 4.2, a 1.2 mH third-harmonic injection inductor is chosen, and the detailed design procedure is not elaborate here.
   The input LC filter is mainly used for filtering the pulse currents generated by the converter and providing the instantaneous energy storage for the rear-end inverter such that three-phase sinusoidal input currents and proper operation of the converter can be achieved. The lower limit of the capacitance of the capacitor C_F is designed such that the voltage ripple of the capacitor can be limited to a certain level in order to prevent a distortion of the output voltage and enable safe operation of the converter. Besides, the upper limit of the capacitance of the capacitor should be selected based on the required input power factor at rated power. By considering the system parameters such as the power rating, the input voltage, the switching frequency, and the desired input power factor, a 6.6 μF capacitor (2 parallel EPCOS MKP B32924 3.3 μF, 305V film capacitors) is chosen. After selecting the capacitor, the input filtering inductor L_F is normally determined by the cutoff frequency of the input filter:

$$L_F = \frac{1}{4\pi^2 f_{cutoff}^2 C_F} \tag{4.25}$$

Table 4.2 System specifications of T^{3}HIMC

Parameters	Value
Power rating	1.5kW
Input line-line voltage	220 Vrms
Input frequency f_n	50 Hz
Switching frequency f_s	20 kHz
Inductor L_F	300 μH
Capacitor C_F	6.6 μF
Inductor L_y	1.2 mH
Inductor L_o	3 mH
Resistor R_L	25 ohm

Generally, the cutoff frequency f_{cutoff} is designed as $f_{cutoff} = (0.1 \sim 0.2)\, f_s$. In this chapter, a 300 µH inductor is selected.

2. *Semiconductor Selection*

   The semiconductor switches are selected based on the voltage and current stresses. Considering a 220 Vrms input voltage, the 650V IGBT devices are chosen here. The detailed analysis of the voltage and current stresses of the two-level 3HIMC has been presented in [25]. Thus this section mainly focuses on the selection of the bidirectional switches.

   As the same as in conventional T-type three-level converters, the bidirectional switches S_{ay}, S_{by}, S_{cy}, S_{ro}, S_{so}, and S_{to} in the T^3HIMC topology need to conduct current in each direction and block voltage in both directions. In this chapter, the bidirectional switches were implemented by an anti-series connection of two reverse-conducting IGBTs with common emitter configuration. On the other hand, the switching behavior and current stress of the switches S_{ay}, S_{by}, S_{cy} and the switches S_{ro}, S_{so}, S_{to} are quite different. Thus these two kinds of bidirectional switches should be selected based on their own features and requirements. As can be known from the operating principles of the T^3HIMC topology, the switches S_{ay}, S_{by}, S_{cy} of the third-harmonic injection circuit commutate at line frequency and their current stresses are only half of the amplitude of the input current, and thus the IGBT IKW30N65EL5 (650V/30A, Infineon) with low saturation voltage is chosen for the switches S_{ay}, S_{by}, S_{cy}. While for the switches S_{ro}, S_{so}, S_{to} of the three-level inverter, they commutate at high-frequency and high-current stress, thus high speed IGBT IKW40N65ES5 (650V/40A, Infineon) with both low switching losses and low saturation voltage is selected.

4.5.2 Simulation Results

Figure 4.7 illustrates the capability of T^3HIMC producing three-phase multilevel outputs with controllable amplitude and frequency. The output parameters were set as $m_i = 0.5$, $f_o = 50$ Hz in Fig. 4.7a, and $m_i = 0.9$, $f_o = 30$ Hz in Fig. 4.7b, respectively, where f_o is the output frequency. The waveforms shown in Fig. 4.7 consist of the input phase voltage u_{sa}, the input current ia, the output line-line voltage u_{rs}, and the output current i_r . As can be seen from Fig. 4.7a, the input current is nearly sinusoidal and in phase with the input phase voltage, except for a slight phase advance caused by the capacitive current drawn by the filtering capacitor. Thus, it is clear that T^3HIMC is able to generate sinusoidal input current and unity power factor at the input side. Of equal interest are the output waveforms, where a three-level output line-line voltage and a sinusoidal output current are observed. It can be seen from the envelope of the output line-line voltage that three distinctive levels including the upper DC voltage u_{pO}, the lower DC voltage u_{On}, and the zero voltage are used for synthesizing the output line-line voltage. This can be explained by the fact that, from the SVM perspective, only small vectors, medium

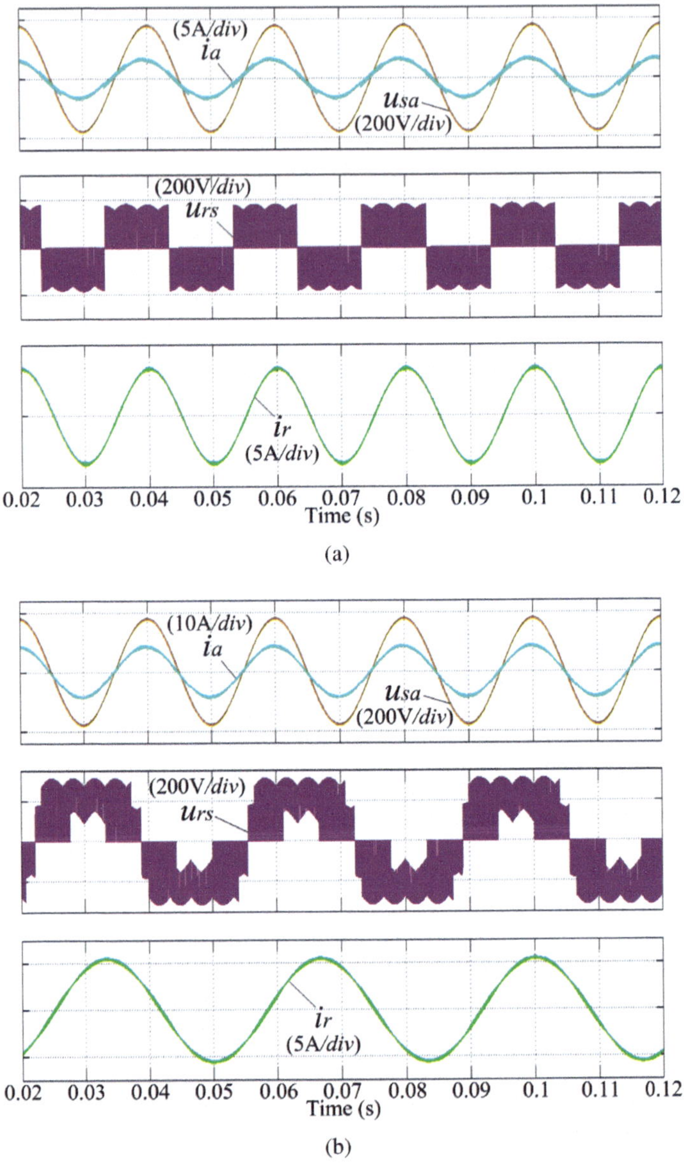

Fig. 4.7 Simulated waveforms of T^{3}HIMC with different m_i and f_o. (**a**) $m_i = 0.5$ and $f_o = 50$ Hz. (**b**) $m_i = 0.9$ and $f_o = 30$ Hz

vectors, and zero vectors are used for producing the expected voltage vector at low modulation index.

As shown in Fig. 4.7b, different from the three-level output line-line voltage in Fig. 4.7a, a noticeable five-level output line-line voltage is generated in T^{3}HIMC.

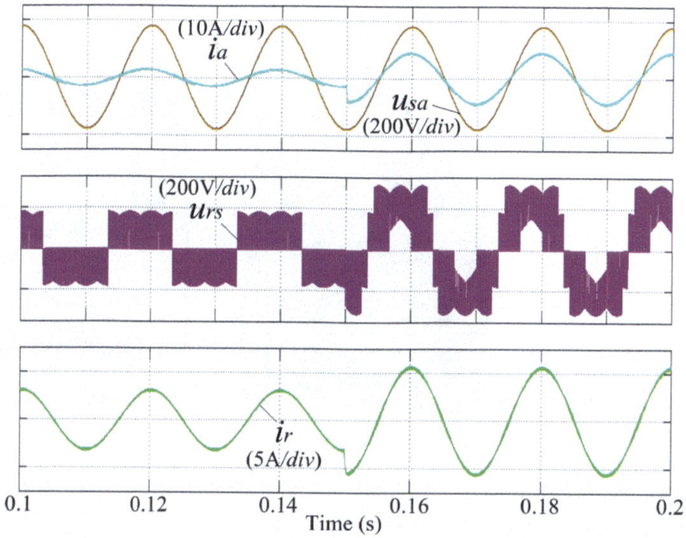

Fig. 4.8 Simulated waveforms of T³HIMC in dynamic condition

This is reasonable since five distinctive levels including the DC-link voltage u_{pn}, the upper DC voltage u_{pO}, the lower DC voltage u_{On}, and the zero voltage are utilized for synthesizing the output voltages at high modulation index. Again, the desired features of sinusoidal input and output currents and unity power factor at the input side are achieved. Thus the results shown in Fig. 4.7 verified the correctness of the T³HIMC topology and modulation strategies.

Figure 4.8 shows the dynamic test of the system. In Fig. 4.8, the output frequency f_o is set as 50Hz, and the modulation index m_i is 0.5 at first. At 0.15s, m_i is step from 0.5 to 0.9. As can be seen from Fig. 4.8, the input and output currents are still sinusoidal, and the dynamic performance is good in the presented T³HIMC system.

4.5.3 Experimental Results

To validate the theoretical analysis and simulation results, a laboratory prototype of T³HIMC with the specifications given in Table 4.2 is built based on a configurable main circuit, as shown in Fig. 4.9. The control platform of T³HIMC is implemented by a combination of floating-point DSP TMS320F28335 and FPGA EP2C8T144C8N.

Figure 4.10 demonstrates the experimental results of T³HIMC generating three-phase multilevel outputs with different amplitudes and frequencies, and Fig. 4.11 shows the experimental results of T³HIMC in dynamic condition. It is worth noting that the experimental results shown in Figs. 4.10 and 4.11 correspond to the

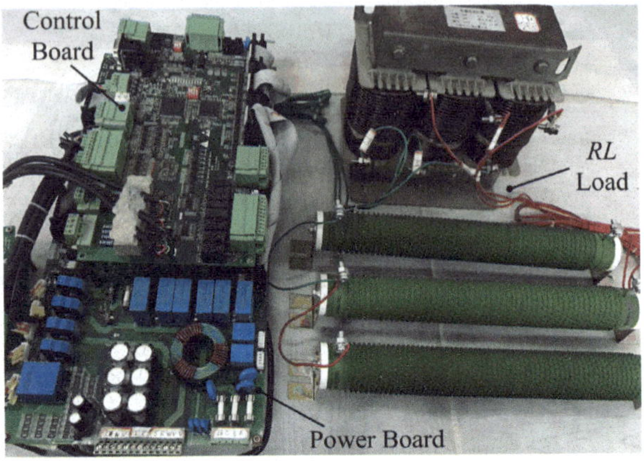

Fig. 4.9 Experimental setup of the T^3HIMC system

simulated results shown in Figs. 4.7 and 4.8, and the experimental conditions and commands are exactly the same as those in the simulation. As can be seen from Figs. 4.10 and 4.11, the experimental results match the simulated results very well, except for a slight reduction in the amplitudes and higher distortions of the input and output currents. The higher distortions of the input and output currents in the experiments are attributed to the nonidealities of the power source (distortions and unbalance), the nonidealities of the power devices, the dead time effects, the narrow switching pulses, the measurement error of the transducers, and the quantization error of the digital control system, etc. The input and output waveforms quality could be improved by optimizing the design of the prototype and the experimental conditions. Besides, it can be seen from Figs. 4.10 and 4.11 that the output voltage in the experiments contains some spikes. The spikes are mainly caused by the EMI noise picked up by the voltage probe and the screen persistence effect of the digital storage oscilloscope. Referring to Figs. 4.10 and 4.11, multilevel output line-line voltage and sinusoidal input and output currents are achieved in the prototype, and thus the validity of T^3HIMC is verified experimentally.

To verify the feasibility of the control algorithm for balancing the neutral-point voltage in the practical converter, the balancing capability under various operation conditions such as different modulation indices, different output frequencies, and different loads has been tested, and the results are shown in Fig. 4.12. The output parameters were set as follows: The modulation index m_i is set as 0.45 in Fig. 4.12a and 0.9 in Fig. 4.12b–d; the output frequency f_o is set as 40 Hz in Fig. 4.12a–b and 60 Hz in Fig. 4.12c–d; the load is a series-connected RL branch ($R = 25$ ohm, $L = 3\,m$H) in Fig. 4.12a–c and an RLC branch composed of an inductor in series with parallel-connected capacitor and resistor (R=25 ohm, $L = 3\,m$H, $C = 10\,\mu$F) in Fig. 4.12d. From the top to the bottom, the upper DC voltage u_{pO}, the lower DC voltage u_{On}, the neutral-point voltage u_{Og}, and the output line-line voltage u_{rs} are shown sequentially in Fig. 4.12. In Fig. 4.12, the control algorithm was inactive at first and then was activated by an activation command. As can be seen from Fig. 4.12, before the control algorithm is activated, the neutral-point voltage

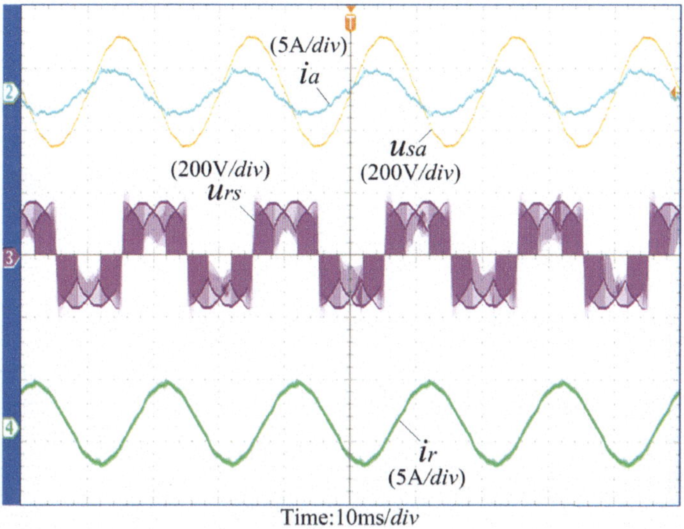

(a)

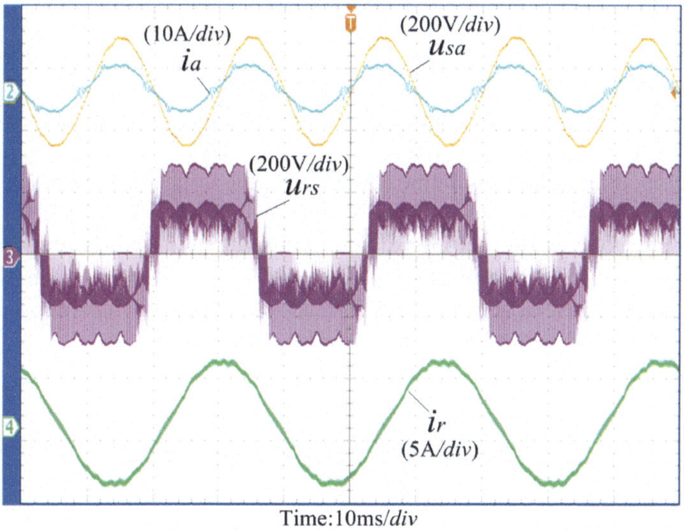

(b)

Fig. 4.10 Experimental waveforms of T[3]HIMC with different m_i and f_o. (**a**) $m_i = 0.5$ and $f_o = 50$ Hz. (**b**) $m_i = 0.9$ and $f_o = 30$ Hz. CH1 is the input phase voltage u_{sa}, CH2 is the input current i_a, CH3 is the output line-line voltage u_{rs}, CH4 is the output current i_r

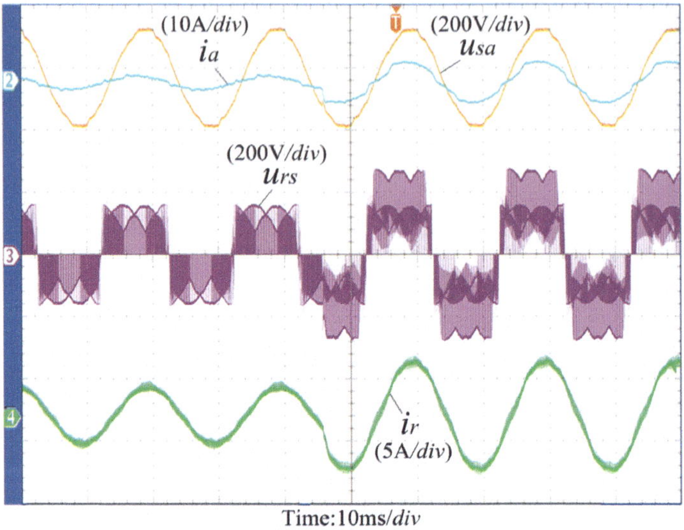

Fig. 4.11 Experimental waveforms of T³HIMC in dynamic condition. CH1 is the input phase voltage u_{sa}, CH2 is the input current i_a, CH3 is the output line-line voltage u_{rs}, CH4 is the output current i_r

of the practical converter is not zero. Instead, it exhibits low-frequency oscillation, although zero averaged neutral-point current under ideal conditions is guaranteed by the developed modulation scheme. As a result, the upper and lower DC voltages are unbalanced, and the output line-line voltage is distorted. This can be seen from the uneven envelope of the output voltage. After activating the control algorithm, the neutral-point voltage is controlled to be almost zero and the upper and lower DC voltages become balanced. Thus the effectiveness of the balancing control algorithm under various operation conditions is verified experimentally.

4.5.4 Comparisons Between T³HIMC *and 3HIMC*

Although it is difficult to make a very fair comparison between two different converter topologies, it is still worth giving a short qualitative assessment considering the basic performance of a converter in order to provide a general guideline for preselecting a suitable topology for dedicated application. In this chapter, the performances of T³HIMC and 3HIMC including the waveforms quality and converter efficiency are evaluated and compared.

The input and output waveforms quality of T³HIMC and 3HIMC at different modulation indices are evaluated, and the results are presented in Fig. 4.13. Figure 4.13a and c shows the input and output waveforms of T³HIMC with the modulation indices being 0.45 and 0.9, respectively. As a comparison, the results of

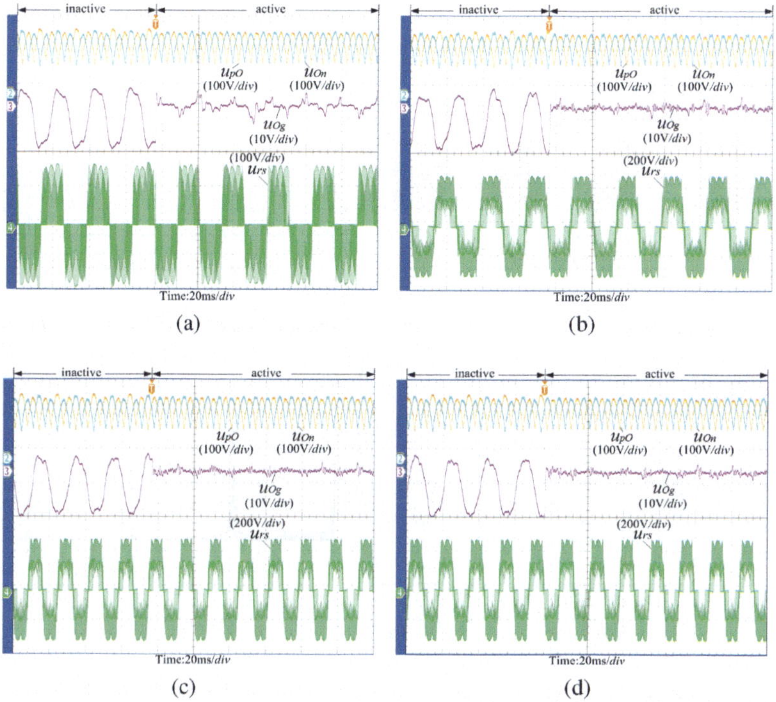

Fig. 4.12 Experimental results of neutral-point voltage balancing control under various operation conditions. (**a**) $m_i = 0.45$ and $f_o = 40$ Hz with RL load. (**b**) $m_i = 0.9$ and $f_o = 40$ Hz with RL load. (**c**) $m_i = 0.9$ and $f_o = 60$ Hz with RL load. (**d**) $m_i = 0.9$ and $f_o = 60$ Hz with RLC load. CH1 is the upper DC voltage u_{pO}, CH2 is the lower DC voltage u_{On}, CH3 is the neutral-point voltage u_{Og}, and CH4 is the output line-line voltage u_{rs}

3HIMC with the modulation indices being 0.45 and 0.9 are depicted in Fig. 4.13b and d, respectively. The output frequency is set as 40 Hz, and an RL load specified in Table 4.2 is used in Fig. 4.13. At a low modulation index, shown in Fig. 4.13a and b, T^{3}HIMC is able to synthesize the output line-line voltage with lower voltage levels. As shown in Fig. 4.13a, the output line-line voltage of T^{3}HIMC is constructed by the input phase-to-neutral voltages instead of the input line-line voltages for 3HIMC, shown in Fig. 4.13b. Consequently, the harmonic components in the output line-line voltage are reduced and a lower THD is achieved. By comparing Fig. 4.13a with Fig. 4.13b, the dominant switching frequency-related harmonics of the output line-line voltage of T^{3}HIMC are reduced from 11.2 to 10.7V (f_s) and from 94.2 to 32.3V ($2f_s$), and the THD of the output line-line voltage (THDu) is reduced significantly from 142.84% to 71.81%, while for the comparison of the input performance, the THD of the input current (THDi) of T^{3}HIMC is increased slightly from 8.87% to 11.61%. For the case of high modulation index, shown in Fig. 4.13c and d, the output line-line voltage of T^{3}HIMC clearly consists of five distinctive voltage levels instead of three voltage levels for 3HIMC. Compared with 3HIMC, the dominant switching

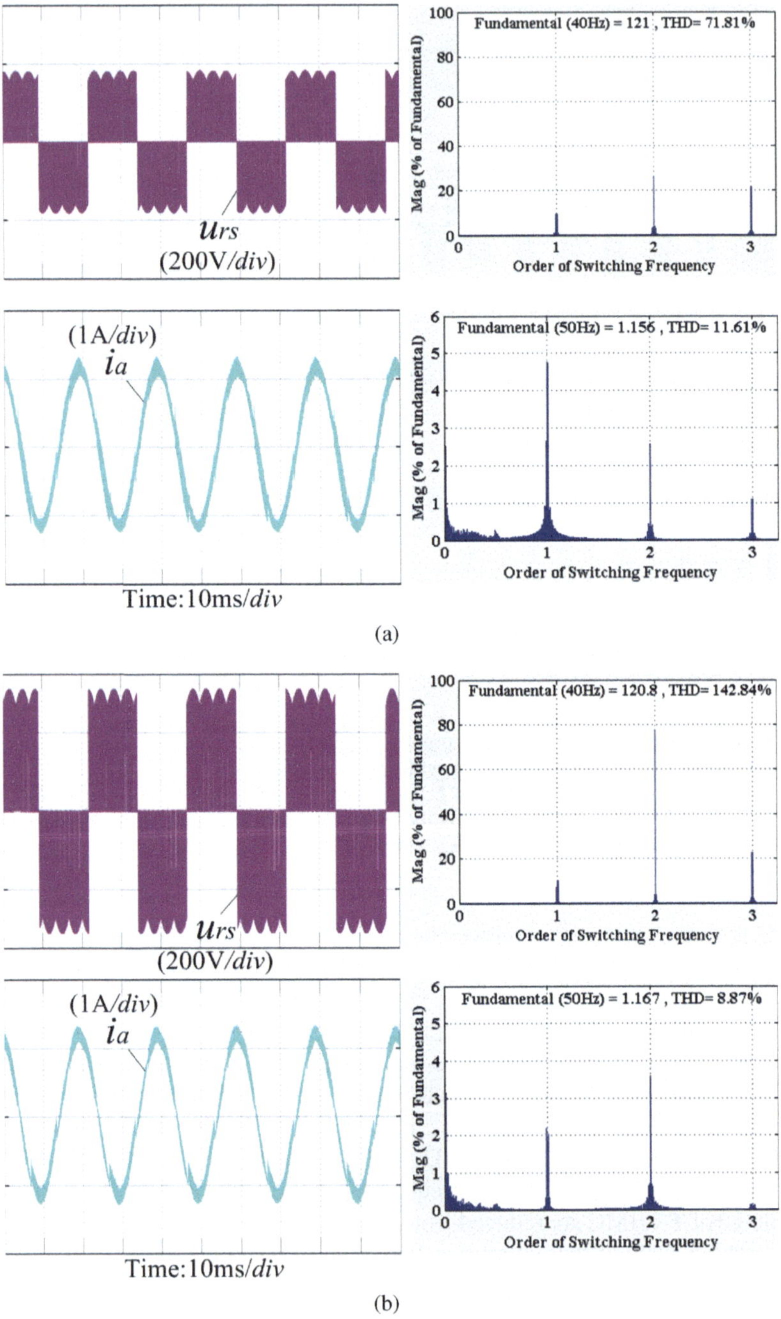

Fig. 4.13 Input and output performance comparisons between T^3HIMC and 3HIMC. (**a**) T^3HIMC with $m_i = 0.45$. (**b**) 3HIMC with $m_i = 0.45$. (**c**) T^3HIMC with $m_i = 0.9$. (**d**) 3HIMC with $m_i = 0.9$

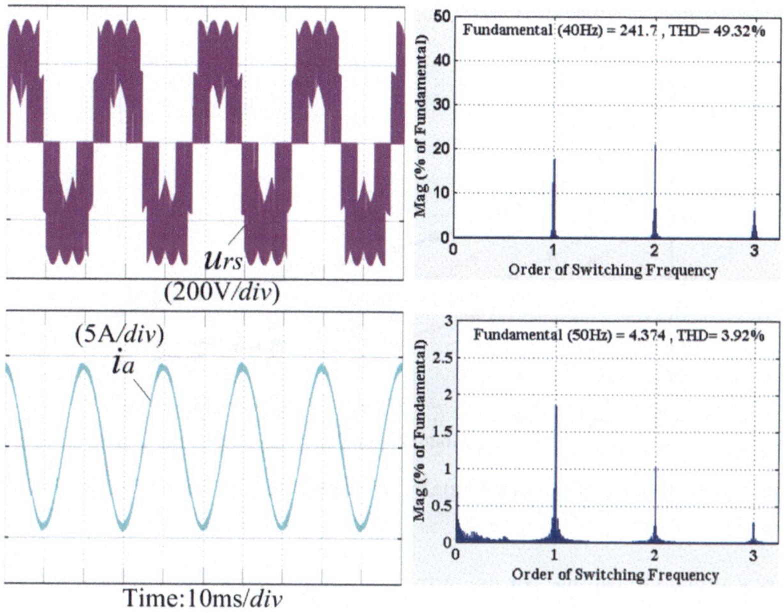

(c)

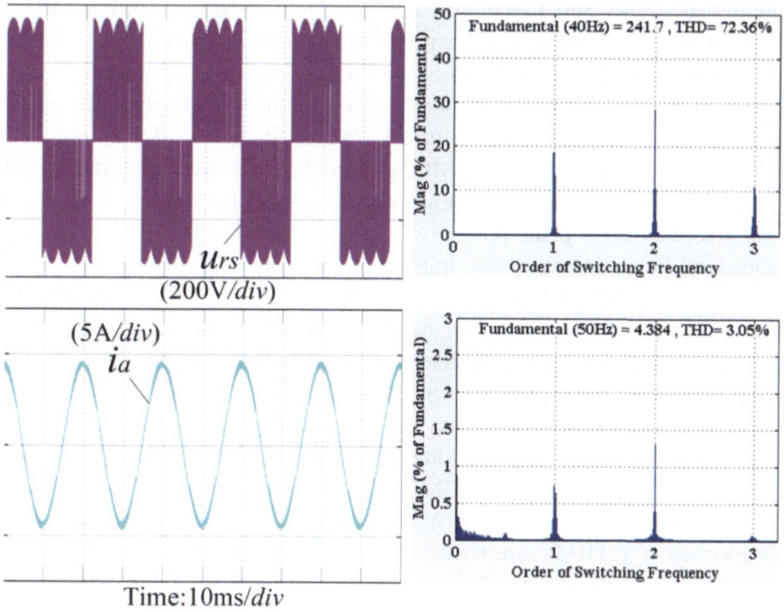

(d)

Fig. 4.13 (continued)

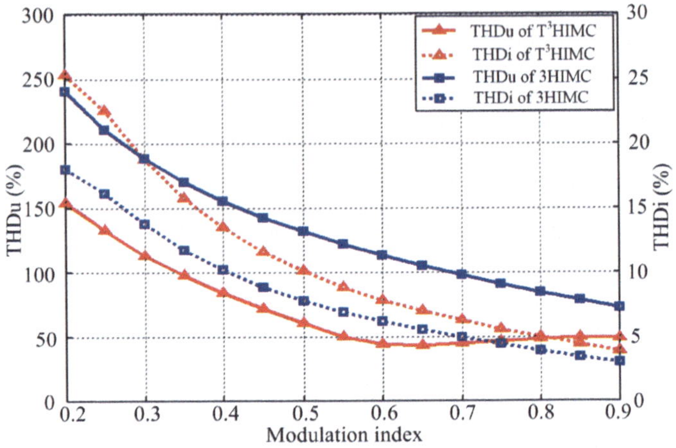

Fig. 4.14 THDs comparison between T³HIMC and 3HIMC

frequency-related harmonics of the output line-line voltage of T³HIMC are reduced from 43.5 to 37.6V (f_s) and from 67.7 to 53.6V ($2f_s$), and the THDu and THDi are decreased from 72.36% to 49.32% and increased from 3.05% to 3.92%, respectively.

For both topologies, the THDu and THDi are measured and compared over the whole modulation indices range. The related results are shown in Fig. 4.14, where f_o is set as 40Hz and the load specified in Table 4.2 is used. As can be seen from Fig. 4.14, T³HIMC has a superior performance than 3HIMC in terms of THDu because of its multilevel output feature. However, in terms of the input current quality, 3HIMC is better than T³HIMC, especially for lower modulation indices. This is mainly attributed to the presence of the neutral-point current in T³HIMC. Based on the operating principles of T³HIMC and 3HIMC, the input currents are synthesized by distributing the impressed DC-link current accordingly to the input phases. From the SVM point of view, the use of small vectors in T³HIMC disconnects the rectifier from one of the DC-link and causes discontinuity in the DC-link current. This increases the harmonic components of the input current and causes degradation of the input performance slightly.

In Fig. 4.15, the converter efficiency comparison between T³HIMC and 3HIMC is carried out. Under the same operation conditions T³HIMC shows higher converter efficiency when compared with 3HIMC. This can be explained by the fact that although the conduct losses of the VSI of T³HIMC are increased slightly, the switching losses of T³HIMC are reduced significantly due to a lower commutation voltage, especially for high switching frequencies. In addition, due to the reduction of the harmonics at the outputs, losses in passive components such as the output inductors and the load machine can also be reduced, which leads to a further increase of the overall efficiency of T³HIMC.

It can be found that although the complexity in circuits and modulation strategies are increased slightly, T³HIMC could provide an enhanced output power quality

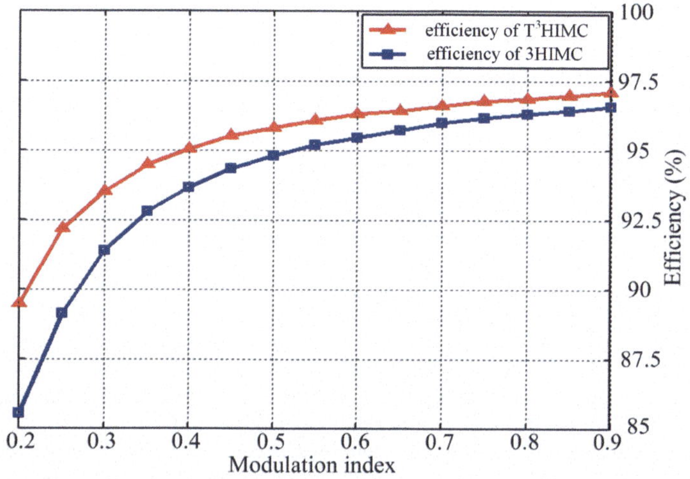

Fig. 4.15 Converter efficiency comparison between T^{3}HIMC and 3HIMC

and higher efficiency when compared with 3HIMC, which makes T^{3}HIMC an attractive choice for the applications where high output power quality and excellent conversion efficiency are essential.

4.6 Conclusions

In this chapter, a three-level T-type indirect matrix converter topology as well as a carrier-based modulation method is proposed. In addition to having the advantages such as extended input reactive power control range and no need for synchronization in modulation, T^{3}HIMC could provide an improved output power quality. To address the issue of neutral-point potential drift caused by the nonlinearities of the practical converter, a control algorithm for balancing the neutral-point voltage is presented. The developed modulation scheme and control algorithm are generalized and can be extended to other three-level IMCs. By applying the developed modulation scheme and control algorithm, T^{3}HIMC is able to synthesize multilevel output voltages and sinusoidal input currents simultaneously. Simulation and experimental results clearly show that T^{3}HIMC could achieve better output performance and higher converter efficiency than that of 3HIMC. By having sinusoidal input/output currents, improved output waveforms quality, high conversion efficiency, and extended input reactive power control range, T^{3}HIMC is an attractive candidate for many applications such as WECS, FACTS, and so on.

References

1. L. Huber, D. Borojevic, Space vector modulated three-phase to three-phase matrix converter with input power factor correction. IEEE Trans. Ind. Appl. **31**(6), 1234–1246 (1995)
2. P. Wheeler, J. Rodriguez, J. Clare, L. Empringham, A. Weinstein, Matrix converters: a technology review. IEEE Trans. Ind. Electron. **49**(2), 276–288 (2002)
3. P. Zwimpfer, H. Stemmler, Modulation and realization of a novel two-stage matrix converter, in *Proceedings of the Brazilian Power Electronics Conference* (2001), pp. 485–490
4. L. Wei, T.A. Lipo, A novel matrix converter topology with simple commutation, in *Proceedings of the IEEE Industry Applications Society Annual Meeting*, vol. 3 (2001), pp. 1749–1754
5. L. Wei, T.A. Lipo, Matrix converter with reduced number of switches, in *ConJ Record of the 2th W'EMPEC Anniversary Meeting, Madison* (2001)
6. J.W. Kolar, M. Baumann, F. Schafmeister, H. Ertl, Novel three-phase AC–DC–AC sparse matrix converter, in Proceedings of the APEC, vol. 2 (2002), pp. 777–791
7. J.W. Kolar, F. Schafmeister , S.D. Round, H. Ertl, Novel three-phase ac-ac sparse matrix converters. IEEE Trans. Power Electron. **22**(5), 1649–1661 (2007)
8. J. Kang, E. Yamamoto, et al., Medium-voltage matrix converter design using cascaded single-phase power cell modules. IEEE Trans. Ind. Electron. **58**(11), 5007–5013 (2011)
9. W. Jiacheng, W. Bin, X. Dewei, N.R. Zargari, Multimodular matrix converters with sinusoidal input and output waveforms. IEEE Trans. Ind. Electron. **59**(1), 17–26 (2012)
10. W. Jiacheng, W. Bin, et al., Phase-shifting transformer-fed multimodular matrix converter operated by a new modulation strategy. IEEE Trans. Ind. Electron. **60**(10), 4329–4338 (2013)
11. S. Yong, Y. Xu, et al., Research on a novel capacitor clamped multilevel matrix converter. IEEE Trans. Power Electron. **20**(5), 1055–1065 (2005)
12. X. Lie, J.C. Clare, et al., Capacitor clamped multilevel matrix converter space vector modulation. IEEE Trans. Ind. Electron. **59**(1), 105–115 (2012)
13. X. Lie, Y. Li, et al., Research on the amplitude coefficient for multilevel matrix converter space vector modulation. IEEE Trans. Power Electron. **27**(8), 3544–3556 (2012)
14. M.Y. Lee, P.W. Wheeler, C. Klumpner, A new modulation method for the three-level-output-stage matrix converter, in *Proceedings of the PCC* (2007), pp. 776–783
15. P.C. Loh, F. Blaabjerg, F. Gao, A. Baby, D.A. C. Tan, Pulsewidth modulation of neutral-point-clamped indirect matrix converter. IEEE Trans. Ind. Appl. **44**(6), 1805–1814 (2008)
16. C. Klumpner, M. Lee, P. Wheeler, A new three-level sparse indirect matrix converter, in *Proceedings of the IECON* (2006), pp. 1902–1907
17. M.Y. Lee, P. Wheeler, C. Klumpner, Space-vector modulated multilevel matrix converter. IEEE Trans. Ind. Electron. **57**(10), 3385–3394 (2010)
18. Y. Sun, W. Xiong, M. Su, X. Li, H. Dan, J. Yang, Topology and modulation for a new multilevel diode-clamped matrix converter. IEEE Trans. Power Electron. **29**(12), 6352–6360 (2014)
19. R. Cardenas, R. Pena, P. Wheeler, J. Clare, G. Asher, Control of the reactive power supplied by a WECS based on an induction generator fed by a matrix converter. IEEE Trans. Ind. Electron. **56**(2), 429–438 (2009)
20. J. Monteiro, J. Silva, S. Pinto, J. Palma, Matrix converter-based unified power-flow controllers: advanced direct power control method. IEEE Trans. Power Del. **26**(1), 420–430 (2011)
21. F. Schafmeister, J.W. Kolar, Novel hybrid modulation schemes significantly extending the reactive power control range of all matrix converter topologies with low computational effort. IEEE Trans. Ind. Electron. **52**(1), 194–210 (2012)
22. M. Rivera, J. Rodriguez, J. Espinoza, T. Friedli, J. Kolar, A. Wilson, C. Rojas, Imposed sinusoidal source and load currents for an indirect matrix converter. IEEE Trans. Ind. Electron. **59**(9), 3427–3435 (2012)
23. S. Yao, S. Mei, W. Hui, et al., Nonlinearity analysis and compensation strategies for two-stage matrix converter. Proc. CSEE **30**(12), 20–27 (2010) (in Chinese)
24. T. Soeiro, F. Vancu, J. W. Kolar, Hybrid active 3rd harmonic current injection mains interface concept for DC distribution systems. IEEE Trans. Power Electron. **28**(1), 7–13 (2013)

25. H. Wang, M. Su, Y. Sun, J. Yang, G. Zhang, W. Gui, J. Feng, Two-stage matrix converter based on third-harmonic injection technique. IEEE Trans. Power Electron. **31**(1), 533–547 (2016)
26. H. Wang, M. Su, et al., Active third-harmonic injection indirect matrix converter with dual three-phase outputs. IET Power Electron. **9**(4), 657–668 (2016)
27. J. Pou, J. Zaragoza, P. Rodríguez, S. Ceballos, V. Sala, R. Burgos, D. Boroyevich, Fast-processing modulation strategy for the neutral-point-clamped converter with total eli_{min}ation of the low-frequency voltage oscillations in the neutral point. IEEE Trans. Ind. Electron. **54**(4), 2288–2294 (2007)

Chapter 5
Three-Level Matrix Converter with Integrated Neutral-Point Potential Balancer

5.1 Introduction

The MC is a direct AC-AC power conversion topology without bulky energy storage elements [1–7]. Due to the advantages such as bidirectional power flow, sinusoidal input and output currents, and high power density, MCs have been found in many applications such as ASDs, power supply, WECS, FACTS, and so on [8–12].

Recently, to improve the output waveform quality and expand the application area of MCs, the multilevel concept has been applied to MCs [13–25]. Generally, the multilevel MC topologies can be mainly divided into three categories: the multi-modular MCs [13–18], the capacitor clamped MCs [19–21], and the NPC MCs [4, 22–25].

The multi-modular MC has the advantages of good output power quality and flexible expansibility, but a bulky transformer is necessary. Generally, the capacitor clamped MC utilizes flying capacitors to provide the middle voltage levels so as to produce multilevel output voltages. A drawback of the capacitor clamped MC is the need of excessive numbers of capacitors. Besides, complicated control methods would be required to balance the flying capacitor voltages.

As for the NPC MC, it is derived from the IMC topology. Therefore, except the ability to generate multilevel output voltages and the possibility of achieving higher conversion efficiency, the NPC MCs also inherit the advantages of IMCs such as compact structure, simple commutation mechanism, and clamp circuit. The advantages described above make the NPC MCs an attractive choice in some applications. However, several problems of the NPC MCs still exist.

First, similar to the conventional NPC converters, neutral-point voltage balancing is a key challenge to normal operation of the NPC MCs. For the topology studied in [23], to achieve zero averaged neutral-point current and thus keep the balance of the neutral-point potential, the nearest three space vector pulse width modulation method is used, which increases the computational burden and causes degradation of the output performance due to the absence of some voltage vectors.

H. Wang, *Advanced Matrix Converters*, Power Systems,
https://doi.org/10.1007/978-3-031-69902-3_5

Besides, although natural-balancing under ideal conditions can be guaranteed, the nonlinearities of the practical converter such as the dead time effects may cause the neutral-point potential drift and distort the input and output waveforms. For the topology presented in [24], the leakage inductance of the isolated transformer and the nonzero averaged neutral-point current would disrupt the self-balance of the neutral-point voltage. Second, the multilevel NPC MC derived from IMC also inherits the drawbacks of IMC, including the limited control range of the input reactive power and the need for synchronization in the modulations of the rectification and inversion stages. The first drawback is more obvious when the NPC MCs are applied in WECS or FACTS [9, 11], while the second one degrades the input and output power quality of the NPC MCs. However, as a new type of IMC, the third-harmonic injection IMC has the merits of enhanced input reactive power capability and independent modulations of the rectification and inversion stages. And the fundamental topology, the derived topology with dual-outputs, and the three-level version have been studied in [26, 27], and [25], respectively.

In this chapter, to improve the output power quality of the conventional IMC and overcome the drawbacks of the NPC MCs including the neutral-point voltage balancing issue, the limited control range of the input reactive power, and the need for strict synchronization in the modulations, the third-harmonic injection concept is extended to multilevel areas, and a T^2HIMC topology and a neutral-point voltage balancing algorithm are proposed. In addition to possessing the advantages such as bidirectional power flow, sinusoidal inputs/outputs, and high power density, it can obtain improved output power quality. Besides, the active power filtering technique is applied to balance the neutral-point voltage, and the third-harmonic injection circuit is utilized to compensate the neutral-point current. Therefore, self-balance of the neutral-point potential is achieved without additional control effort. Moreover, the control range of the input reactive power is extended significantly, and the rectification and inversion stages can be controlled independently. The remainder of this chapter is organized as follows: Sect. 5.2 introduces the topology and operating principles of the T^2HIMC, Sect. 5.3 presents the carrier-based modulation scheme in detail, followed by the neutral-point voltage balancing algorithm, the control of the third-harmonic injection circuit, and the analysis of the input currents, Sect. 5.4 shows the simulation and experimental results to verify the correctness of the presented methods, Sect. 5.5 presents the performance comparisons with other related methods to demonstrate the effectiveness of the developed methods, and Sect. 5.6 draws the final conclusion.

5.2 Topology and Operating Principles of the T^2HIMC

5.2.1 Topology

The T^2HIMC topology is shown in Fig. 5.1, which consists of an input filter, an IVS, an active third-harmonic injection circuit, and a three-level T-type inverter. The input filter consisting of the inductor L_F and the film capacitor C_F is mainly

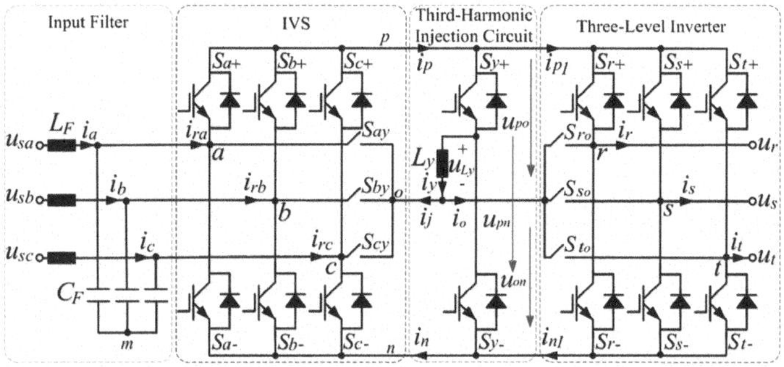

Fig. 5.1 Schematic diagram of the T²HIMC topology

used for filtering the pulse currents generated by the converter. The IVS consists of a line-commutated bidirectional rectifier and three low-frequency bidirectional switches, which acts as a three-pole three-throw switch and provides the DC source voltages for the rear-end three-level inverter. The active third-harmonic injection circuit is composed of a single fast-commutated bridge leg and a third-harmonic injection inductor. The positive, neutral-point, and negative terminals of the three-level inverter connect to nodes p, o, and n of the IVS, respectively. It should be noted that different from the case in the conventional NPC MCs, the neutral-point of the inverter of the T²HIMC is not connected to the star point of the input filtering capacitors. Therefore, star-type connection scheme of the input filtering capacitors is not necessary.

5.2.2 Operating Principles

The operating principles of the T²HIMC are explained as follows: For the IVS, the switches of the rectifier antiparalleled to the conducting diodes, and the bidirectional switch that connected to the input phase with the medium instantaneous voltage, are turned on so as to impose two of the six input line-line voltages across the intermediate DC-link. For the third-harmonic injection circuit, switches S_{y+} and S_{y-} are controlled to generate the desired current i_y injecting into the input phase with the medium instantaneous voltage and the neutral-point of the rear-end inverter. In this manner, the input PFC and neutral-point voltage balancing are achieved. Table 5.1 shows the switching states of the IVS, where θ_{sa} is the phase of the input phase voltage u_{sa}. For example, when the three-phase input voltages satisfy $u_{sa} > u_{sb} > u_{sc}$ (denoted as sector 1), switches S_{a+}, S_{c-}, and S_{by} of the IVS are turned on, and nodes a, b, and c connect to the positive terminal p, neutral-point terminal o, and negative terminal n, respectively. As a result, the input line-line voltages u_{ab} and u_{bc} provide the upper DC source voltage u_{po} and the lower DC source voltage u_{on} of the split DC source, respectively, and so forth. Consequently, the split DC source

Table 5.1 Switching states of the IVS

θ_{sa}	Sector	S_{ay}	S_{by}	S_{cy}	S_{a+}	S_{a-}	S_{b+}	S_{b-}	S_{c+}	S_{c-}
0-π/3	1	0	1	0	1	0	0	0	0	1
π/3-2π/3	2	1	0	0	0	0	1	0	0	1
2π/3-π	3	0	0	1	0	1	1	0	0	0
π-4π/3	4	0	1	0	0	1	0	0	1	0
4π/3-5π/3	5	1	0	0	0	0	0	1	1	0
5π/3-2π	6	0	0	1	1	0	0	1	0	0

voltage is always formed by two different input line-line voltages. According to the requirements of the load, the inverter provides three-phase and three-level output voltages with variable frequency and amplitude. In this manner, sinusoidal three-phase input/output currents and controllable input power factor are attainable.

It can be found from the operating principles that the split DC source is featured by unequal upper and lower DC voltages, since the upper and lower DC voltages in each sector are now formed by two different input line-line voltages. As a result, proper control strategies should be developed for achieving sinusoidal input and output waveforms.

5.3 Control Scheme of the T^{2}HIMC

As can be seen from the operating principles, the switches of the IVS are commutated with line frequency, and thus the DC-link voltage is featured by a six-pulse shape. This implies that different from the conventional NPC IMCs, the IVS and inverter of the T^{2}HIMC can be controlled independently. Since the switching states of the IVS are determined only by the input voltages, the control of the IVS is relatively simple, and the switches only need to switch according to Table 5.1. Therefore, this section mainly focuses on the modulation scheme of the inverter and the control of the third-harmonic injection circuit.

5.3.1 Carrier-Based Modulation Scheme

For the inverter, both the carrier-based modulation method and the SVM method could be adopted. Compared with the SVM method, the carrier-based modulation method has the advantage of relatively low computational burden and hence is chosen here. The concrete analysis is given as follows.

Assume that the three-phase output reference voltages, also known as the modulation signals, are given by

$$\begin{cases} u_{rN} = U_{om} \cos(\omega_o t + \phi) \\ u_{sN} = U_{om} \cos(\omega_o t - 2\pi/3 + \phi) \\ u_{tN} = U_{om} \cos(\omega_o t + 2\pi/3 + \phi) \end{cases} \tag{5.1}$$

where u_{rN}, u_{sN}, and u_{tN} are the desired output phase-to-neutral voltages referenced to the neutral-point of the star-connected load, and U_{om}, ω_o, and ϕ represent the magnitude, angular frequency, and initial phase of the modulation signals, respectively.

To maximize the utilization of the DC-link voltage, the initial modulation signals are modified by adding a zero-sequence voltage as follows:

$$\begin{cases} u_{iO} = u_{iN} + u_{NO} \\ u_{NO} = -0.5[\max(u_{iN}) + \min(u_{iN})] \end{cases}, \ i \in \{r, s, t\} \tag{5.2}$$

where u_{iO} represent the desired output phase voltages referenced to the virtual midpoint of the DC-link, u_{NO} is the zero-sequence voltage, and max() and min() are the operators of the maximum and minimum values, respectively.

The modulation process described in (5.1) and (5.2) is the same as that of the third-harmonic injection IMC [26]. However, since the upper and lower DC voltages of the inverter are unequal, it is necessary to compensate this difference by introducing an extra zero-sequence voltage for the purpose of maximizing the utilization of the DC-link voltage:

$$\begin{cases} u_{io} = u_{iO} + u_{Oo}, \ i \in \{r, s, t\} \\ u_{Oo} = 0.5(u_{po} - u_{on}) \end{cases} \tag{5.3}$$

where u_{io} are the desired output phase voltages referenced to the neutral-point of the split DC source, and u_{Oo} is the extra zero-sequence voltage.

For the convenience of digital implementation, the desired output phase voltages are normalized according to the upper and lower DC voltages, respectively. The normalized modulation signals and the duty ratios are determined by

$$\overline{u}_{io} = \begin{cases} u_{io}/u_{po}, & \text{when } u_{io} \geq 0 \\ u_{io}/u_{on}, & \text{when } u_{io} < 0 \end{cases}, \ i \in \{r, s, t\} \tag{5.4}$$

$$d_{ip} = \begin{cases} \overline{u}_{io}, & \text{when } \overline{u}_{io} \geq 0 \\ 0, & \text{when } \overline{u}_{io} < 0 \end{cases}, \ i \in \{r, s, t\} \tag{5.5}$$

$$d_{in} = \begin{cases} 0, & \text{when } \overline{u}_{io} \geq 0 \\ -\overline{u}_{io}, & \text{when } \overline{u}_{io} < 0 \end{cases}, \ i \in \{r, s, t\} \tag{5.6}$$

where $\overline{u}_{io}$ are the normalized modulation signals, and d_{ip} and d_{in} represent the duty ratios connected to the positive and negative terminals of the DC-link, respectively.

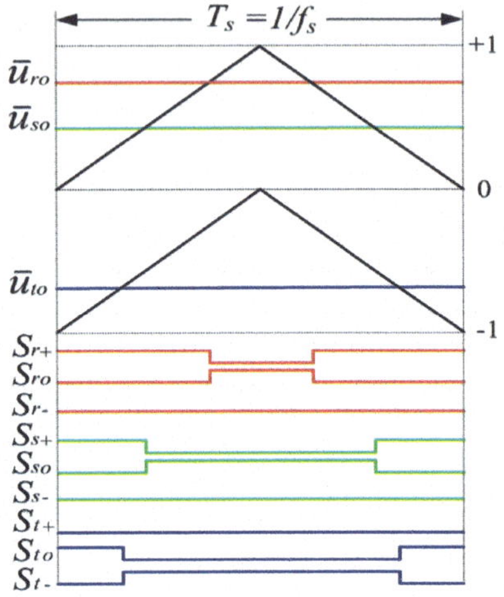

Fig. 5.2 Schematic diagram of the PD modulation scheme

For the carrier-based modulation strategies of the three-level NPC inverters, there exist two carriers. Usually, the phase disposition (PD) scheme, the phase opposition disposition scheme, and the alternative phase opposition disposition scheme are commonly used [28]. In general, among the aforementioned schemes, the PD scheme has the advantage of lower output voltage harmonic distortion and hence is selected here. Under the condition of $\overline{u}_{ro} > \overline{u}_{so} > \overline{u}_{to}$, Fig. 5.2 shows the schematic diagram and switching pattern of the PD modulation scheme, where T_s and f_s are the switching period and switching frequency, respectively.

5.3.2 *Neutral-Point Voltage Balancing*

Neutral-point voltage balancing is one of the main concerns for the NPC converters [29]. Similar to other three-level IMCs, the neutral-point potential of the T^2HIMC is more susceptive to the disturbance of the neutral-point current due to the small capacitance of the input capacitor. Thus appropriate measures should be taken to guarantee normal operation of the T^2HIMC.

For the T-type three-level inverter, the averaged neutral-point current can be derived as

$$\overline{i}_o = \sum (1 - d_{ip} - d_{in})\, i_i, \;\; i \in \{r, s, t\} \tag{5.7}$$

It can be found from (5.7) that zero averaged neutral-point current within a switching period cannot be guaranteed by the developed PD scheme. Thus the T^2HIMC would suffer from oscillation of the neutral-point voltage and distortions of the input currents. Although natural-balancing of the neutral-point potential can be achieved by applying some alternate modulation strategies such as the DSPWM [30], a drawback of these methods is degradation of the output power quality and increase of the switching frequency.

To address the neutral-point voltage balancing issue without degrading the output power quality and increasing the switching frequency of the converter, the active power filtering technique is applied here to balance the neutral-point voltage. In particular, a current equal to the averaged neutral-point current is generated by the third-harmonic injection circuit so as to compensate the neutral-point current. By doing this, the averaged neutral-point current flowing into the IVS is zero. Therefore, neutral-point voltage balancing is achieved, and extra control effort of the capacitor voltages is not required.

5.3.3 Controller Design

Similar to the third-harmonic injection IMC, proper control and design of the third-harmonic injection circuit are the key challenges to implement the converter [26]. For the T^{2}HIMC, this issue will become more crucial because both the PFC and neutral-point voltage balancing are realized by synthesizing the proper third-harmonic injection inductor current.

According to Fig. 5.1, the third-harmonic injection inductor current i_y can be expressed as

$$i_y = i_o + i_j \tag{5.8}$$

It can be found from (5.8) that the third-harmonic injection inductor current can be divided into two parts: the neutral-point current i_o and the injection current i_j. Thus to achieve balancing of the neutral-point voltage and input PFC, the third-harmonic injection inductor current must be controlled to match the required neutral-point current and injection current exactly.

Referring to Fig. 5.1, the mathematical model of the third-harmonic injection circuit can be described as follows:

$$u_{Ly} = L_y \frac{di_y}{dt} \tag{5.9}$$

where u_{Ly} is the voltage imposed across the third-harmonic injection inductor.

It can be seen from (5.9) that the third-harmonic injection inductor current can be controlled by changing u_{Ly}. Similar to the case in [26], a PI controller is used for controlling the third-harmonic injection inductor current, and the detailed design procedure using the conventional frequency-domain analysis tools is not given here for the sake of brevity. To improve the dynamic tracking speed, duty ratio of the switch S_{y+} in steady state is used as a feed-forward term.

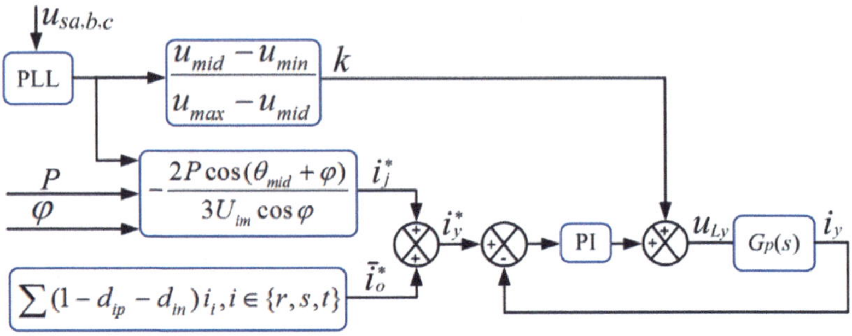

Fig. 5.3 Control scheme of the third-harmonic injection circuit

The overall control scheme of the third-harmonic injection circuit is shown in Fig. 5.3, where u_{max}, u_{mid}, and u_{min} represent the maximum, medium, and minimum values among the three-phase input voltages, θ_{mid} is the phase angle of the input phase voltage having the medium value, P is the active power of the converter, φ is the desired input displacement angle, U_{im} is the amplitude of the input phase voltage, i_j^*, $\overline{i}_o^*$, and i_y^* denote the expected injection current, the expected averaged neutral-point current, and the expected third-harmonic injection inductor current, respectively, and $G_p(s) = 1/L_y s$ is the transfer function of the plant. First, the expected injection current i_j^* is calculated based on the input phase voltages, the active power of the converter, and the expected input displacement angle, and the expected averaged neutral-point current $\overline{i}_o^*$ is calculated according to (5.7). Then the reference third-harmonic injection inductor current i_y^* is obtained by adding the expected injection current and the expected averaged neutral-point current. The combination of the PI controller and feed-forward term ensures good control performance of the system.

For the third-harmonic circuit, the inductor should be selected properly, and the voltage and current stresses of the components should also be evaluated.

According to the design criteria in [26], the selection of the third-harmonic injection inductor should consider both the current ripple and current tracking performance, and the lower and upper limits of the inductance depend on specific third-harmonic injection inductor current. As can be seen from Fig. 5.3, i_y^* is composed of the averaged neutral-point current and the injection current. The injection current depends on the active power of the converter and the input power factor [26], while the neutral-point current is decided by the output/input frequency ratio, the initial phase of the output voltage, the modulation index, and the load condition. As a result, it is difficult to obtain an explicit expression of i_y^*, and the range of the inductance for each operating condition is different. Thus the inductor should be determined based on a specific condition. In this chapter, the inductor is designed based on the rated operating condition and the system parameters listed in Table 5.2. Besides, unity one input and output power factors, a modulation index

Table 5.2 System parameters of T²HIMC

Parameters	Value
Power rating	1.5 kW
Input line-line voltage	220 Vrms, 50 Hz
Switching frequency of the inverter	20 kHz
Switching frequency of the third-harmonic injection circuit	40 kHz
Inductor L_F	300 μH
Capacitor C_F	6.6 μF
Inductor L_y	1.2 mH
Load	25 ohm + 3 mH

m_i of 1, the same input and output frequency/phase, a ripple index γ of 0.4, and a width index ζ Of 0.06 are considered. By considering the above system parameters, a 1.2 mH (0.011686 pu) inductor is selected.

On the other hand, the current rating of the components in the third-harmonic injection circuit, including the third-harmonic injection inductor L_y and the switches S_{y+}/S_{y-}, is selected based on the worst operating condition. It can be deduced from (5.7–5.8) that the amplitude of the third-harmonic injection inductor current is given by

$$I_y = I_{om} - 0.5I_{im} = I_{om}\left[1 - \sqrt{3}m_i \cos\varphi_o/4\right] \tag{5.10}$$

where i_y, I_{im}, and I_{om} are the amplitude of the third-harmonic injection inductor current, the input current, and the output current, respectively, and φ_o is the output power factor angle. It can be seen from (5.10) that the worst case occurs at the maximum output current and zero output power factor. Therefore, the current rating of the components of the third-harmonic injection circuit is determined by the maximum amplitude of the output current. According to the operating principles, the maximum voltage stress on switches S_{y+} and S_{y-} is $\sqrt{3}U_{im}$.

5.3.4 Analysis of the Input Currents

For the T²HIMC, sinusoidal input and output currents should be guaranteed simultaneously. In general, the desired sinusoidal and symmetry output currents can be obtained easily by properly synthesizing the balanced output voltages, as shown in part A of Sect. 5.3. However, for the T²HIMC with the developed modulation strategy and neutral-point balancing algorithm, whether sinusoidal input currents and the controllable input power factor could be achieved should be reexamined, although the mathematical proof of sinusoidal input currents and controllable input power factor of the third-harmonic injection IMC has been conducted in [26].

Assume that the input voltages are given by

$$\begin{cases} u_{sa} &= U_{im}\cos(\theta_{sa}) \\ u_{sb} &= U_{im}\cos(\theta_{sa} - 2\pi/3) \\ u_{sc} &= U_{im}\cos(\theta_{sa} + 2\pi/3) \end{cases} \tag{5.11}$$

Sort the input voltages according to the relationships of the instantaneous values as follows:

$$\begin{cases} u_{max} = \max(u_{sa}, u_{sb}, u_{sc}) \\ u_{mid} = \mathrm{mid}(u_{sa}, u_{sb}, u_{sc}) \\ u_{min} = \min(u_{sa}, u_{sb}, u_{sc}) \end{cases} \tag{5.12}$$

where mid() is the operator of the medium value.

In each switching period, according to the voltage-second balance of L_y, the duty ratio equation can be obtained as

$$\begin{cases} u_{mid} = k u_{max} + k' u_{min} \\ k + k' = 1 \end{cases} \tag{5.13}$$

where k and k' are the steady-state duty ratios of switches S_{y+} and S_{y-}. And the duty ratios are solved as

$$\begin{cases} k = \dfrac{u_{mid} - u_{min}}{u_{max} - u_{min}} \\ \\ k' = \dfrac{u_{max} - u_{mid}}{u_{max} - u_{min}} \end{cases} \tag{5.14}$$

In sector 1, it is found from Table 5.1 that the current i_{rb} is equal to the minus value of the injection current i_j. Suppose that the third-harmonic injection inductor current is given by

$$\begin{cases} i_y = -I_{pm}\ \cos(\theta_{sa} - 2\pi/3) - I_{qm}\ \sin(\theta_{sa} - 2\pi/3) + \bar{i}_o \\ \quad\ = -[GU_{im}\cos(\theta_{sa} - 2\pi/3 + \varphi)]/\cos\varphi + \bar{i}_o \\ \tan\varphi = I_{qm}/I_{pm} \end{cases} \tag{5.15}$$

where I_{pm} and I_{qm} are the amplitudes of the active and reactive components of the input current, respectively, and $G = 2P/3U_{im}^2$ is the equivalent input conductance.

According to Kirchhoff's law, the averaged input current of phase b, denoted as $\bar{i}_{rb}$, can be obtained as

$$\bar{i}_{rb} = -\bar{i}_j = \bar{i}_o - i_y = [GU_{im}\cos(\theta_{sa} - 2\pi/3 + \varphi)]/\cos\varphi \tag{5.16}$$

Using the topology shown in Fig. 5.1, the averaged input current of phase a can be expressed as

$$\bar{i}_{ra} = ki_y + \bar{i}_{p1} = ki_y + (P - u_{on}\bar{i}_o)/u_{pn} \tag{5.17}$$

where $\bar{i}_{p1}$ is the averaged positive DC-link current.

In sector 1, the DC-link voltage u_{pn} is equal to the input line-line voltage u_{ac}, and the lower DC source voltage u_{on} is equal to the input line-line voltage u_{bc}. According to (5.14), the duty ratio k is equal to u_{bc}/u_{ac}. Substituting (5.14) and (5.15) into (5.17), $\bar{i}_{ra}$ is deduced further as

$$\begin{aligned}\bar{i}_{ra} &= \frac{u_{bc}}{u_{ac}}\left[\bar{i}_o - \frac{GU_{im}\cos(\theta_{sa} - 2\pi/3 + \varphi)}{\cos\varphi}\right] + \frac{P - u_{bc}\bar{i}_o}{u_{ac}} \\ &= [GU_{im}\cos(\theta_{sa} + \varphi)]/\cos\varphi\end{aligned} \tag{5.18}$$

As the three-phase input currents satisfy $\bar{i}_{ra} + \bar{i}_{rb} + \bar{i}_{rc} = 0$, the averaged input current of phase c is obtained as

$$\bar{i}_{rc} = GU_{im}\cos(\theta_{sa} + 2\pi/3 + \varphi)\,/\,\cos\varphi \tag{5.19}$$

According to Eqs. (5.16), (5.18), and (5.19), the three-phase input currents are symmetry and sinusoidal with a desired displacement angle φ. It should be noted that the active and reactive components of the input currents in (5.15) can be controlled independently. Thus it is easy to find that the range of the input displacement angle is from $-\pi/2$ to $\pi/2$, and the input reactive power control range is theoretically from $-\infty$ to ∞. For the remaining sectors, similar results can be obtained, and the derivation process is not repeated here.

5.4 Simulation and Experimental Results

In this section, the functionality and performance of the T^2HIMC were firstly evaluated by simulation using Matlab/Simulink software and then were validated experimentally. A simulation study was conducted by using the same parameters as the final laboratory prototype.

5.4.1 Simulation Results

Figure 5.4 shows the waveforms of T^2HIMC with different modulation indices and output frequencies. The output parameters are $m_i = 0.5$, $f_o = 50\,\text{Hz}$ in

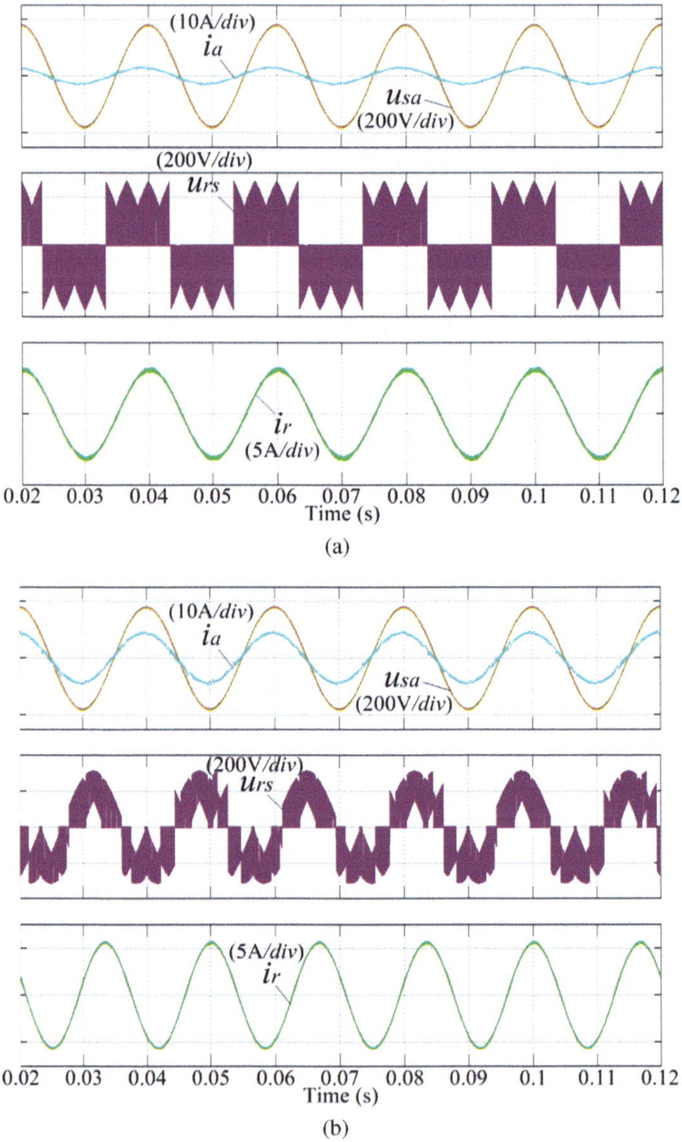

Fig. 5.4 Simulated waveforms of the T^2HIMC with different m_i and f_o. (**a**) $m_i = 0.5$ and $f_o = 50\,\text{Hz}$. (**b**) $m_i = 0.9$ and $f_o = 60\,\text{Hz}$

Fig. 5.4a and $m_i = 0.9$, $f_o = 60$ Hz in Fig. 5.4b, respectively, and the desired input displacement angle is 0, where f_o is the output frequency. The waveforms shown in Fig. 5.4 consist of the input voltage u_{sa}, the input current i_a, the output voltage u_{rs}, and the output current i_r. As can be seen from Fig. 5.4a, the input current is nearly sinusoidal and in phase with the input voltage. It can be found also that the three-level output voltage and sinusoidal output current are obtained. This is because from the SVM point of view, only small, medium, and zero vectors are used for synthesizing the voltage vector at a low modulation index.

As shown in Fig. 5.4b, different from the three-level output voltage in Fig. 5.4a, a noticeable five-level output voltage is generated. This is reasonable since five distinctive levels are utilized for synthesizing the output voltages at high modulation indices. Again, the desired features of sinusoidal input/output currents and unity input power factor are achieved.

Figure 5.5 presents the simulation results with non-unity input power factors. The amplitude of the input reactive current reference is 4A, and the expected input displacement angles are $-\pi/2$ to $\pi/2$ in Fig. 5.5a and b, respectively. It can be found that pure input reactive currents with the desired amplitude are generated. The results in Fig. 5.5 demonstrate the wide input reactive power control range of the T^2HIMC.

To verify the effectiveness of the developed algorithm for balancing the neutral-point voltage, the capability of balancing the neutral-point potential is tested, as shown in Fig. 5.6. The output parameters are $m_i = 0.9$, $f_o = 50$ Hz, and the input displacement angles are $\pi/12$ to $-\pi/12$ in Fig. 5.6a and b, respectively. In Fig. 5.6, the balance algorithm is inactive at first. After 0.1 s, the algorithm is enabled. As can be seen from Fig. 5.6, the upper and lower DC source voltages are uneven, and the input currents are distorted seriously when the balance algorithm is inactive. This is

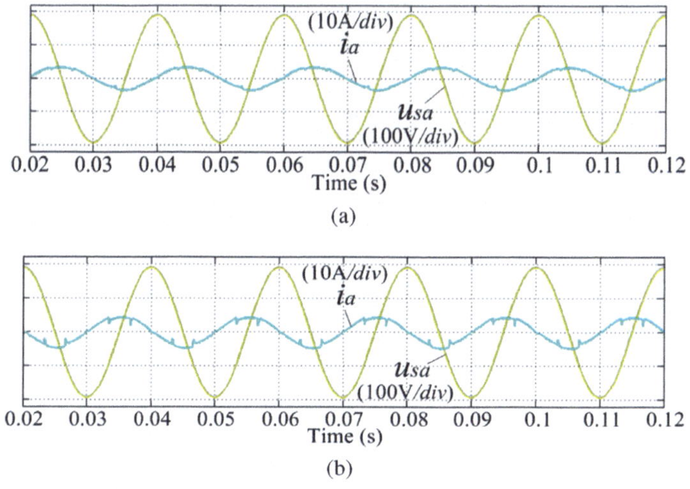

Fig. 5.5 Simulated waveforms of the T^2HIMC with non-unity input power factors. (**a**) $\varphi = \pi/2$. (**b**) $\varphi = -\pi/2$

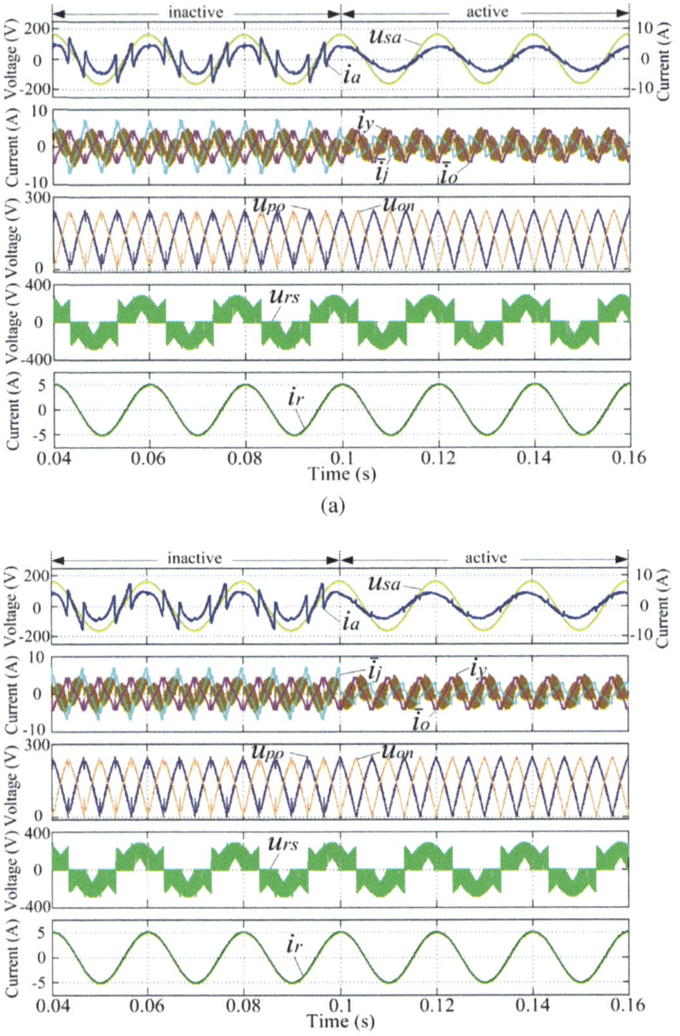

Fig. 5.6 Simulated results of the algorithm for balancing the neutral-point voltage under different input power factors. **(a)** $\varphi = \pi/12$. **(b)** $\varphi = -\pi/12$

because the neutral-point current generated by the inverter flows into the input side of the converter directly. However, sinusoidal outputs are still guaranteed because of compensation of the modulation signals according to the real-time upper and lower DC source voltages. After activating the balance algorithm, the neutral-point current is compensated completely, and thus sinusoidal input currents, and balanced upper and lower DC source voltages are achieved. Besides, the amplitude of the averaged injection current $\overline{i}_j$ is reduced, which is beneficial to reduce the power losses.

It can be seen from the simulation results that sinusoidal input/output currents, multilevel output voltage, and extended control range of the input reactive power are achieved in T^2HIMC. Besides, the sinusoidal input currents and balanced upper and lower DC source voltages clearly demonstrate the effectiveness of the neutral-point voltage balancing algorithm and the control scheme of the third-harmonic injection circuit. Therefore, the correctness and feasibility of the proposed methods are verified by simulation.

5.4.2 Experimental Results

To validate the theoretical analysis and simulation results, a 1.5 kW prototype with the parameters listed in Table 5.2 is built. For the switches of the IVS which commutate at line frequency, the IGBT IKW30N65EL5 is chosen due to its low saturation voltage. For the rest of the switches, high speed IGBT IKW40N65ES5 is used. Besides, the dead time is set as 1 μs.

The experimental results shown in Figs. 5.7, 5.8, and 5.9 correspond to the simulated results shown in Figs. 5.4, 5.5, and 5.6, and the experimental conditions are the same as these in the simulation. As can be seen from Figs. 5.7, 5.8, and 5.9, the experimental results match the simulated results well, except for a slight reduction in the amplitudes and higher distortions of the input and output currents. The higher distortion of the currents in the experiments is mainly attributed to the following factors. Firstly, sinusoidal input currents are achieved only if the third-harmonic injection inductor current is controlled to exactly track the reference current with high di/dt rate. However, as a single-phase shunt active power filter in essence, the practical current tracking performance is greatly affected by the control delay, the input filtering capacitor voltage ripple, the dead time, and so on. Secondly, although the averaged value of the neutral-point current flowing into the input side is zero with the developed control method, its instantaneous value is not zero. Instead, it is featured by high peak value and contains many high-order harmonic currents, which would increase the THD of the input currents inevitably. Thirdly, similar to the situation in many other IMCs, the input currents of the T^2HIMC are not controlled forwardly but are formed passively and indirectly under the together action of the input voltages and the modulation of the inversion stage. As a result, the input current waveforms are susceptive to the power quality of the grid, and a tiny distortion and unbalance of the grid voltage would cause a considerable distortion of the input currents. Referring to Figs. 5.7, 5.8, and 5.9, multilevel output voltage and sinusoidal input/output currents are achieved; thus the validity of the presented methods is verified experimentally.

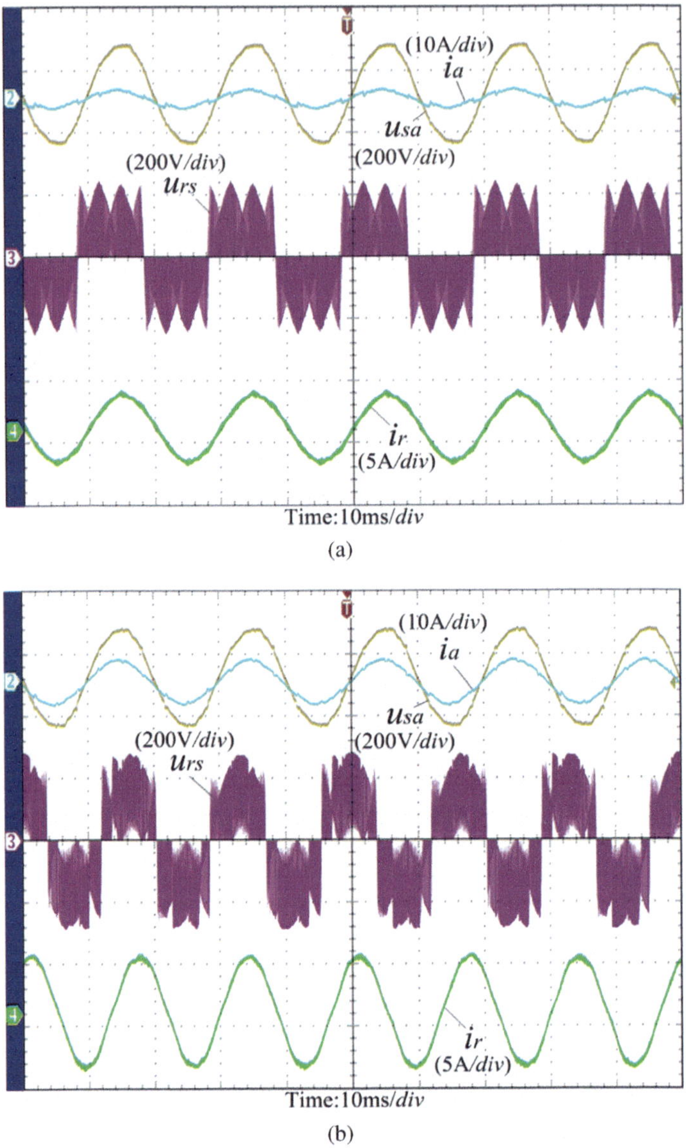

Fig. 5.7 Experimental waveforms of the T^2HIMC with different m_i and f_o. (**a**) $m_i = 0.5$ and $f_o = 50\,\text{Hz}$. (**b**) $m_i = 0.9$ and $f_o = 60\,\text{Hz}$

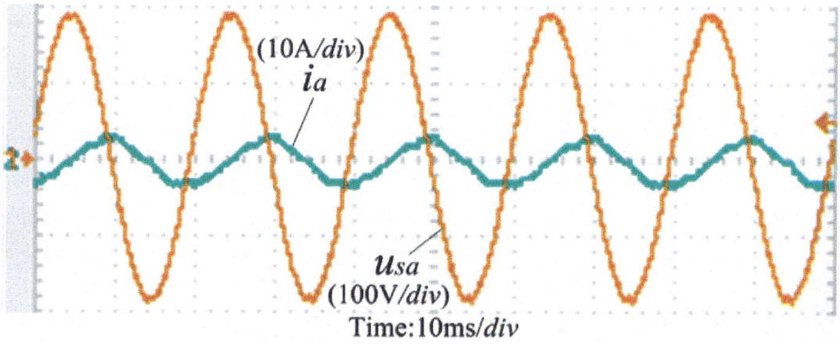

(a)

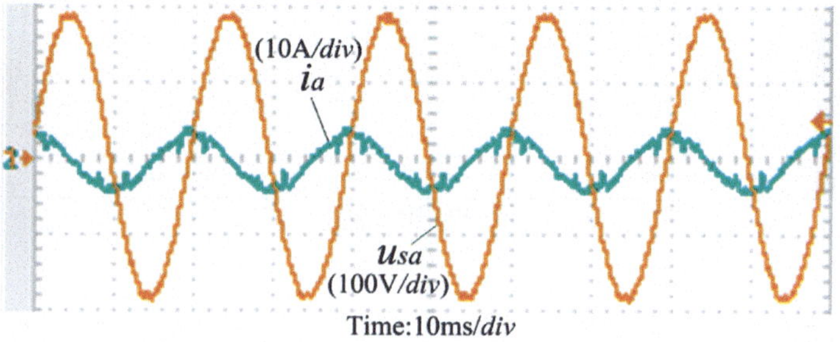

(b)

Fig. 5.8 Experimental waveforms of the T^2HIMC with non-unity input power factors. (**a**) $\varphi = \pi/2$. (**b**) $\varphi = -\pi/2$

5.5 Performance Comparisons

To evaluate both the advantages and disadvantages of the proposed methods, first the waveform quality, as well as the overall efficiency under different modulation schemes, is compared. Then a comparison of the output waveform quality among different topologies is conducted.

Figures 5.10 and 5.11 present the comparisons of the input and output waveform quality between DSPWM and the developed methods. Figure 5.10a and c shows the input and output waveforms of the T^2HIMC under the DSPWM scheme with the modulation indices being 0.45 and 0.9, respectively. As a comparison, the results of the proposed method are depicted in Fig. 5.10b and d, and $f_o = 50$ Hz. At a low modulation index, shown in Fig. 5.10a and b, the THD of the output line-line voltage under the proposed scheme is reduced from 85.13% to 80.94% when compared with the DSPWM method, while for the comparison of the input performance, the THD of the input current under the presented scheme is reduced significantly from 11.28% to 6.15%. For the case of high modulation index, shown in Fig. 5.10c and d, the THDs of the output voltage and input current under the proposed scheme are decreased from 46.86% to 30.97% and from 4.88% to 4.59%, respectively.

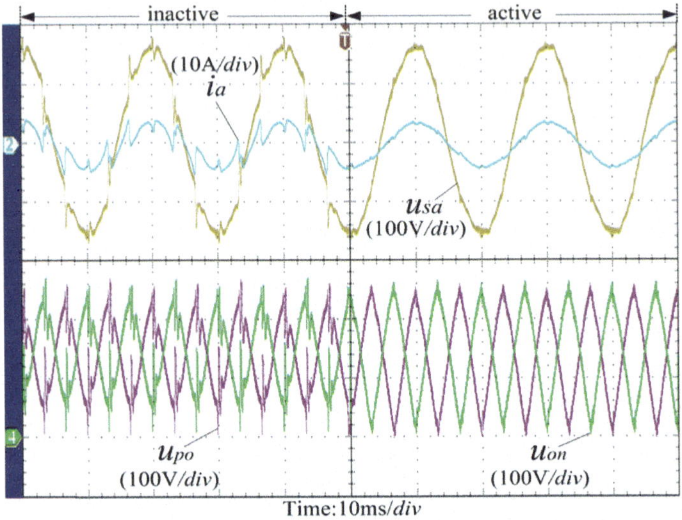

(a)

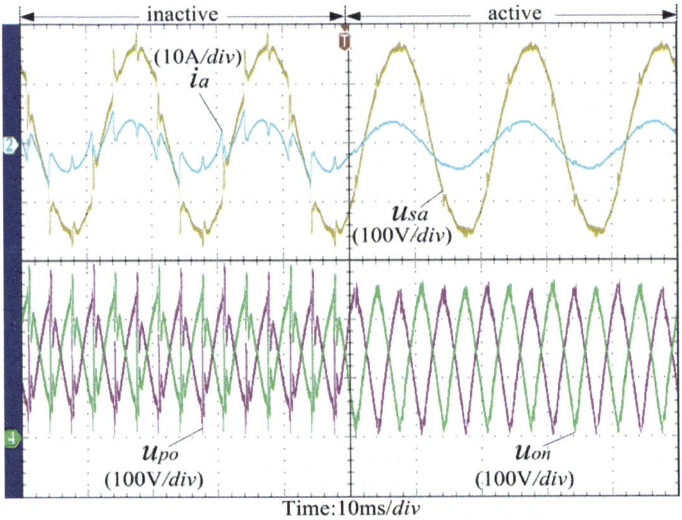

(b)

Fig. 5.9 Experimental results of the algorithm for balancing the neutral-point voltage under different input power factors. (**a**) $\varphi = \pi/12$. (**b**) $\varphi = -\pi/12$

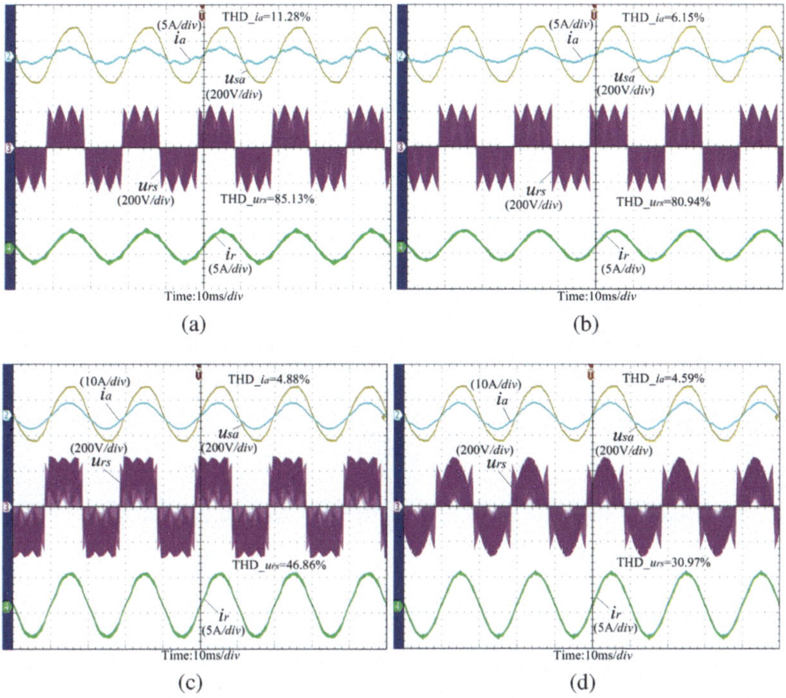

Fig. 5.10 Experimental waveforms of T^{2}HIMC under (**a**) DSPWM method with m_i=0.45, (**b**) The proposed method with m_i =0.45, (**c**) DSPWM method with m_i=0.9, and (**d**) The proposed method with m_i =0.9

Figure 5.11 depicts the Fast Fourier Transform analysis of the input and output currents corresponding to the cases in Fig. 5.10. As can be seen from Fig. 5.11, the input and output currents in the experiments contain some low-order harmonics because of the nonidealities of the practical converter. Compared with the DSPWM method, the harmonics (especially the 5th and 7th harmonics) of the input current under the developed method are reduced significantly, while for the comparison of the output current harmonics, the proposed method is also superior to the DSPWM method.

The results in Figs. 5.10 and 5.11 show that the proposed method has a superior performance to the DSPWM method in terms of THDs of the output voltage and input current. This is because some vectors are not available under the DSPWM method, which would cause degradation of the output power quality. In addition, the harmonic components in the injection current under the developed method are lower than these of the DSPWM method, which leads to reduction in the THD of the input current to some extent.

To compare the efficiency for both modulation methods, the converter efficiency is measured, as shown in Fig. 5.12. Compared with DSPWM, the efficiency under

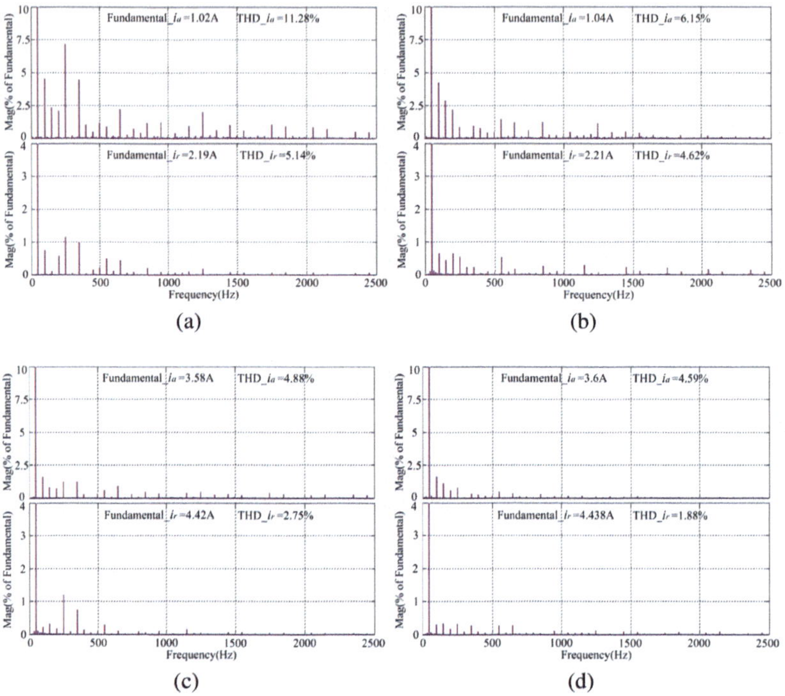

Fig. 5.11 Spectra of the input and output currents under (**a**) DSPWM method with m_i=0.45, (**b**) The proposed method with m_i =0.45, (**c**) DSPWM method with m_i=0.9, and (**d**) The proposed method with m_i =0.9

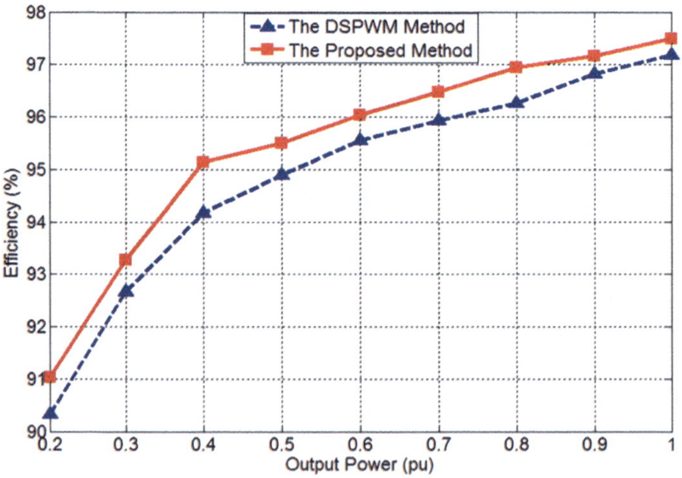

Fig. 5.12 Efficiency comparison of the T^2HIMC

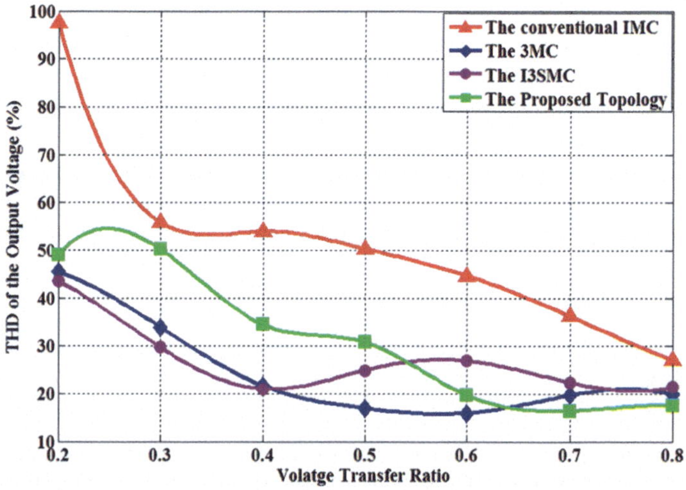

Fig. 5.13 THD of the output voltage for different topologies

the developed method is obviously higher due to a reduction of the equivalent switching frequency by 25% and reduced harmonics losses.

In Fig. 5.13, the output waveform quality of the T^2HIMC is compared with the conventional IMC, the so-called 3MC [22] and I3SMC [23], and the THDs of the output voltage are calculated. It is clear that all of the three-level IMCs clearly have a superior output performance to the conventional IMC in terms of THDs of the output voltage. Comparing the T^2HIMC with the 3MC and I3SMC, the THD of the output voltage for both the 3MC and I3SMC is slightly lower than that of the T^2HIMC in low VTR region, while in high VTR region, the THD performance of the T^2HIMC is better. This is a remarkable merit of T^2HIMC because usually the converter is expected to operate in high VTR region for the purpose of maximizing the utilization of the power devices' voltage rating.

It can be found from Figs. 5.10, 5.11, 5.12, and 5.13 that for the T^2HIMC, the developed modulation method is superior to DSPWM in terms of input/output power quality and efficiency. In addition, the three-level IMCs show better output power quality than the conventional IMC, and the T^2HIMC has the best output THD performance among the three three-level IMCs in high VTR region at the cost of a slightly more complicated circuit.

5.6 Conclusions

In this chapter, a three-level third-harmonic injection IMC topology and a carrier-based modulation strategy are proposed. Moreover, a neutral-point voltage balancing algorithm based on active power filtering technique is developed. Compared with the conventional NPC MCs, self-balance of the neutral-point potential without additional control effort and extended control range of the input reactive power are achieved. Besides, independent control of the rectification and inversion stages enhances the expansibility and implementation flexibility of the topology. Simulation and experimental results verify the correctness of the proposed methods. Because the T^2HIMC has the advantages of sinusoidal input/output currents and enhanced input reactive power characteristic and output power quality, the T^2HIMC belongs to the family of the NPC MCs and is an attractive choice for ASDs, WECS, and other applications.

References

1. L. Huber, D. Borojevic, Space vector modulated three-phase to three-phase matrix converter with input power factor correction. IEEE Trans. Ind. Appl. **31**(6), 1234–1246 (1995)
2. L. Wei, T.A. Lipo, A novel matrix converter topology with simple commutation. Proc. IEEE Ind. Appl. Soc. Annu. Meeting **3**, 1749–1754 (2001)
3. D. Casadei, G. Serra, et al., Matrix converter modulation strategies: a new general approach based on space-vector representation of the switch state. IEEE Trans. Ind. Electron. **49**(2), 370–381 (2002)
4. J.W. Kolar, M. Baumann, et al., Novel three-phase AC–DC–AC sparse matrix converter, in *Proceedings of the APEC*, vol. 2 (2002), pp. 777–791
5. P. Wheeler, J. Rodriguez, et al., Matrix converters: a technology review. IEEE Trans. Ind. Electron. **49**(2), 276–288 (2002)
6. J.W. Kolar, F. Schafmeister, et al., Novel three-phase ac-ac sparse matrix converters. IEEE Trans. Power Electron. **22**(5), 1649–1661 (2007)
7. J. Rodriguez, M. Rivera, et al., A review of control and modulation methods for matrix converters. IEEE Trans. Ind. Electron. **59**(1), 58–70 (2012)
8. R. Pena, R. Cardenas, et al., A topology for multiple generation system with doubly fed induction machines and indirect matrix converter. IEEE Trans. Ind. Electron. **56**(10), 4181–4193 (2009)
9. R. Cardenas, R. Pena, et al., Control of the reactive power supplied by a WECS based on an induction generator fed by a matrix converter. IEEE Trans. Ind. Electron. **56**(2), 429–438 (2009)
10. Y. Sun, M. Su, et al., Indirect four-leg matrix converter based on robust adaptive back-stepping control. IEEE Trans. Ind. Electron. **58**(9), 4288–4298 (2011)
11. J. Monteiro, J. Silva, et al., Matrix converter-based unified power-flow controllers: advanced direct power control method. IEEE Trans. Power Del. **26**(1), 420–430 (2011)
12. X. Li, Y. Sun, et al., Modulation methods for indirect matrix converter extending the input reactive power range. IEEE Trans. Power Electron. **32**(6), 4852–4863 (2017)
13. J. Change, Modular AC-AC variable voltage and variable frequency power converter system and control, U.S. Patent 5 909 367 (1999)
14. J. Kang, E. Yamamoto, et al., Medium-voltage matrix converter design using cascaded single-phase power cell modules. IEEE Trans. Ind. Electron. **58**(11), 5007–5013 (2011)
15. W. Jiacheng, W. Bin, et al., Multimodular matrix converters with sinusoidal input and output waveforms. IEEE Trans. Ind. Electron. **59**(1), 17–26 (2012)
16. W. Jiacheng, W. Bin, et al., Phase-shifting transformer-fed multimodular matrix converter operated by a new modulation strategy. IEEE Trans. Ind. Electron. **60**(10), 4329–4338 (2013)
17. Y. Sun, W. Xiong, et al., Modulation strategies based on mathematical construction method for multi-modular matrix converter. IEEE Trans. Power Electron. **31**(8), 5423–5434 (2016)

18. Y. Sun, W. Xiong, et al., Carrier-based modulation strategies for multi-modular matrix converters. IEEE Trans. Ind. Electron. **63**(3), 1350–1361 (2016)
19. S. Yong, Y. Xu, et al., Research on a novel capacitor clamped multilevel matrix converter. IEEE Trans. Power Electron. **20**(5), 1055–1065 (2005)
20. X. Lie, J.C. Clare, et al., Capacitor clamped multilevel matrix converter space vector modulation. IEEE Trans. Ind. Electron. **59**(1), 105–115 (2012)
21. X. Lie, Y. Li, et al., Research on the amplitude coefficient for multilevel matrix converter space vector modulation. IEEE Trans. Power Electron. **27**(8), 3544–3556 (2012)
22. P.C. Loh, F. Blaabjerg, et al., Pulsewidth modulation of neutral-point-clamped indirect matrix converter. IEEE Trans. Ind. Appl. **44**(6), 1805–1814 (2008)
23. M. Y. Lee, P. Wheeler, et al., Space-vector modulated multilevel matrix converter. IEEE Trans. Ind. Electron. **57**(10), 3385–3394 (2010)
24. Y. Sun, W. Xiong, et al., Topology and modulation for a new multilevel diode-clamped matrix converter. IEEE Trans. Power Electron. **29**(12), 6352–6360 (2014)
25. L. Wang, H. Wang, et al., A three-level T-type indirect matrix converter based on the third-harmonic injection technique. IEEE J. Emerg. Sel. Topics Power Electron **5**(2), 841–853 (2017)
26. H. Wang, M. Su, et al., Two-stage matrix converter based on third-harmonic injection technique. IEEE Trans. Power Electron. **31**(1), 533–547 (2016)
27. H. Wang, M. Su, et al., Active third-harmonic injection indirect matrix converter with dual three-phase outputs. IET Power Electron. **9**(4), 657–668 (2016)
28. R. Naderi, A. Rahmati, Phase-shifted carrier PWM technique for general cascaded inverters. IEEE Trans. Power Electron. **23**(3), 1257–1269 (2008)
29. R. Maheshwari, S. Munk-Nielsen, et al., Design of neutral-point voltage controller of a three-level NPC inverter with small DC-link capacitors. IEEE Trans. Ind. Electron. **60**(5), 1861–1871 (2013)
30. J. Pou, J. Zaragoza, et al., Fast-processing modulation strategy for the neutral-point-clamped converter with total elimination of the low-frequency voltage oscillations in the neutral point. IEEE Trans. Ind. Electron. **54**(4), 2288–2294 (2007)

Chapter 6
Three-Level Indirect Matrix Converter for Adjustable Speed Drives

6.1 Introduction

Different from conventional back-to-back converters, MC aims for high power density and achieves direct AC-AC power conversion without the DC-link energy storage element [1]. Besides, MC also possesses some attractive characteristics such as bidirectional power flow, sinusoidal input and output currents, and high reliability [2–13]. In many fields such as ASDs [14, 15], Marine Electric Propulsion [16, 17], WECS [18], and FACTS [19, 20], it is becoming obvious that MCs can be applied as alternatives to conventional inverters.

To ameliorate the output waveform quality and expand the application area of MCs, the multilevel MCs are gradually in the spotlight [21–23]. By generating stepped multilevel output voltages, the multilevel MCs, compared with the conventional two-level MCs, can achieve better output power quality, the reduced voltage stress on the power switches, smaller common-mode voltage, and lower du/dt stress. Multilevel MCs can be classified into three categories: the multi-modular MCs [21–25], the capacitor clamped MCs [26, 27], and the NPC MCs [28–32].

The multi-modular MC, which can be constructed from cascaded three-phase to single-phase MC modules as shown in Fig. 6.1a, possesses some merits such as flexible expansibility and good output power quality [25]. However, the essential bulky and expensive multi-winding transformer is an inherent drawback. The capacitor clamped MC is shown in Fig. 6.1b that utilizes flying capacitors to produce multilevel output voltages [26]. However, excessive numbers of capacitors and the high complexity of balancing the flying capacitor voltages underline that capacitor clamped MCs cannot be fully embraced by the industry.

The NPC MC usually stems from the conventional IMC. In addition to the ability to synthesize multilevel output voltages and potentially higher conversion efficiency, some intrinsic merits of IMCs remain in NPC MCs. Nevertheless, several issues still need to be solved. As the key of NPC converters, the significance of neutral-point potential balancing is self-evident, because the capacitance of the input capacitors

H. Wang, *Advanced Matrix Converters*, Power Systems,
https://doi.org/10.1007/978-3-031-69902-3_6

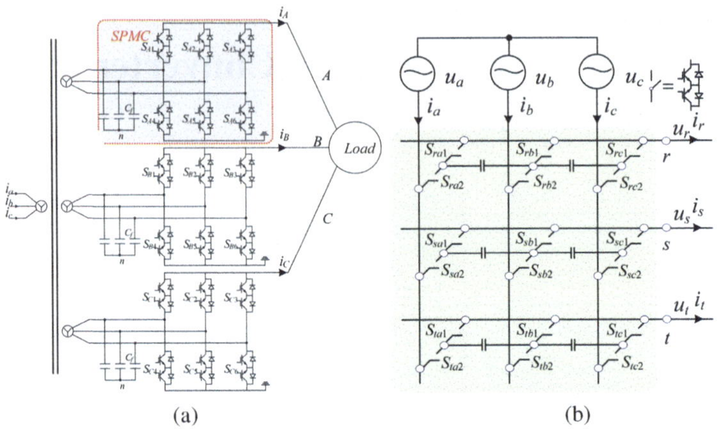

Fig. 6.1 Circuits of (**a**) multi-modular MC and (**b**) capacitor clamped MC

in the NPC MCs is relatively small compared to the bulky DC-link capacitors in conventional NPC converters. The nonzero average neutral-point current caused by the modulation scheme may disrupt the balance of the neutral-point voltage [29]. Literature [30] utilizes the nearest three space vectors pulse width modulation method to acquire a zero average neutral-point current and the balanced neutral-point potential, but there is no denying that the complexity and computational burden increase.

The leakage inductance of the isolated transformer and the nonzero average neutral-point current can cause distortions of the input currents [29]. In [32] a T^2HIMC topology is developed, and the active power filtering concept is applied to solve the neutral-point potential balance issue, and the third-harmonic injection circuit is utilized to compensate for the neutral-point current generated by the inverter stage. However, similar to the T^3HIMC topology presented in [31], the input current performance of the T^2HIMC may be poor because the input currents are partially synthesized by the modulation of the third-harmonic injection circuit. Thus, the input current quality is directly affected by the tracking performance of the third-harmonic injection inductor current and the injected high-frequency harmonic currents flowing from the neutral-point of the inverter.

In this chapter, to ameliorate the power quality and surmount the foregoing defects of the NPC MCs, a T3LIMC topology derived from the SWISS Rectifier [33] is presented, which also inherits advantages of the conventional two-level IMCs such as bidirectional power flow capability, sinusoidal inputs and outputs, controllable input power factor, no DC-link energy storage elements, and compact structure. A simple carrier-based modulation scheme is developed for the T3LIMC. By properly choosing the zero-sequence voltage and constructing the desired neutral-point current of the NPC inverter, the ZCS operation of the rectifier and the balanced neutral-point potential are achieved simultaneously without extra control efforts. Thereby, T3LIMC improves the quality of output waveforms by realizing three-level output and increases efficiency with low complexity and high power density. The remainder of this chapter is organized as follows: Sect. 6.2 introduces the topology and operating principles of the T3LIMC, Sect. 6.3 presents the carrier-based modulation schemes in detail, followed by the mathematical proof of the sinusoidal input currents and balanced neutral-point potential, Sect. 6.4 shows the experimental results to verify the validity of the presented methods, followed by the efficiency evaluation and performance comparisons with other related methods to demonstrate the effectiveness of the developed methods, and Sect. 6.5 draws the conclusion.

6.2 Topology and Operating Principles

In this section, the topology structure and operating principles of the presented T3LIMC are discussed briefly. First, the construction details of the input filter, rectifier, and T-type three-level inverter of the T3LIMC are presented. Afterward, the operating principles of the T3LIMC are analyzed, from which the requirements for the modulation scheme with the goals of balanced neutral-point potential and symmetric sinusoidal input currents are determined.

6.2.1 Topology

As shown in Fig. 6.2, the T3LIMC is composed of an LC input filter, a bidirectional current-source-type rectifier, and a T-type three-level inverter. The pulsed currents produced by the rectifier can be filtered by the LC input filter to yield three-phase sinusoidal input currents. The rectifier consists of six line-commutated switches ($S_{i+/-}$, $i = a, b, c$), three line- frequency bidirectional switches (S_{iy}, $i = a, b, c$),

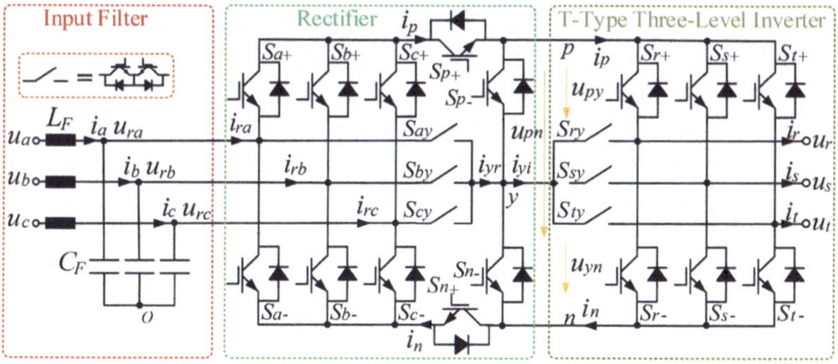

Fig. 6.2 Topology of the presented T3LIMC

and four high-frequency switched switches ($S_{i+/-}$, $i = p, n$), which provides the split DC-link voltages for the load-side T-type three-level inverter. In addition, p, y, and n represent the junctions of load-side three-level inverter and front-end rectifier, and it is apparent that there are some differences between the presented T3LIMC and conventional NPC MCs.

6.2.2 Operating Principles

The operating principles of the T3LIMC are given as follows:

For the line-commutated switches $S_{i+/-}$ and S_{iy} ($i = a, b, c$) of the front-end rectifier, the upper switch that connected to the input phase with the maximum instantaneous voltage, the bottom switch that connected to the input phase with the minimum instantaneous voltage, and the bidirectional switch that connected to the input phase with the medium instantaneous voltage are always turned on.

The four high-frequency switched switches of the rectifier $S_{k+/-}$ ($k = p, n$) are controlled to operate complementarily. S_{p+} or S_{p-} and S_{n+} or S_{n-} are turned on at an arbitrary time to synthesize the required intermediate split DC-link voltages. When the switches S_{a+}, S_c and S_{by} are turned on, the neutral point terminal y connects to input phase b, the positive terminal p connects to input phase a or b according to the switching states of the switches S_{p+} and S_p, and the upper DC source voltage u_{py} consists of the input line-line voltage u_{ab} and zero voltage. Accordingly, the negative terminal n connects to input phase c or b as S_{n+} or S_{n-} is turned on, and the lower DC source voltage u_{yn} is constituted by the input line-line voltage u_{bc} and zero voltage, respectively, and so forth. Thus, the split DC source voltages are featured by unequal upper and lower voltages.

The T-type three-level inverter of T3LIMC is modulated to provide three-phase output voltages with variable frequency and amplitude. However, since the neutral-point current generated by the inverter injects into one of the three input phases and the split DC-link voltages are directly supported by the input filtering capacitor voltages, the modulation of the inverter may disturb the normal operation of the rectifier, which will lead to unbalanced input filtering capacitor voltages and distorted input currents. Section 6.3 provides the elaboration of the developed modulation scheme that obtains symmetric sinusoidal three-phase input/output currents and the controllable input power factor.

6.2.3 Analysis of Input Power Factor

Usually, in the modulation process of the input current, the phase shift of the input current caused by the input filter is ignored, and the phase angle between the input voltage u_i ($i = a, b, c$) on the grid side, and the input current (before filtering) i_{ri} ($i = a, b, c$) on the main circuit side is directly used as the input power factor angle. However, the actual input power factor angle is the angle between the grid-side input voltage u_i $(i = a, b, c)$ and the grid-side input current (filtered) i_i $(i = a, b, c)$. Due to the input filter, there is an inevitable phase difference between input current before and after filtering. Next comes the analysis.

For phase A, assuming that u_a is in phase with i_a, i.e., $u_a = U_a e^{j0}$ and $i_a = I_a e^{j0}$. Then u_{ra} and i_{ra} are expressed as

$$\begin{aligned} u_{ra} &= u_a - j\omega L_F i_a = U_a - j\omega L_F I_a \\ i_{ra} &= i_a - j\omega C_F u_{ra} \\ &= \left(1 - \omega^2 L_F C_F\right) I_a - j\omega C_F U_a \\ &= \sqrt{\left[\left(1 - \omega^2 L_F C_F\right) I_a\right]^2 + (\omega C_F U_a)^2}\, e^{-j\tan^{-1}\left[\frac{\omega C_F U_a}{\left(1 - \omega^2 L_F C_F\right) I_a}\right]} \end{aligned} \tag{6.1}$$

From (6.1), i_{ra} always lags behind i_a at a certain angle δ:

$$\delta = \tan^{-1}\left[\frac{\omega C_F U_a}{\left(1 - \omega^2 L_F C_F\right) I_a}\right] \tag{6.2}$$

For the other phases, similar analyses can be done. The objective δ is not only related to the parameters of the input filter but also the amplitude of the grid-side voltage and current. An appropriate modulation scheme, which is introduced in Sect. 6.3, can compensate for the phase difference.

6.3 Modulation for Neutral-Point Potential Balance

The presented T3LIMC topology is modulated by synchronous modulations of the front-end rectifier and the load-side inverter, where the carrier-based modulation scheme or the SVM scheme can be adopted. Compared with the SVM scheme, the carrier-based modulation scheme has the advantages of relatively low computational burden and implementation effort [34]; thus it is used for both the rectifier and the inverter.

6.3.1 Carrier-Based Modulation Scheme for the Rectifier

The rectifier uses a dual-carrier-based modulation scheme to synthesize the expected symmetric sinusoidal input currents with desired input power factors, as well as the controllable DC-link voltage.

Assume that the three-phase input voltages are

$$\begin{cases} u_a = U_{im}\cos(\omega_i t) \\ u_b = U_{im}\cos(\omega_i t - 2\pi/3) \\ u_c = U_{im}\cos(\omega_i t + 2\pi/3) \end{cases} \tag{6.3}$$

where U_{im} and ω_i represent the amplitude and angular frequency of the input voltages, respectively.

According to the operating principles of the T3LIMC, when $u_a > u_b > u_c$ (defined as sector 1), the average values of the input currents i_{ra} and i_{rc} (denoted as $\overline{i}_{ra}$ and $\overline{i}_{rc}$) are adjusted by designing the switch-on duration of the switches S_{p+} and S_{n+}, respectively. The duty ratios of the switches S_{p+} and S_{n+}, denoted as d_1 and d_2, are given by

$$\begin{cases} d_1 = m_r\cos(\omega_i t - \varphi) \\ d_2 = -m_r\cos(\omega_i t + 2\pi/3 - \varphi) \end{cases} \tag{6.4}$$

where m_r ($0 \le m_r \le 1$) and φ are the modulation index of the rectifier and input power factor angle, respectively.

As shown in (6.4), to guarantee the duty ratios d_1 and d_2 within the ranges of [0, 1], the input power factor angle of the T3LIMC satisfies

$$-\pi/6 \le \varphi \le \pi/6 \tag{6.5}$$

Then, the average DC-link voltage $\overline{u}_{pn}$ during each switching period is calculated as

$$\overline{u}_{pn} = 1.5 m_r U_{im} \cos\varphi \tag{6.6}$$

It can be found from (6.5) and (6.6) that the maximum available VTR of the T3LIMC is $(\sqrt{3}\cos\varphi)/2$. Thus, m_r is set as 1 to maximize the VTR of the T3LIMC.

The analytical derivations for other sectors are similar to the above. Generally, the duty ratios for the rectifier are derived as

$$\begin{cases} d_1 = m_r\cos(\theta_{max} - \varphi) \\ d_2 = -m_r\cos(\theta_{min} - \varphi) \end{cases} \tag{6.7}$$

where θ_{max} and θ_{min} are the phase angles of the input phases with the maximum and the minimum instantaneous voltages, respectively.

The switching pattern of the rectifier in sector 1 is shown in Fig. 6.3, where T_s and f_s are the switching period and switching frequency, respectively.

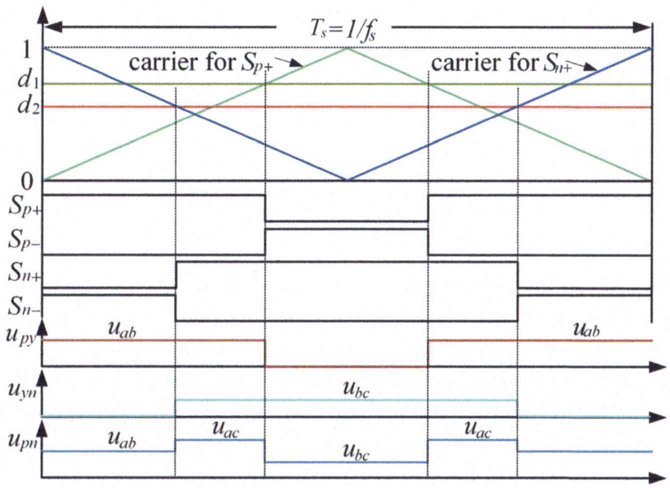

Fig. 6.3 Schematic diagram of the modulation scheme for the rectifier

6.3.2 *Implementation of the Inverter Modulation*

The multi-carrier-based modulation of the inverter is discussed in this section. Since strict synchronization between the rectifier and the inverter modulation of the T3LIMC is required, the conventional multicarrier-based modulation scheme for multilevel inverters cannot be directly applied here. The derivation of the multi-carrier-based DSPWM [35] for the T3LIMC is therefore described as follows.

Assume that the three-phase expected output voltages u_s^*, u_r^*, and u_t^*, denoted as the modulation signals, are given by

$$\begin{cases} u_r^* = U_{om}\cos\left(\omega_o t + \phi\right) \\ u_s^* = U_{om}\cos\left(\omega_o t - 2\pi/3 + \phi\right) \\ u_t^* = U_{om}\cos\left(\omega_o t + 2\pi/3 + \phi\right) \end{cases} \tag{6.8}$$

where $U_{om} = (\sqrt{3}/2)m_i m_r \cos(\varphi)\, U_{im}$, ω_o, and ϕ represent the magnitude, angular frequency, and initial phase of the expected output voltages, respectively, and $m_i (0 \le m_i \le 1)$ is the modulation index of the inverter.

First, sort the original modulation signals by instantaneous values as

$$\begin{cases} u_{max}^* = \max(u_r^*, u_s^*, u_t^*) \\ u_{mid}^* = \mathrm{mid}\,(u_r^*, u_s^*, u_t^*) \\ u_{min}^* = \min(u_r^*, u_s^*, u_t^*) \end{cases} \tag{6.9}$$

where u_{max}^*, u_{mid}^*, and u_{min}^* represent the maximum value, medium value, and minimum value of the modulation signals, respectively, and max(), mid(), and min() are the operators.

The modulation signals in (6.9) are modified by adding a common zero-sequence voltage to achieve maximum utilization of the DC-link voltage, and they are expressed as

$$\begin{cases} u_{max}^{*'} = u_{max}^{*} + u_0 \\ u_{mid}^{*'} = u_{mid}^{*} + u_0 \\ u_{min}^{*'} = u_{min}^{*} + u_0 \end{cases} \tag{6.10}$$

where u_0 is the common zero-sequence voltage, and $u_{max}^{*'}$, $u_{mid}^{*'}$, and $u_{min}^{*'}$ are the modified modulation signals.

To control the average neutral-point current and thereby balance the neutral-point potential, one degree of freedom of the modified modulation signals is introduced by decomposing them into the positive parts and the negative parts, which are

$$\begin{cases} u_{max}^{*'} = u_{maxp}^{*} + u_{maxn}^{*} \\ u_{mid}^{*'} = u_{midp}^{*} + u_{midn}^{*} \\ u_{minn}^{*'} = u_{minp}^{*} + u_{minn}^{*} \end{cases} \tag{6.11}$$

where the subscripts "p" and "n" represent the positive modulation signal and the negative modulation signal of the modified modulation signals, respectively.

From (6.10) and (6.11), the modified modulation signals are designed as

$$\begin{cases} u_{maxp}^{*} = u_{max}^{*} + u_0 \\ u_{maxn}^{*} = 0 \\ u_{midp}^{*} = 0.5u_{mid}^{*} + u_{01} \\ u_{midn}^{*} = 0.5u_{mid}^{*} + u_{02} \\ u_{minp}^{*} = 0 \\ u_{minn}^{*} = u_{min}^{*} + u_0 \end{cases} \tag{6.12}$$

where u_{01} and u_{02} are the zero-sequence voltages of the positive and negative modulation signals, respectively.

For the convenience of digital implementation, the positive and negative modulation signals are normalized as $\overline{u}_{jp}^{*}$ and $\overline{u}_{jn}^{*}$ according to the split upper and lower DC-link voltages, and

$$\begin{cases} \overline{u}_{jp}^{*} = u_{jp}^{*}/\overline{u}_{py} \\ \overline{u}_{jn}^{*} = u_{jn}^{*}/\overline{u}_{yn} \end{cases}, \quad j \in \{max, mid, min\} \tag{6.13}$$

where $\overline{u}_{py}$ and $\overline{u}_{yn}$ denote the average upper and lower DC-link voltages during a switching period, respectively.

Then, the corresponding upper and lower duty ratios of the switches, denoted as d_{jp} and d_{jn}, are given as

$$\begin{cases} d_{jp} = \overline{u}^*_{jp} \\ d_{jn} = -\overline{u}^*_{jn} \end{cases}, \ j \in \{max, mid, min\} \tag{6.14}$$

As discussed in Sect. 6.3.1, the upper and lower DC-link voltages are formed from different input line-line voltages and zero voltage dominating by d_1 and d_2. Consequently, the switching period T_s in the inverter is divided into two time-varying subperiods. The conventional carrier-based modulation process occurs twice within one switching period, and the carrier frequency varies with time. The duty ratios of the inverter are calculated as

$$\begin{cases} d_{1jp} = d_1 d_{jp} \\ d_{2jn} = d_2 d_{jn} \end{cases}, \ j \in \{max, mid, min\} \tag{6.15}$$

where d_{1jp} and d_{2jn} are the final upper and lower duty ratios, respectively.

To find feasible zero-sequence voltages for DSPWM, the zero-sequence voltages are developed from

$$\begin{cases} \overline{i}_{yi} = \sum d_{jy} i_j, \ j \in \{max, mid, min\} \\ i_{max} + i_{mid} + i_{min} = 0 \\ u^*_{max} + u^*_{mid} + u^*_{min} = 0 \\ u_0 = u_{01} + u_{02} \end{cases} \tag{6.16}$$

with the constraints

$$\begin{cases} 0 \leq \overline{u}^*_{jp} \leq 1 \\ -1 \leq \overline{u}^*_{jn} \leq 0 \end{cases}, \ j \in \{max, mid, min\} \tag{6.17}$$

where i_{max}, i_{mid}, and i_{min} represent the output currents corresponding to u^*_{max}, u^*_{mid}, and u^*_{min}, respectively; $d_{jy} = 1 - d_{1jp} - d_{2jn}$ are the duty ratios connected to the neutral terminal of the DC-link, and $\overline{i}_{yi}$ is the average neutral-point current within each switching period.

From (6.16) and (6.17), one set of the zero-sequence voltages, to keep balanced neutral-point potential and sinusoidal input currents, is given by

$$\begin{cases} u_{01} = [-0.5 u^*_{mid} (\overline{u}_{yn} - \overline{u}_{py}) - \overline{u}_{py} u^*_{min}]/\overline{u}_{pn} \\ u_{02} = [0.5 u^*_{mid} (\overline{u}_{yn} - \overline{u}_{py}) - \overline{u}_{yn} u^*_{max}]/\overline{u}_{pn} \\ u_0 = u_{01} + u_{02} \end{cases} \tag{6.18}$$

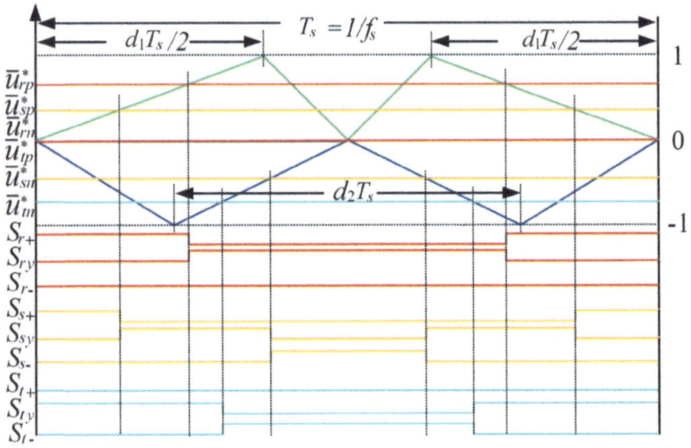

Fig. 6.4 Schematic diagram and switching pattern of the modulation scheme of the inverter of the T3LIMC

It should be noted that the average neutral-point current in each switching period is usually kept at zero for conventional NPC MCs [6], but it does not hold for the presented T3LIMC. The nonzero average neutral-point current of the inverter of the T3LIMC is constructed from (6.16), where the original discontinuous input current of the input phase with the medium instantaneous voltage is completely compensated. As a result, sinusoidal input currents and balanced neutral-point potential are achieved, and the detailed proof is presented in Sect. 6.3.3. Figure 6.4 shows the schematic diagram and switching pattern of the modulation scheme of the inverter of the T3LIMC, when $u_r^* = u_{max}^*$, $u_s^* = u_{mid}^*$, $u_t^* = u_{min}^*$.

6.3.3 Proof of Sinusoidal Input Currents

It is necessary for T3LIMC to realize sinusoidal input/output currents and balanced neutral-point potential simultaneously. According to Sect. 6.3.2, the desired sinusoidal and symmetric output currents are easy to acquire as long as balanced output voltages can be synthesized. Sinusoidal input currents, controllable input power factor, and balanced neutral-point potential are reexamined as follows.

Without loss of generality, the case that $u_a > u_b > u_c$ and $u_r^* = u_{max}^*$, $u_s^* = u_{mid}^*$, $u_t^* = u_{min}^*$ is used to prove the sinusoidal input currents and balanced neutral-point potential. Assume that the output currents are

$$\begin{cases} i_r = I_{om}\cos(\omega_o t + \phi + \varphi_L) \\ i_s = I_{om}\cos(\omega_o t - 2\pi/3 + \phi + \varphi_L) \\ i_t = I_{om}\cos(\omega_o t + 2\pi/3 + \phi + \varphi_L) \end{cases} \tag{6.19}$$

where i_r, i_s, and i_t are the three-phase output currents, and I_{om} and φ_L represent the magnitude of the output currents and the load angle, respectively.

When $u_a > u_b > u_c$ and $u_r^* = u_{max}^*$, $u_s^* = u_{mid}^*$, $u_t^* = u_{min}^*$, it is found from the previous modulation schemes that the average input current $\bar{i}_{ra}$ of phase a and $\bar{i}_{rc}$ of phase a are calculated as

$$\begin{cases} \bar{i}_{ra} = \bar{i}_p = d_{1rp}\, i_r + d_{1sp}\, i_s \\ \bar{i}_{rc} = -\bar{i}_n = d_{2sn}\, i_s + d_{2tn}\, i_t \end{cases} \tag{6.20}$$

where $\bar{i}_p$ and $\bar{i}_n$ are the averaged upper and lower DC-link currents, respectively.

The average DC-link voltages are given as

$$\begin{cases} \bar{u}_{py} = \sqrt{3}d_1\, U_{im} \cos(\omega_i t + \pi/6) \\ \bar{u}_{yn} = \sqrt{3}d_2\, U_{im} \cos(\omega_i t - \pi/2) \end{cases} \tag{6.21}$$

Substituting (6.9)–(6.15), (6.18), (6.19), and (6.21) into (6.20), $\bar{i}_{ra}$ and $\bar{i}_{rc}$ are deduced further as

$$\begin{cases} \bar{i}_{ra} = \dfrac{U_{om}\, I_{om} \cos(\varphi_L) \cos(\omega_i t - \varphi)}{U_{im} \cos(\varphi)} \\ \\ \bar{i}_{rc} = \dfrac{U_{om}\, I_{om} \cos(\varphi_L) \cos(\omega_i t - \varphi + 2\pi/3)}{U_{im} \cos(\varphi)} \end{cases} \tag{6.22}$$

As the three-phase input currents satisfy $\bar{i}_{ra} + \bar{i}_{rb} + \bar{i}_{rc} = 0$, the average input current of phase b is obtained as

$$\bar{i}_{rb} = \frac{U_{om}\, I_{om} \cos(\varphi_L) \cos(\omega_i t - \varphi - 2\pi/3)}{U_{im} \cos(\varphi)} \tag{6.23}$$

Also, the average neutral-point current during a switching period, denoted as $\bar{i}_{yi}$, is obtained from (6.16) as

$$\bar{i}_{yi} = \frac{U_{om}\, I_{om} \cos(\varphi_L) \cos(\omega_i t - \varphi - 2\pi/3)}{U_{im} \cos(\varphi)} = \bar{i}_{rb} \tag{6.24}$$

From (6.24), it clearly shows that, different from the zero average neutral-point current in the conventional multilevel IMCs [20], the average neutral-point current of the T3LIMC is exactly the average input current corresponding to the input phase with the medium instantaneous voltage with the developed scheme, and the input currents will only then be symmetric and sinusoidal.

It can be seen from (6.22) and (6.23) that symmetric and sinusoidal three-phase input currents with the desired input power factor angle are obtained. Therefore, the input filtering capacitor voltages that support the split DC-link voltages of the inverter are symmetric and sinusoidal. Since the input filtering capacitors of the T3LIMC play the same role as the split DC-link capacitors of conventional NPC inverter, it can be concluded that balanced neutral-point potential is achieved in T3LIMC. In other cases, the similar analysis can be done.

6.3.4 Commutation Issue

For the rectifier, from the T3LIMC topology and modulation scheme, switches $S_{i+/-}$ and S_{iy} $(i = a, b, c)$ are line-commutated, and the analysis focuses on high-frequency switches $S_{k+/-}$ $(k = p, n)$. Switches S_{x+} $(x = r, s, t)$ of the inverter are off when S_{p+} is switching, which is the same as S_{n+}. The DC-link currents i_p and i_n are zero during commutation so that the ZCS of S_{p+} and S_{n+} can be guaranteed. Besides, no currents flow through the S_{k-} $(k = p, n)$ due to the balanced neutral-point potential. Thereby, the rectifier realizes ZCS without considering line-commutated switches. Compared with conventional MC, these features greatly simplify the commutating processes. For bidirectional switches S_{iy} $(i = a, b, c)$, the current-based four-step commutation strategy is utilized to guarantee the normal operation of the converter [5]. Besides, the commutation mechanism of the load-side inverter is the same as the conventional T-type three-level inverter. When the load side inverter stopped, the front end rectifier has to keep working until the load inductors is deenergized completely.

6.4 Experimental Studies

To validate the theoretical analyses and effectiveness, the presented topology and modulation schemes are tested on a laboratory prototype shown in Fig. 6.5. The main circuit is shown in Fig. 6.1, where fast commutated and line frequency-commutated switches are constructed by IGBT IKW40N65ES5 and IKW30N65EL5, respectively. The T3LIMC drives an IM with the delta-winding connection. A PMSM, which is equipped with rated power $P_{rate} = 1.1$ kW and rated speed $n_{rate} = 1390$ rpm, is used as the load. The specific system (T3LIMC and IM) parameters are listed in Table 6.1, and the dead time is set to 1 μs. The vector control is adopted to control the IM [36], as shown in Fig. 6.6. When the input voltage is low, the field-weakening control is used to reach the desired speed. All control algorithms are executed on a floating-point DSP TMS320F28335. Currents and voltages are measured through Hall-effect current sensors and voltage sensors severally, which are transmitted to the DSP by the signal processing circuit. To facilitate the analysis, all experimental data are acquired by external probes and an oscilloscope (Tektronix MDO 3014) and then imported into MATLAB for redrawing.

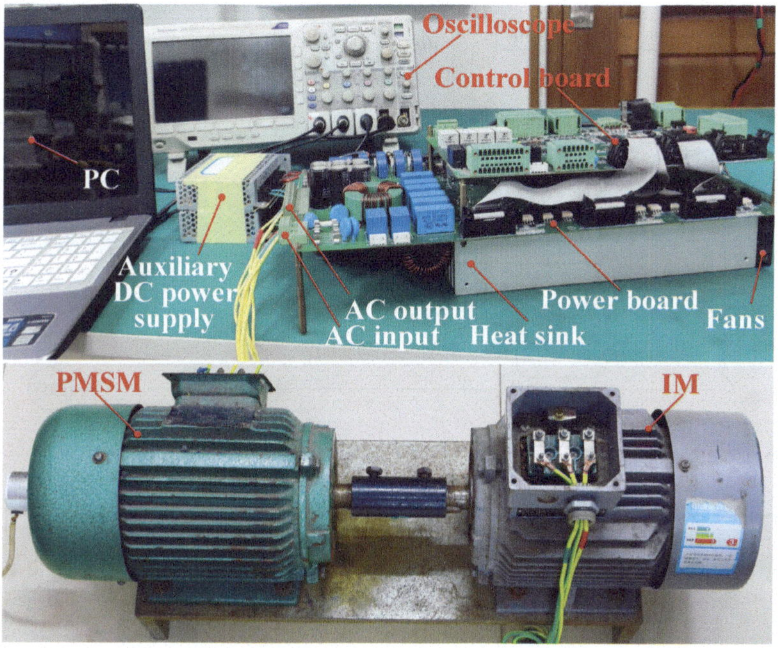

Fig. 6.5 Experimental setup of T3LIMC-IM system

Table 6.1 Parameters of T3LIMC-IM system

Parameters	Value
Input line-line voltage, input frequency	220 Vrms, 50 Hz
Switching frequency f_s	8 kHz
Inductor L_F	200 μH
Capacitor C_F	6.6 μF
Rated power	1.1 kW
Rated current	2.89 A
Rated speed n_{rate}	1390 rpm
Rated torque	6 N·m
Flux amplitude reference ψ_n	0.5 Wb
Stator resistance R_s	6.4 Ω
Stator resistance R_r	4.8 Ω
Stator self-inductance L_s	0.364 H
Rotor self-inductance L_r	0.364 H
Magnetizing inductance L_m	0.336 H
Moment of inertia J	0.02 kg $\cdot\ m^2$
Number of pole pairs n_p	2

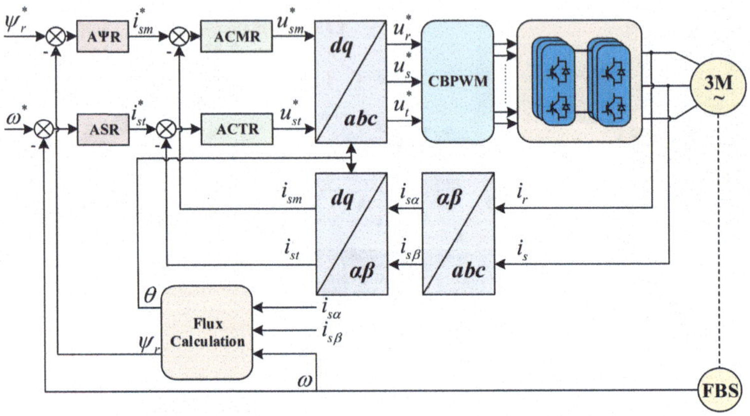

Fig. 6.6 Control diagram of the T3LIMC-based IM system

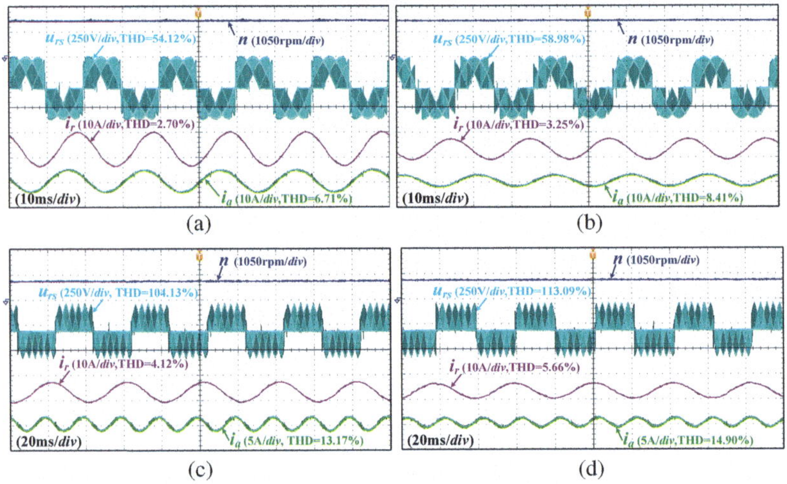

Fig. 6.7 Experimental waveforms of the T3LIMC-IM system with different n_r and P. (**a**) n_r =1390 rpm and P = 1.0 kW. (**b**) n_r =1390 rpm and P = 0.55 kW. (**c**) n_r =700 rpm and P = 0.3 kW. (**d**) n_r =700 rpm and P = 0.2 kW

6.4.1 *Steady-State Performance*

Figure 6.7 shows the waveforms of the rotor speed n, the output line-line voltage u_{rs}, the output current i_r and the input current i_a of the T3LIMC-IM system with different rotor speed reference n_r and load power P, and the desired input power factor angle is 0. It is apparent that the rotor speed tracks the reference well and sinusoidal input/output currents are acquired. Note that the balanced neutral-point potential is necessary for sinusoidal input currents. Therefore, the experimental results also indirectly verify the neutral-point potential balancing ability of the developed modulation schemes. As expected, when the modulation index (MI) that proportional to the rotor speed changes from high to low, the output line-line voltage also transitioned from the five-level waveforms in Fig. 6.7a and b to the three-level waveforms in Fig. 6.7c and d accordingly. Only small, medium, and zero vectors are used for synthesizing the expected output voltage vector when m_i is low, while five distinctive levels are required at a high modulation index of the inverter.

In the experiment, the input currents are also affected by many external factors such as sampling precision, ADC accuracy, fluctuations in actual parameters, tiny error of PLL, the added dead time, the voltage ripple of the input filtering capacitors, etc. Moreover, the input currents of the T3LIMC are indirectly formed by the synchronous modulations of the rectifier and the inverter; thus they have a low tolerance for the power quality of the grid. In other words, the distortion of the grid will be amplified in the input currents. Figure 6.8a and b displays the harmonic spectra of u_{rs} corresponding to Fig. 6.7a and c. It serves to show that the five-

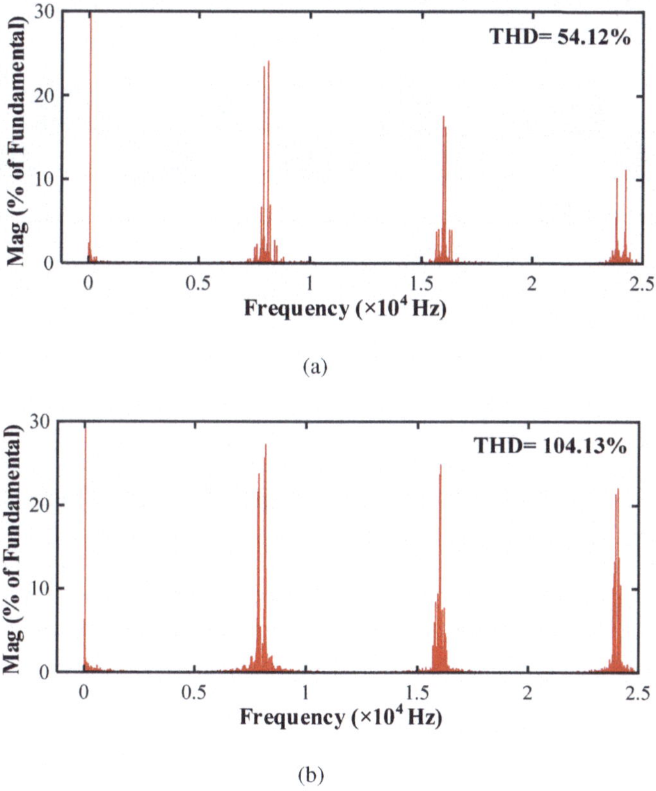

Fig. 6.8 Harmonic spectra of output line-line voltage u_{rs} with (**a**) n_r =1390 rpm, P = 1.0 kW and (**b**) n_r =700 rpm, P = 0.3 kW

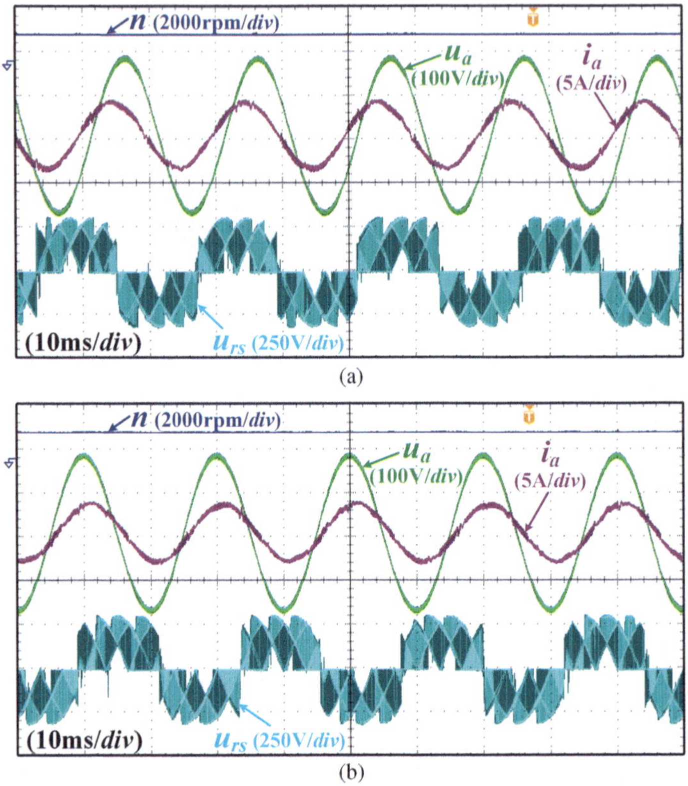

Fig. 6.9 Experimental waveforms of the T3LIMC-IM system with non-unity input power factors. (**a**) $\varphi = -\pi/6$. (**b**) $\varphi = \pi/6$

level output line-line voltage (corresponding to three-level output phase voltage) possesses the lower THD, and the concentrated harmonics are easy to filter.

When non-unity input power factors occur, the experimental results at $n_r = 1200$ rpm are shown in Fig. 6.9, the expected input power factor angles in Fig. 6.9a and b are $-\pi/6$ and $\pi/6$, respectively. It can be observed that the desired input power factors are obtained, and sinusoidal input and output currents are also achieved, which demonstrates the controllable input power factor ability of the T3LIMC. The harmonic spectra of u_{rs} in Fig. 6.9 are shown in Fig. 6.10, which indicates that good output performance remains. Besides, the maximum of fundamental output line-line voltage decreases because of the reduced average DC-link voltage $\overline{u}_{pn}$ according to (6.6).

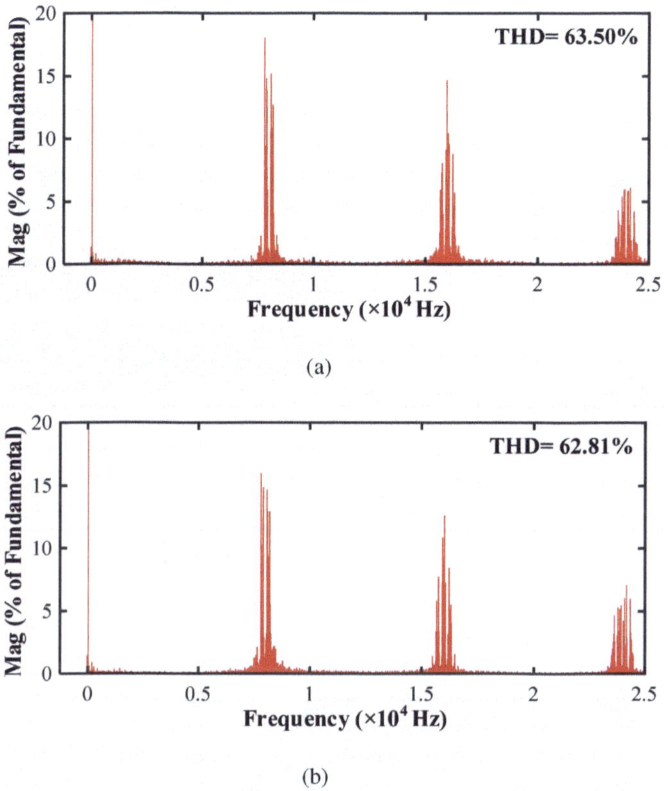

Fig. 6.10 Harmonic spectra of output line-line voltage u_{rs} with non-unity input power factors. (**a**) $\varphi = -\pi/6$. (**b**) $\varphi = \pi/6$

6.4.2 Dynamic Performance

Figure 6.11 shows the dynamic responses of n, u_{rs}, i_r, and i_a of the T3LIMC-IM system when the induction motor starts from a standstill to 700 rpm and then to 1390 rpm at full load ($P = 1.0\,\text{kW}$). Quick and smooth acceleration processes and strong performance of tracking reference come up.

Experimental results of responses during speed reversal at rated speed and full load are presented in Fig. 6.12. It is obvious that the magnitude of rotor flux ψ_r can always be maintained as a constant regardless of the change in speed. Meanwhile, there are no excessive fluctuations in the output current and voltage. The stable and smooth transition waveforms are apparent.

Figure 6.13 displays the dynamic waveforms with the sudden increase of external load, where t_1 is the turning instant. The rotor speed is quickly restored to 1390 rpm without fluctuations, which demonstrates good speed tracking performance. Above all, the good dynamic performance is verified in the presented T3LIMC.

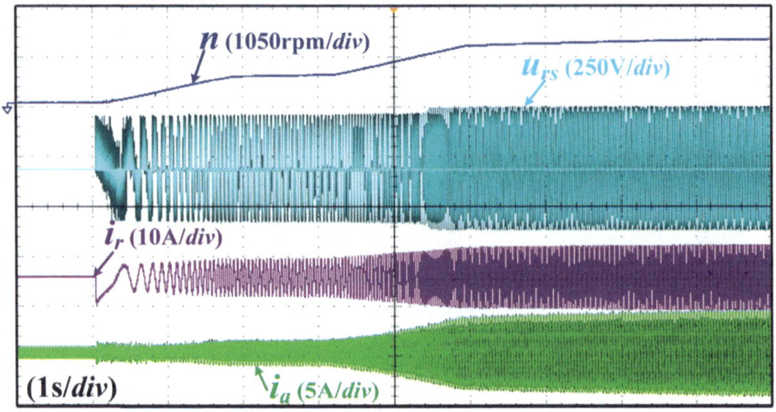

Fig. 6.11 Experimental waveforms of two-stage accelerating from a standstill

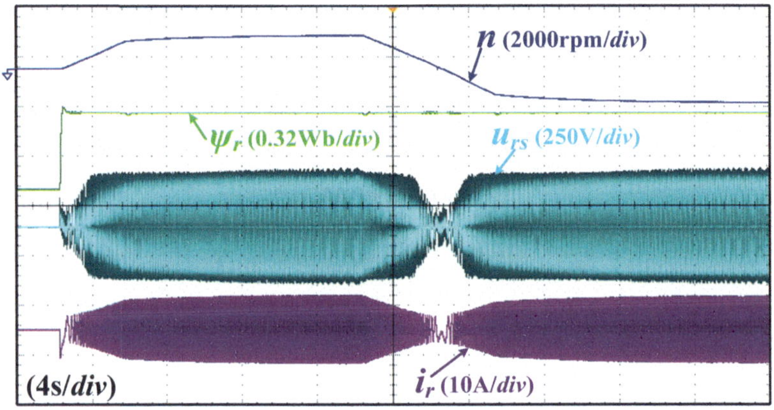

Fig. 6.12 Experimental waveforms of responses during speed reversal at 1390 rpm

Fig. 6.13 Experimental dynamic waveforms with the sudden increase of external load

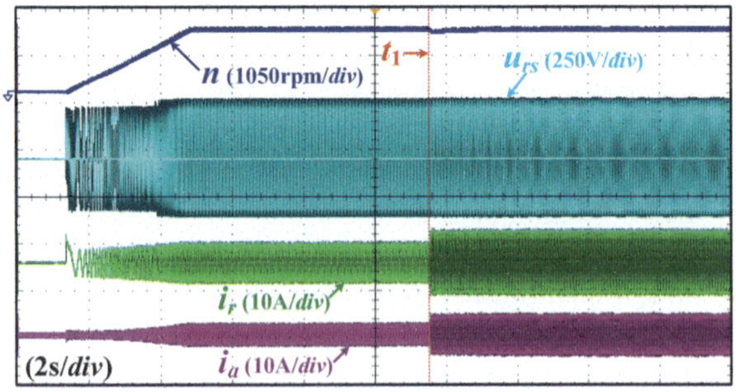

6.4.3 Efficiency Evaluation

To evaluate the conversion efficiency of the T3LIMC, first, the power losses and system efficiency with forward power flow are calculated. From the topology and modulation scheme of T3LIMC, the loss analysis of the inverter is the same as that of a conventional T-type three-level inverter, and the loss calculation process of the rectifier is described as follows.

(1) *Conduction Losses:* It can be seen that no currents flow through the switches $S_{i+/-}$ $(i = a, b, c)$ and S_{k-} $(k = p, n)$ due to the balanced neutral-point potential. The conduction losses of the rectifier occur only in the diodes and switches S_{k+} $(k = p, n)$ and S_{iy} $(i = a, b, c)$. For the diodes, the conduction losses of phase A are calculated as

$$P_{cD\_A} = \frac{1}{2\pi}\int_0^{2\pi} V_F(|i_a|)\cdot|i_a|\,d\theta \tag{6.25}$$

where $V_F(i)$ is acquired from the datasheet and I_{im} is the amplitude of i_a.

For switch S_{p+}, the conduction loss is calculated as

$$P_{cS\_p} = \frac{1}{2\pi}\int_0^{2\pi} V_{CE1}(|i_p|)\cdot|i_p|\,d\theta \tag{6.26}$$

where $V_{CE1}(i)$ is acquired from the datasheet.

For switches S_{iy} $(i = a, b, c)$, i_{yr} must flow through one and only one of these at the same time, and the conduction losses are calculated as

$$P_{cS\_y} = \frac{1}{2\pi}\int_0^{2\pi} \left(V_{CE2}(|i_{yr}|)\right)\cdot|i_{yr}|\,d\theta \tag{6.27}$$

where $V_{CE2}(i)$ is acquired from the datasheet.

(2) *Switching Losses:* According to the analysis of commutation in Sect. 6.3.4, the switching loss of the rectifier is approximately zero.

(3) *Overall Losses:* According to the symmetry of the topology, the overall losses of the rectifier are calculated as

$$P_{loss\_all} = 3P_{cD\_A} + 2P_{cS\_p} + P_{cS\_y} \tag{6.28}$$

Then the practical system efficiency is measured by the HIOKI 3390 three-phase wattmeter. The calculated efficiency and the measured one are shown in Fig. 6.14. It should be noted that in the efficiency diagram, the auxiliary power consumed by the control electronics, the IGBT drivers, and the cooling fans are included.

As can be seen from Fig. 6.14, the peak calculated efficiency of the T3LIMC is 96.74%. In contrast, the peak measured efficiency reaches 96.65%, which is a bit lower than the calculated one but is still satisfactory for a small power rating

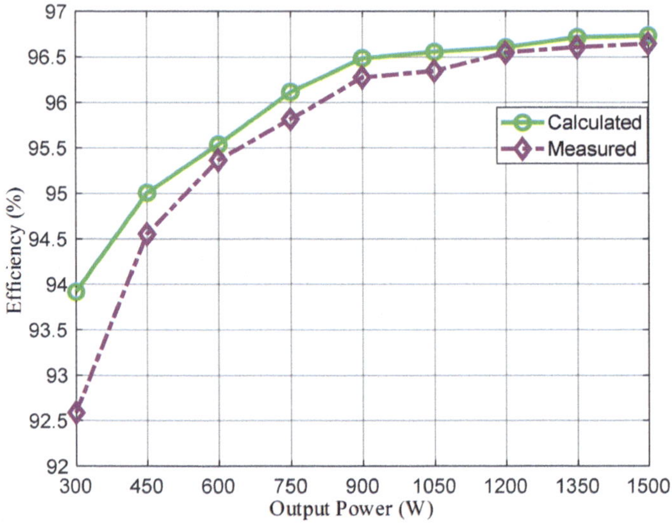

Fig. 6.14 Calculated and measured efficiency of the T3LIMC prototype

converter. This can be explained by the fact that the losses dissipated on input filtering capacitors, print-circuit board traces, and current sensors are not considered in theoretical calculation. The good efficiency of the T3LIMC is attributed to the low switching losses of the converter. As can be known from the operating principles and modulation schemes of the T3LIMC, the four high-frequency switched switches of the rectifier switch at zero current, and the rest switches of the rectifier commutate at line frequency. Consequently, the switching losses of the rectifier can be ignored. Moreover, due to the lower commutating voltage, the switching losses of the inverter are also decreased significantly when compared to the conventional two-level IMCs, especially for the high switching frequency condition.

6.4.4 Performance Comparisons

In this section, the input and output performance of the T3LIMC and other three-level IMCs are evaluated and compared, where comprehensive comparisons of the THDs of the input and output waveforms with different three-level IMCs topologies are carried out.

From the perspective of topological structure, the T3LIMC is similar to other three-level IMCs since they are all featured by the combination of a current-source-type rectifier and a three-level inverter. To evaluate the input and output waveform performance of the developed T3LIMC, the THDs of the input current and the output voltage of the T3LIMC are compared with the so-called T^3HIMC and the T^2HIMC. The comparison results are given in Fig. 6.15.

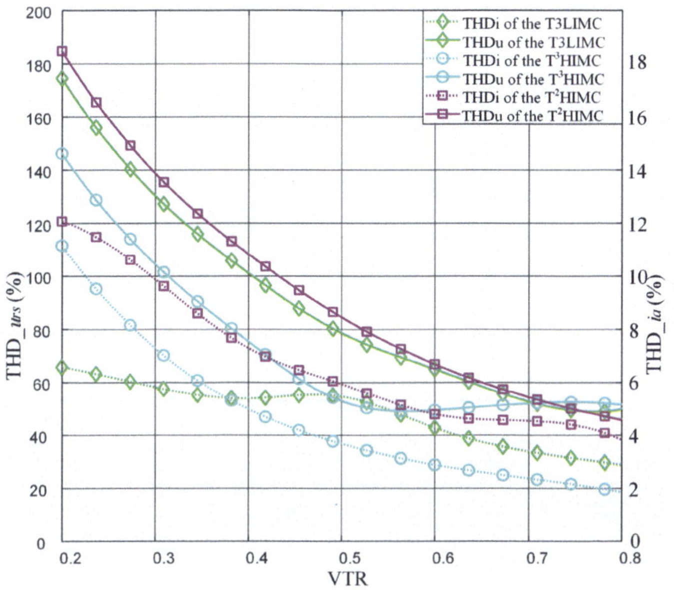

Fig. 6.15 THD comparisons of the input current and the output voltage for different three-level IMCs

For the aspect of output waveform performance, THDs of the output voltage of the T3LIMC are superior to those of the T^2HIMC almost throughout the VTR region, while they are inferior to those of the T^3HIMC when VTR <0.71 and slightly better than that of the T^3HIMC within high VTR region. This is because output voltages of the T3LIMC are synthesized by the input phase to phase voltages and zero voltage, while they are input phase to neutral voltages and the input phase to phase voltages for T^3HIMC and the input phase to phase voltages for T^2HIMC, respectively. This is one of the advantages of T3LIMC in ASDs application because the converter usually operates within a high speed and VTR region.

T3LIMC is better than T^2HIMC in regard to the input current performance during the whole VTR range. This can be attributed to the fact that the input currents, which are synthesized by the inverter instead of both the third-harmonic injection circuit and the inverter for the T^2HIMC, are not compromised by the control and tracking performance of the third-harmonic injection circuit. Besides, the third-harmonic injection circuit of the T^2HIMC also undertakes the task of synthesizing the appropriate third-harmonic injection inductor current to compensate for the neutral-point current generated by the inverter, which will introduce more high-frequency components of the input currents.

Therefore, the T3LIMC has decent input current THD performance and moderate output voltage THD performance among the three candidates. Moreover, compared with the T^3HIMC and the T^2HIMC, the advantages, such as good input/output waveforms and natural neutral-point potential balancing ability, will make the

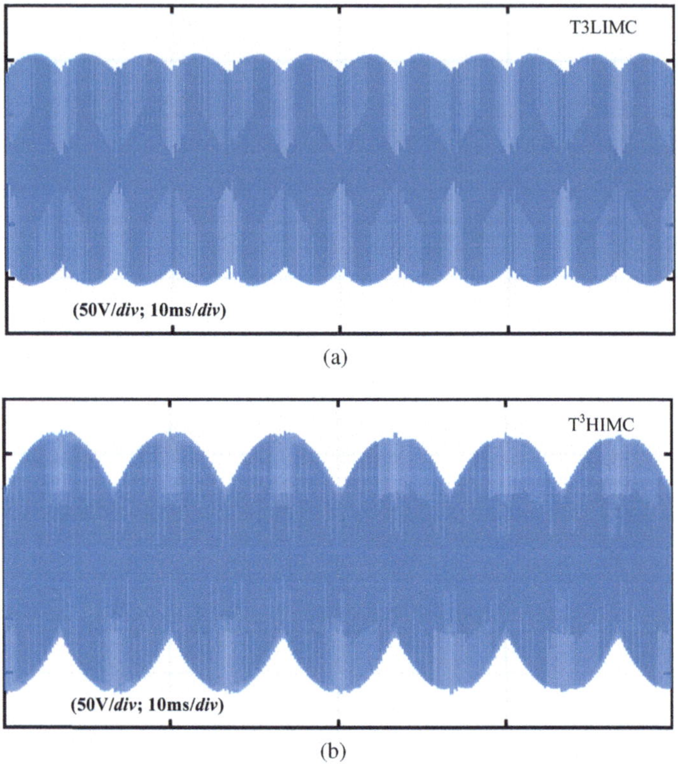

Fig. 6.16 Common-mode voltage waveforms of (**a**) T3LIMC and (**b**) T^3HIMC

T3LIMC be a good candidate in some applications like ASDs. Besides, the common-mode voltage waveforms of T3LIMC and T^3HIMC are given in Fig. 6.16. The results show that T3LIMC is slightly superior to T^3HIMC in the common-mode voltage amplitude.

6.5 Conclusions

In this chapter, a T-type three-level IMC topology and its modified multicarrier-based modulation schemes are presented. By designing a specific zero-sequence voltage in the carrier-based modulation scheme and thereby constructing the desired neutral-point current of the inverter, the balancing ability of neutral-point potential, ZCS operation of rectifier, controllable input power factor, and multilevel output are simultaneously achieved with low complexity. Experimental results validate the developed topology and its modulation schemes. Meanwhile, the good steady-state and dynamic performance of T3LIMC is demonstrated in the motor drive system.

Due to its merits such as good output performance, low complexity, high efficiency, and high power density, T3LIMC has some advantages and a good prospect in AC-AC power conversion applications such as WECS, electric airplanes, and marine electric propulsion.

References

1. J.W. Kolar, T. Friedli, J. Rodriguez, P.W. Wheeler, Review of Three-Phase PWM AC–AC Converter Topologies. IEEE Trans. Ind. Electron. **58**(11), 4988–5006 (2011)
2. L. Huber, D. Borojevic, Space vector modulated three-phase to three-phase matrix converter with input power factor correction. IEEE Trans. Ind. Appl. **31**(6), 1234–1246 (1995)
3. L. Wei, T. A. Lipo, A novel matrix converter topology with simple commutation, in *Conference Record of the 2001 IEEE Industry Applications Conference. 36th IAS Annual Meeting*, vol. 3 (2001), pp. 1749–1754
4. D. Casadei, G. Serra, et al., Matrix converter modulation strategies: a new general approach based on space-vector representation of the switch state. IEEE Trans. Ind. Electron. **49**(2), 370–381 (2002)
5. P. Wheeler, J. Rodriguez, et al., Matrix converters: a technology review. IEEE Trans. Ind. Electron. **49**(2), 276–288 (2002)
6. J.W. Kolar, F. Schafmeister, et al., Novel three-phase AC-AC sparse matrix converters. IEEE Trans. Power Electron. **22**(5), 1649–1661 (2007)
7. J. Rodriguez, M. Rivera, et al., A review of control and modulation methods for matrix converters. IEEE Trans. Ind. Electron. **59**(1), 58–70 (2012)
8. H. Wang, M. Su, et al., Two-stage matrix converter based on third-harmonic injection technique. IEEE Trans. Power Electron. **31**(1), 533–547 (2016)
9. H. Wang, M. Su, et al., Active third-harmonic injection indirect matrix converter with dual three-phase outputs. IET Power Electron. **9**(4), 657–668 (2016)
10. H. Nguyen, H. Lee, et al., A modulation scheme for matrix converters with perfect zero common-mode voltage. IEEE Trans. Power Electron. **31**(8), 5411–5422 (2016)
11. H. Wang, Y. Zhang, et al., Control method for the two-stage matrix converter to enhance the linear voltage transfer ratio. IET Power Electron. **11**(14), 2295–2301 (2018)
12. M. Su, J. Lin, et al., A new modulation strategy to reduce common mode current of indirect matrix converter. IEEE Trans. Ind. Electron. **66**(9), 7447–7452 (2019)
13. W. Xiong, Y. Sun, et al., A cost-effective and low-complexity predictive control for matrix converters under unbalanced grid voltage conditions. IEEE Access **7**, 43895–43905 (2019)
14. J. Lei, S. Feng, B. Zhou, H.-N. Nguyen, J. Zhao, W. Chen, A simple modulation scheme with zero common-mode voltage and improved efficiency for direct matrix converter-fed PMSM drives. IEEE J. Emerg. Sel. Topics Power Electron. **8**(4), 3712–3722 (2020)
15. T. Mir, B. Singh, A.H. Bhat, FS-MPC based speed sensorless control of matrix converter fed induction motor drive with zero common mode voltage. IEEE Trans. Ind. Electron. **68**(10), 9185–9195 (2021)
16. R.W.G. Bucknall, K.M. Ciaramella, On the conceptual design and performance of a matrix converter for marine electric propulsion. IEEE Trans. Power Electron. **25**(6), 1497–1508 (2010)
17. S. Kuznetsov, Advanced AC to AC power conversion techniques for naval ship propulsion drives, in *2011 IEEE Electric Ship Technologies Symposium, Alexandria, VA* (2011), pp. 420–423
18. R. Cardenas, R. Pena, et al., Control of the reactive power supplied by a WECS based on an induction generator fed by a matrix converter. IEEE Trans. Ind. Electron. **56**(2), 429–438 (2009)

19. J. Monteiro, J. Silva, et al., Matrix converter-based unified power-flow controllers: advanced direct power control method. IEEE Trans. Power Del. **26**(1), 420–430 (2011)
20. X. Li, Y. Sun, et al., Modulation methods for indirect matrix converter extending the input reactive power range. IEEE Trans. Power Electron. **32**(6), 4852–4863 (2017)
21. J. Kang, E. Yamamoto, et al., Medium-voltage matrix converter design using cascaded single-phase power cell modules. IEEE Trans. Ind. Electron. **58**(11), 5007–5013 (2011)
22. W. Jiacheng, W. Bin, et al., Multimodular matrix converters with sinusoidal input and output waveforms. IEEE Trans. Ind. Electron. **59**(1), 17–26 (2012)
23. W. Jiacheng, W. Bin, et al., Phase-shifting transformer-fed multimodular matrix converter operated by a new modulation strategy. IEEE Trans. Ind. Electron. **60**(10), 4329–4338 (2013)
24. Y. Sun, W. Xiong, et al., Modulation strategies based on mathematical construction method for multi-modular matrix converter. IEEE Trans. Power Electron. **31**(8), 5423–5434 (2016)
25. Y. Sun, W. Xiong, et al., Carrier-based modulation strategies for multi-modular matrix converters. IEEE Trans. Ind. Electron. **63**(3), 1350–1361 (2016)
26. S. Yong, Y. Xu, et al., Research on a novel capacitor clamped multilevel matrix converter. IEEE Trans. Power Electron. **20**(5), 1055–1065 (2005)
27. X. Lie, J.C. Clare, et al., Capacitor clamped multilevel matrix converter space vector modulation. IEEE Trans. Ind. Electron. **59**(1), 105–115 (2012)
28. P.C. Loh, F. Blaabjerg, et al., Pulse width modulation of neutral-point-clamped indirect matrix converter. IEEE Trans. Ind. Appl. **44**(6), 1805–1814 (2008)
29. Y. Sun, W. Xiong, et al., Topology and modulation for a new multilevel diode-clamped matrix converter. IEEE Trans. Power Electron. **29**(12), 6352–6360 (2014)
30. M.Y. Lee, P. Wheeler, et al., Space-vector modulated multilevel matrix converter. IEEE Trans. Ind. Electron. **57**(10), 3385–3394 (2010)
31. L. Wang, H. Wang, et al., A three-level T-type indirect matrix converter based on the third-harmonic injection technique. IEEE J. Emerg. Sel. Topics Power Electron. **5**(2), 841–853 (2017)
32. H. Wang, M. Su, et al., Topology and modulation scheme of a three-level third-harmonic injection indirect matrix converter. IEEE Trans. Ind. Electron. **64**(10), 7612–7622 (2017)
33. M.F. Vancu, T. Soeiro, et al., Comparative evaluation of bidirectional buck-type PFC converter systems for interfacing residential DC distribution systems to the smart grid, in *Proceedings of the 38th Annual Conference IEEE Industrial Electronics Society* (2012), pp. 5153–5160
34. Y.-D. Yoon, S.-K. Sul, Carrier-based modulation technique for matrix converter. IEEE Trans. Power Electron. **21**(6), 1691–1703 (2006)
35. J. Pou, J. Zaragoza, et al., Fast-processing modulation strategy for the neutral-point-clamped converter with total elimination of the low-frequency voltage oscillations in the neutral point. IEEE Trans. Ind. Electron. **54**(4), 2288–2294 (2007)
36. P. Vas, *Vector Control of AC Machines* (Oxford University Press, New York, 1990)

Chapter 7
Third-Harmonic Injection Micro-inverter for AC Photovoltaic Module Applications

7.1 Introduction

In recent years, the issues of electrical energy shortage and environmental pollution have resulted in the increasing demand of renewable energy sources, such as wind power, hydrogen energy, solar energy, and so on. Among these renewable energy sources, the PV sources have the advantages of no supply limitations, emission-free, and high reliability, and thus the number of installations over the world has experienced a steep rise [1, 2]. In 2016, the total installed capacity of PV has reached to about 300 GW, with an annual increased capacity of 70 GW and an annual growth rate of 30%. Meanwhile, the average price per watt of the PV module in 2016 was \$0.36/Wp, and the accumulative reduction in the price is roughly 80% during 2010–2016 [3–5]. Moreover, the PV sources are expected to be the biggest contributors to power generation among all renewable energy sources by 2040 [2, 6–12].

Generally, the grid-interactive PV systems can be divided into three categories: the centralized inverter system, the string inverter system, and the AC module system [13–18]. Among these, the centralized and string inverter systems are superior in terms of conversion efficiency, unit cost, power density, and system monitoring/maintenance and thus become the main form in the practical applications with a great commercial success. In respect to the residential and commercial applications, the centralized and string inverter systems are used, where the PV panels are usually connected in series so as to obtain a sufficiently high DC voltage required by the central inverters. In these cases, the output currents of all PV panels are equal to the string current, and therefore it is difficult to achieve global MPPT when there are parameters mismatch between PV modules. Consequently the efficiency of energy harvesting significantly reduces. Besides, the centralized and string inverter systems also have some other disadvantages such as the safety problem due to DC wires, no expandability, and so on [16, 18].

Different from the centralized and string inverter systems, the AC module system is a kind of PV system with small capacity, which integrates the inverter into the

H. Wang, *Advanced Matrix Converters*, Power Systems,
https://doi.org/10.1007/978-3-031-69902-3_7

PV panel and directly converts the low DC voltage to high AC voltage compatible with the grid. By equipping each PV panel with a micro-inverter, the PV panels are operated independently in their respective MPP, and hence the issue of the power generation reduction caused by module mismatch is eliminated. Compared with the centralized and string inverter systems, the PV micro-inverter has the advantages of improved energy harvesting, friendly "plug-and-play" operation, enhanced flexibility and expandability, excellent system redundancy and no DC cabling/safety issue, although the challenges of reducing unit cost and enhancing conversion efficiency in the PV micro-inverter system remain. Because of these advantages, the PV micro-inverters have attracted an increasing attention and became competitive alternatives to the centralized and string inverters for PV system, and many PV micro-inverter topologies and control schemes have been proposed and discussed [16–33]. However, several problems of the conventional micro-inverters still exist.

Firstly, the power density and reliability issues caused by the energy storage elements are the main concerns for most of the existing PV micro-inverters. For the conventional single-phase micro-inverters, the imbalance of the instantaneous power between the PV side and the grid side causes power pulsation at twice the line frequency, which would affect the MPPT of the PV panel; therefore the bulky electrolytic capacitors are usually equipped at the PV input side instead of the active power decoupling techniques [18, 20, 30–34], which served as the energy buffering elements in many commercial micro-inverters. Although it has the advantage of the low cost, the electrolytic capacitors inevitably introduce challenges on the power density and the reliability, which is sensitive to panel-integrated micro-inverters [21–26]. Secondly, most of the micro-inverters consist of a quasi-single-stage architecture and utilize an unfolding bridge to realize DC/AC conversion [17, 21–26]. As a result, the micro-inverters could only be operated in unity output power factor condition and cannot provide reactive power compensation, which makes the PV micro-inverters fail to meet the requirements of some electric power regulations [35].

In this chapter, to solve the power density/reliability issues caused by the bulky energy storage elements and improve the output reactive power control range, a three-phase micro-inverter topology without bulky energy storage capacitors is presented and investigated systematically. The proposed micro-inverter mainly consists of a DC/DC stage including four flyback converters, a third-harmonic injection circuit derived from [36–40], and a line-commutated CSI. The flyback converters are used to realize independent MPPT of each PV panel, and the third-harmonic injection circuit is responsible for output PFC, while the CSI is commutated at line frequency so as to accomplish the DC/AC inversion. In addition to possessing the advantageous features of improved energy harvesting, "plug-and-play" operation, enhanced modularity/flexibility, and excellent system redundancy, the developed micro-inverter has the potential of achieving high efficiency/reliability, extended output reactive power control range, and three-phase balanced output. This chapter is organized as follows: Sect. 7.2 introduces the topology and operating principles of the micro-inverter, followed by the mathematical verification of sinusoidal output currents, Sect. 7.3 presents the control algorithm for tracking MPP and the control scheme of the third-harmonic injection circuit, Sect. 7.4 shows the simulation and experimental results to verify the presented methods, and Sect. 7.5 draws the final conclusion of this chapter.

7.2 Circuit Configuration and Operating Principles

7.2.1 Circuit Configuration

The proposed 1000 W three-phase micro-inverter directly connects four adjacent 250 W PV panels, and the system configuration is shown in Fig. 7.1, which consists of a DC/DC conversion stage, an active third-harmonic current injection circuit, a three-phase line-commutated CSI, and an output *LC* filter. The DC/DC conversion stage is composed of four flyback converters, and each flyback converter consists of two interleaved sub-converters.

It should be noted that different from the single PV input situation in [21–26], the developed micro-inverter has four PV input terminals and is installed directly on the back of one PV panel, and each PV panel is connected to one of the PV input terminals of the micro-inverter. Thus, each PV panel is operated independently with its own Mpp. Usually, in the conventional micro-inverters, a bulky PV input capacitor is set to decouple the power pulsation at twice the line frequency. Differently, the input capacitor C_i of the suggested micro-inverter is used only for providing high-frequency input pulse current of the flyback converter. Therefore, ceramic capacitors with small capacitance, long lifespan, and high reliability are adopted instead of the electrolytic capacitors.

The line-commutated CSI consists of three arms and three bidirectional switches. Compared with the conventional VSIs, the selected CSI has the following advantages: (1) The switches of the CSI are commutated with line frequency here, the switching losses of power switches are ignored, therefore the power losses are relatively low, and high conversion efficiency is achieved, and (2) the CSI realizes

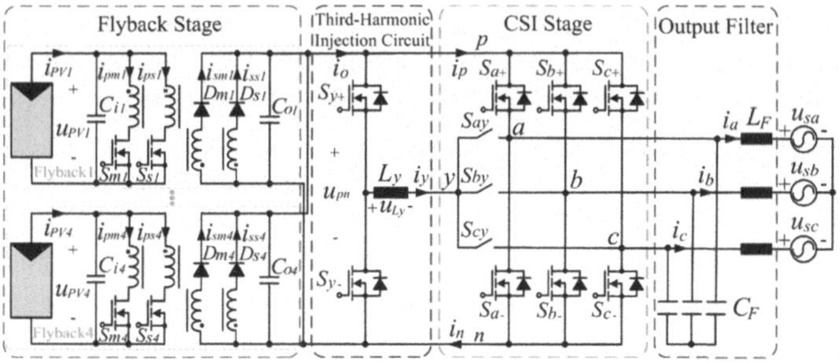

Fig. 7.1 Schematic diagram of the proposed micro-inverter topology

DC/AC inversion without any energy storage capacitors, and consequently high power density and reliability of the converter are attainable.

The output filter consisting of inductor L_F and film capacitor C_F is mainly used for smoothing the pulse currents generated by the inverter so as to produce three-phase sinusoidal output currents. The active third-harmonic current injection circuit is composed of a third-harmonic injection inductor and a bridge leg, which constitutes the topology used for AC PV module applications. By proper control of the third-harmonic injection circuit, extended output reactive power control range is achieved.

The DC/DC stage and the CSI work together to accomplish the energy conversion process and cannot work independently. Therefore, the presented topology is similar to the conventional single-phase micro-inverters in terms of construction approach of the topology, operating principles, and functionality and may be regarded as the generalization of the conventional micro-inverter from single-phase to three-phase.

7.2.2 Operating Principles

Assuming that the three-phase grid voltages are symmetrical and sinusoidal, the operating principles of the suggested three-phase micro-inverter are described as follows. For the flyback stage, each flyback converter tracks the respective MPP of the corresponding PV module, and the two sub-converters of each flyback converter are placed in parallel and interleaved with 180° phase shift. By interleaving the two sub-converters, the input/output current ripples are reduced, and the power range of each flyback converter is extended. The third-harmonic injection circuit is essentially a time-sharing multiplexing single-phase shunt active power filter, and the switches S_{y+} and S_{y-} working in high frequency are switched to generate the desired quasi-third-harmonic current i_y, which injects into the output phase with the minimum absolute voltage, and then the PFC of the output side is achieved.

The line-commutated CSI can be regarded as a grid voltage selector, which connects output phase a, b, c to either node p, y, or n periodically according to the instantaneous values of the output phase voltages. Take the case of $u_{sa}>u_{sb}>u_{sc}$, for example, only the switch S_{a+} in the upper arms, the switch S_{c-} in the lower arms, and the bidirectional switch S_{by} are in the on-state, in order to impose the piecewise six-pulse shape voltage u_{pn} across the intermediate DC-link. In conclusion, the switches of the CSI are worked according to Table 7.1, where θ_{sa} is the phase angle of the output phase voltage u_{sa}.

As described above, the switches of the CSI are commutated at line frequency, and the DC-link voltage exhibits a six-pulse shape waveform, which implies that the flyback stage, the third-harmonic injection circuit, and the CSI stage can be controlled independently. Therefore, this section mainly focuses on the analysis of the operating principle of the flyback stage. Assuming that the three-phase output voltages satisfy $u_{sa}>u_{sb}>u_{sc}$ (denoted as sector 1) and only one sub-converter

Table 7.1 Switching states of the CSI

θ_{sa}	Sector	S_{ay}	S_{by}	S_{cy}	S_{a+}	S_{a-}	S_{b+}	S_{b-}	S_{c+}	S_{c-}
0-$\pi/3$	1	0	1	0	1	0	0	0	0	1
$\pi/3$-$2\pi/3$	2	1	0	0	0	0	1	0	0	1
$2\pi/3$-π	3	0	0	1	0	1	1	0	0	0
π-$4\pi/3$	4	0	1	0	0	1	0	0	1	0
$4\pi/3$-$5\pi/3$	5	1	0	0	0	0	0	1	1	0
$5\pi/3$-2π	6	0	0	1	1	0	0	1	0	0

operating in CCM is included in the topology, the detailed analysis of the micro-inverter in a switching period is explained as follows.

Stage I ($S_{m1} = 1$): Figure 7.2a shows the equivalent circuit of the first stage. In this stage, the switch S_{m1} of flyback converter1 is ON. At the same time, the rectifier diode D_{m1} is in the reverse blocking state, and the secondary current i_{sm1} is zero. Then, the input current i_{pm1}, as well as the magnetizing current i_{mm1}, linearly increases, and the magnetizing current is expressed as

$$L_{m1}di_{mm}/dt = u_{PV1} \tag{7.1}$$

where L_{m1} is the magnetizing inductance of the transformer, and u_{PV1} is the PV voltage.

Stage II ($S_{m1} = 0$): The equivalent circuit of this stage is shown in Fig. 7.2b. In this stage, the switch S_{m1} is turned off, and the input current i_{pm1} is zero. The magnetizing current i_{mm1} linearly decreases, and the magnetizing current is expressed as

$$L_{m1}di_{mm1}/dt = -u_{pn}/N = -(u_{sa} - u_{sc})/N \tag{7.2}$$

where N is the turns ratio of the flyback transformer. Besides, the rectifier diode D_{m1} is in conduction state, and the energy stored in the primary winding is released to the output side. Omitting the effects of the snubber capacitor C_{o1}, the secondary current i_{sm1} and the output current i_o could be obtained by

$$i_o = i_{sm1} = i_{mm1}/N \tag{7.3}$$

Figure 7.2c shows the partial key waveforms of the micro-inverter in a switching period T_s, where f_s is the switching frequency, and k and d are the duty ratios of the switches S_{y+} and S_{m1}.

In this manner, sinusoidal output currents, controllable output power factor, and global MPPT are achieved in the developed micro-inverter. The key waveforms of the micro-inverter are shown in Fig. 7.3.

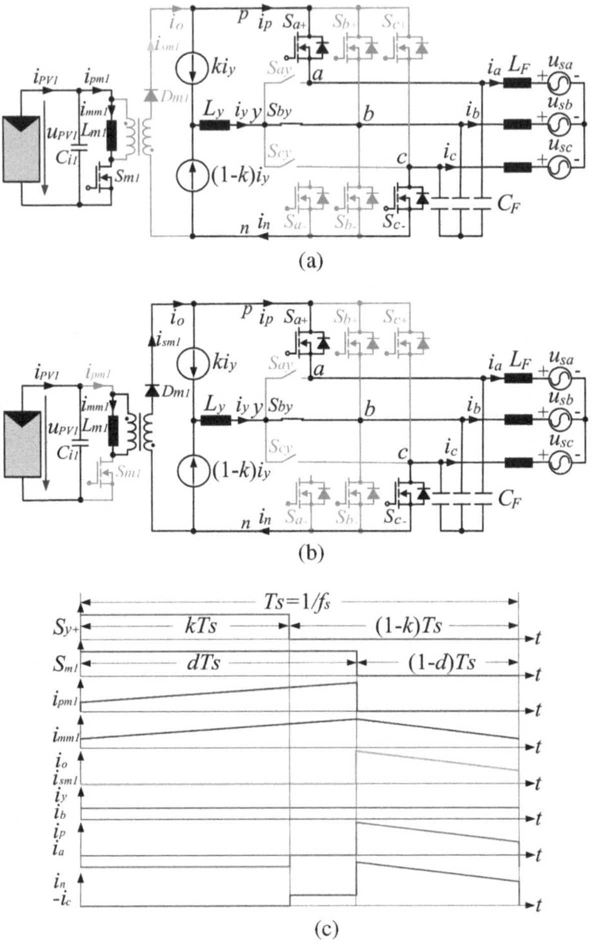

Fig. 7.2 Equivalent circuits of the proposed AC module micro-inverter (assuming $u_{sa}>u_{sb}>u_{sc}$). (**a**) Operating stage I. (**b**) Operating stage II. (**c**) Partial key waveforms of the inverter in a switching period

7.2.3 Analysis of the Input and Output Currents

For grid-connected PV inverters, sinusoidal grid-connected currents, constant output power, and MPPT of each PV panel should be guaranteed simultaneously in steady state. For the developed micro-inverter, the realizability of sinusoidal grid-connected currents and MPPT are carefully verified in the following discussion. Since each sub-converter is operated independently and the three-phase output voltages are symmetrical, one specific case, where the output voltage vector is located in sector 1 and only one sub-converter is included in the flyback stage, is used for analyzing the input and output currents in detail.

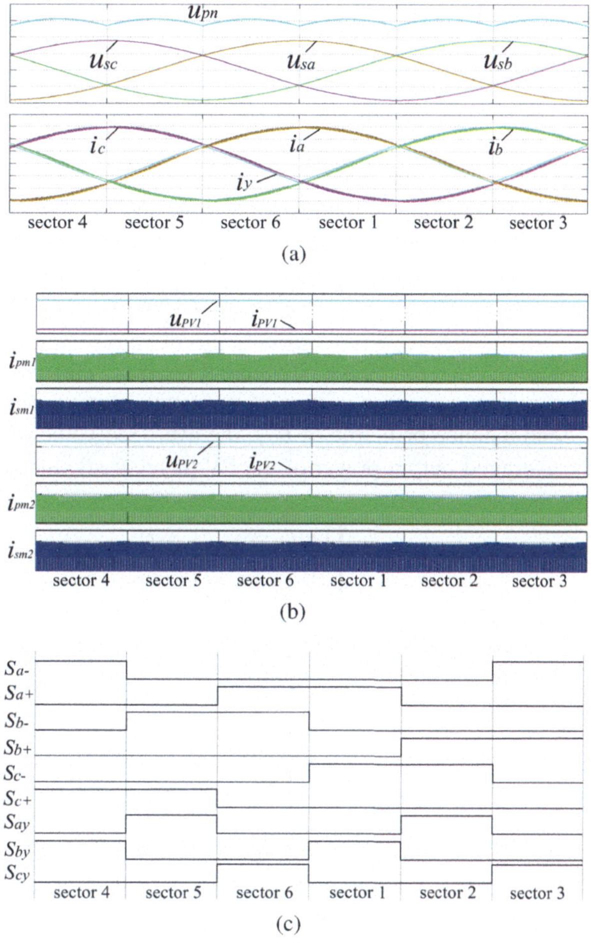

Fig. 7.3 Key waveforms of the proposed micro-inverter. (**a**) Output voltages, output currents, DC-link voltage, and third-harmonic injection current. (**b**) Voltages and currents of the flyback converters. (**c**) Gating signals

Suppose that the three-phase output voltages are given by

$$\begin{cases} u_{sa} = U_{om}\cos(\theta_{sa}) \\ u_{sb} = U_{om}\cos(\theta_{sa} - 2\pi/3) \\ u_{sc} = U_{om}\cos(\theta_{sa} + 2\pi/3) \end{cases} \tag{7.4}$$

where U_{om} is the magnitude of the output phase voltage.

Sort the output voltages according to the relationships of the instantaneous values as follows:

$$\begin{cases} u_{max} = \max(u_{sa}, u_{sb}, u_{sc}) \\ u_{mid} = \text{mid}(u_{sa}, u_{sb}, u_{sc}) \\ u_{min} = \min(u_{sa}, u_{sb}, u_{sc}) \end{cases} \tag{7.5}$$

where max(), mid(), and min() are the operator of the maximum, medium, and minimum values, respectively. As can be known from the operating principles, the DC-link voltage in sector 1 is expressed as

$$u_{pn} = u_{max} - u_{min} = u_{sa} - u_{sc} \tag{7.6}$$

Referring to the equivalent circuits in Fig. 7.3, the duty ratio d is obtained as (7.7) according to the voltage-second balance principle of the magnetizing inductor L_{m1}:

$$d = u_{pn}/\left(Nu_{PV1} + u_{pn}\right) \tag{7.7}$$

Assume that the averaged value of the input current i_{pm1}, denoted as $\overline{i}_{pm1}$, tracks the constant reference current i^*_{pm1} in steady state. Then, according to the ampere-second balance principle of C_{i1}, the averaged value of the output current i_{PV1} and the averaged output power of the PV panel are deduced as

$$\begin{cases} \overline{i}_{PV1} = \overline{i}_{pm1} = i^*_{pm1} \\ \overline{P}_{PV1} = u_{PV1}\overline{i}_{PV1} = u_{PV1}i^*_{pm1} \end{cases} \tag{7.8}$$

The averaged value of the secondary current i_{sm1} and the averaged output power of flyback converter1 are expressed as

$$\begin{cases} \overline{i}_{sm1} = \overline{P}_{PV1}/u_{pn} \\ \overline{P}_o = u_{pn}\overline{i}_{sm1} = \overline{P}_{PV1} \end{cases} \tag{7.9}$$

It can be known from (7.8)–(7.9) that the averaged output power of the PV panel and the averaged output power of flyback converter1, denoted as $\overline{P}_{PV1}$ and $\overline{P}_o$, both are constant in steady state.

In each switching period, according to the voltage-second balance principle of L_y, the duty ratio k is obtained as

$$k = (u_{mid} - u_{min})/(u_{max} - u_{min}) \tag{7.10}$$

The third-harmonic injection inductor current i_y is given by

$$\begin{cases} i_y = I_{pm}\cos(\theta_{sa} - 2\pi/3) + I_{qm}\sin(\theta_{sa} - 2\pi/3) \\ \quad = [GU_{om}\cos(\theta_{sa} - 2\pi/3 + \varphi)]/\cos\varphi \\ \tan\varphi = I_{qm}/I_{pm} \end{cases} \tag{7.11}$$

where I_{pm} and I_{qm} are the amplitudes of the active and reactive components of the grid-connected current, respectively, $G = 2\overline{P}_{PV1}/3U^2_{om}$ is the equivalent output conductance, and φ is the output displacement angle.

Neglecting the effects of the output filtering capacitor C_F, it is found from Table 7.1 that the output phase current i_b in sector 1 is equal to the third-harmonic injection inductor current i_y

$$i_b = i_y \tag{7.12}$$

Using the equivalent circuits shown in Fig. 7.2, the output current of phase a can be expressed as

$$i_a = \bar{i}_{sm1} - ki_y \tag{7.13}$$

In sector 1, the DC-link voltage u_{pn} is equal to the input line-line voltage u_{ac}. According to (7.7), the duty ratio k is equal to u_{bc}/u_{ac}. Substitute (7.9) and (7.11) into (7.13), i_a is deduced as

$$i_a = [GU_{om}\cos(\theta_{sa} + \varphi)]/\cos\varphi \tag{7.14}$$

As the three-phase output currents satisfy $i_a + i_b + i_c = 0$, the output current of phase c is obtained as

$$i_c = GU_{om}\cos(\theta_{sa} + 2\pi/3 + \varphi)/\cos\varphi \tag{7.15}$$

According to Eqs. (7.8), (7.9), (7.11), (7.12, (7.14), and (7.15), the output power of the PV panel is constant, and each PV panel achieves MPPT by adjusting its respective reference current i^*_{pm1}. Besides, the three-phase output currents are symmetrical and sinusoidal with a desired displacement angle. It is easy to find that the range of the output displacement angle of the suggested micro-inverter is from $-\pi/2$ to $\pi/2$. For the remaining sectors and multiple flyback converters configuration, similar results can be obtained, and the derivation process is not repeated here.

7.3 Control Scheme of the Micro-inverter

As discussed in the previous section, the CSI of the micro-inverter commutates at line frequency and hence can be treated as a passive part. Therefore, only the flyback stage and the third-harmonic injection circuit need to be properly controlled so as to realize MPPT and PFC at the output side. In this chapter, the flyback stage is utilized to achieve MPPT, and each flyback converter is controlled to track its Mpp. Furthermore, by adopting shunt active power filtering and time-sharing multiplexing injection techniques, the third-harmonic injection circuit achieves sinusoidal three-phase grid-connected current and the expected output side power factor. The concrete analysis is given as follows.

7.3.1 MPPT Control of the PV System

In general, at the certain irradiance level and ambient temperature, the PV panel has one operating point that results in a maximum output power. In order to maximally extract the output power of the PV panel, the panel-integrated micro-inverter needs to continuously track this operating point even when it changes due to the moving Sun and the partial shading. For a PV panel, it can be known from the current-voltage characteristics that the PV current is determined by the PV voltage. The current and voltage together determine the output power of the PV panel, and either the PV voltage or PV current can represent the MPP of a PV panel. Therefore, the MPP tracking can be realized by regulating either the PV voltage or the PV current to a value that represents the MPP.

In this work, a PO MPPT algorithm based on PV current regulation is adopted for the three-phase micro-inverter due to its advantages of good performance and low implementation efforts [12].

As can be seen from Fig. 7.1, the PV current and the PV voltage can be changed by regulating the averaged input current of the corresponding flyback converter. Therefore, the MPPT of each PV panel is achieved by controlling the averaged input current of the flyback converter to track the reference PV current generated by its MPP tracker. In this manner, global MPPT of the micro-inverter is achieved.

Figure 7.4 shows the control block of the MPPT controller of the micro-inverter, where $G_{f1}(s)$ and $G_{f2}(s)$ are the transfer functions of the sub-converter1 and sub-converter2 of the flyback converter1, respectively. The MPPT controller is a cascaded structure, where the outer control loop is the MPP tracker, and the inner current control loop adopts a conventional PI controller. In the output loop, The MPP tracker measures the PV voltage and PV current and periodically adjusts the current reference and the direction of the perturbation based on the previous operation and the changing of the PV panel power. For the inner loop, the PI controller controls the practical PV current to track the current reference, and the output of the inner control loop is the duty ratio of the switch in flyback converter1. In order to improve the dynamic tracking speed, the steady-state duty ratio d^* is fed forward to the output of current controller, which is obtained by the DC-link voltage and the PV voltage.

As can be seen from (7.8), the averaged value of the primary input current in a switching period is equal to the PV current. Therefore, instead of the PV current, the averaged primary input currents are used to implement the MPPT controller. Although the complexity of the current signals processing and the implementation efforts increase, the latter method obviously has the advantages of low cost of

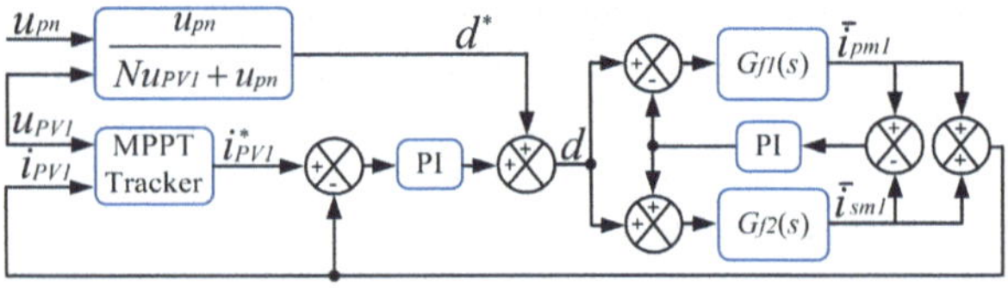

Fig. 7.4 Control block of the MPPT controller of the PV system

current sampling, good dynamic performance, and fast response to over-current fault. In order to guarantee the sinusoidal grid-connected current, the frequency of the outer-loop controller should be relatively low. Moreover, the band width of the inner loop is comparably high so as to decouple the outer and inner loops. The procedure of the controller is designed by conventional frequency-domain analysis tools, and it is not given here for brevity.

Besides, it should be noted that the mismatched parameters between the two sub-converters, such as the magnetizing inductances of the transformers, lead to undesirable current/power imbalance between the two sub-converters. Therefore, extra power balancing control loop is required. In this work, a simple PI controller is adopted to balance the power among the two sub-converters. The difference between the averaged input current of sub-converter1 and sub-converter2 is feed into the power balancing PI regulator, and the output of the PI regulator is added to the original duty ratio of sub-converter1 and subtracted from the original duty ratio of sub-converter2, respectively. As a result, current/power balance between the two sub-converters is achieved.

7.3.2 *Control of the Grid-Connected Currents*

Similar to the third-harmonic injection matrix converter [37], proper control and design of the third-harmonic injection circuit are the key challenges to implement the suggested micro-inverter, because PFC at the output side of the micro-inverter is realized by proper design and control of the third-harmonic injection circuit. The detailed analysis of the controller and design procedure of the third-harmonic injection circuit has been discussed in [37] and thus is not elaborated here.

Figure 7.5 depicts the control block of the whole micro-inverter system, where θ_{mid} is the phase angle of the output phase voltage with the medium value, P_o^* is the total output power reference, and $G_p(s)$ is the system transfer functions of the third-harmonic injection circuit. First, the output power references, generated from the MPP trackers, are added to obtain the total output power reference. Then the third-harmonic injection inductor current reference i_y^* is calculated by the output phase voltages, the total output power reference, and the expected output displacement angle. A PI controller with a feed-forward term is used to control the practical third-harmonic injection inductor current.

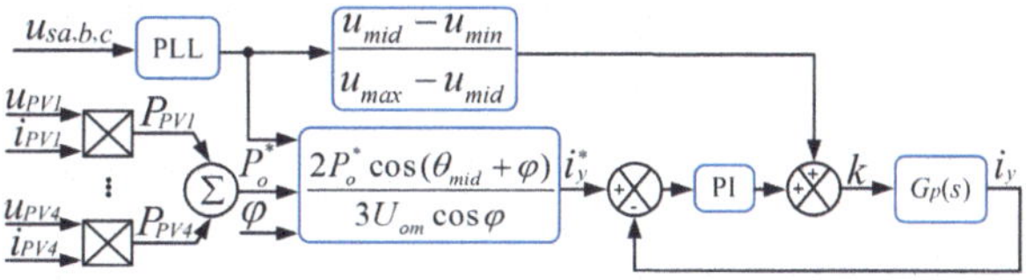

Fig. 7.5 Control block of the micro-inverter system

7.4 Simulation and Experimental Results

In this section, the functionality and performance of the three-phase micro-inverter were firstly evaluated by simulation using Matlab/Simulink software and then were validated experimentally. Simulation study was conducted by using the same parameters as the final laboratory prototype, and the parameters of the micro-inverter are listed in Table 7.2.

7.4.1 Simulation Results

In the following simulation results, in order to accelerate the simulation process, the micro-inverter is studied without of the MPPT control, and the PV panels of the micro-inverter are replaced by constant DC voltage sources. Therefore, the output current reference of each PV panel is directly from the command instead of the MPP tracker.

Figure 7.6 depicts the waveforms of the micro-inverter working with half of the rated power. In this scenario, the output current references of the PV panels were set such that the corresponding PV output powers were $P_{PV1} = 250\,\text{W}$, $P_{PV2} = 125\,\text{W}$, $P_{PV3} = 125\,\text{W}$, and $P_{PV4} = 0\,\text{W}$, respectively. Besides, the expected output displacement angle φ was set as 0. The waveforms shown in Fig. 7.6 consist of the output phase voltage $u_s a$, the output current i_a, the output current i_b, and the output current i_c. As can be seen from Fig. 7.6, the three-phase output currents are nearly sinusoidal and in phase with the output phase voltages, except for a small phase lag caused by the capacitive currents drawn by the output filtering capacitors. Therefore, it is clear that three-phase sinusoidal output currents and unity output power factor are achieved for the developed micro-inverter. Moreover, the output power of the micro-inverter is nearly 500 W, which indicates that each flyback converter has been controlled to effectively track its individual MPP independently.

Figure 7.7 shows the waveforms of the micro-inverter working in full-load condition. In this case, the output current references of the PV panels were set such

Table 7.2 System specifications of the micro-inverter

Parameters	Value
Power rating	1000 W (4 × 250 W)
Output line-line voltage	220 Vrms
Output frequency f_n	50 Hz
Switching frequency f_s	56 kHz
Magnetizing inductance	54 μH
Turns ratio N	6.5
Inductor L_F	300 μH
Capacitor C_F	3.3 μF
Inductor L_y	1.2 mH

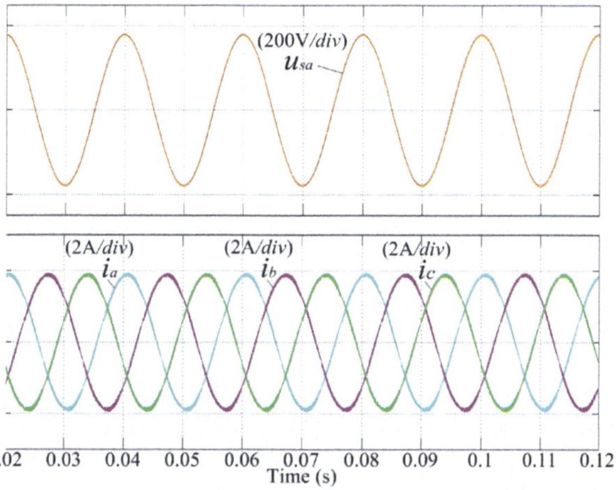

Fig. 7.6 Simulated waveforms of the micro-inverter working with half of the rated power

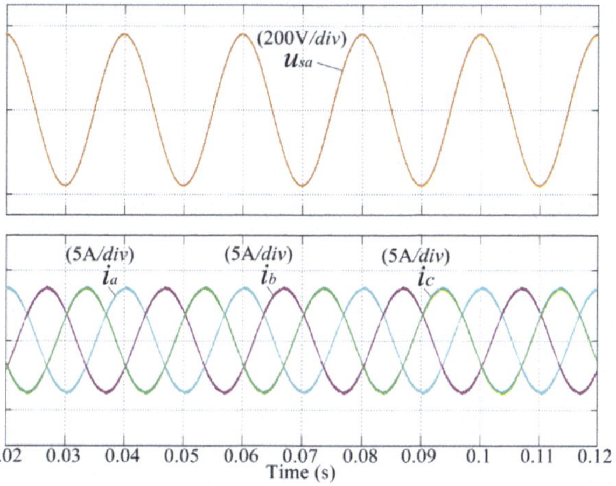

Fig. 7.7 Simulated waveforms of the micro-inverter working in full-load condition

that the corresponding PV output powers were $P_{PV1} = 250\,\text{W}$, $P_{PV2} = 250\,\text{W}$, $P_{PV3} = 250\,\text{W}$, $P_{PV1} = 250\,\text{W}$, and $\varphi = 0$. Again, the desired features of sinusoidal output currents, unity output power factor, and independent MPPT are achieved in the PV system. Thus, the results shown in Figs. 7.6 and 7.7 verify the correctness of the micro-inverter topology and control schemes.

Figure 7.8 illustrates the reactive power capability of the micro-inverter. The micro-inverter runs under full-load condition (1000 W), and the expected output reactive powers are 268 Var (capacitive) in Fig. 7.8a and 577 Var (capacitive) in Fig. 7.8b, and the equivalent output displacement angles are 15 ° and 30 °, respectively. As can be found in Fig. 7.8, three-phase sinusoidal output currents with the desired active and reactive components are generated at the output side. Therefore, the results in Fig. 7.8 demonstrate the wide reactive power control range of the micro-inverter.

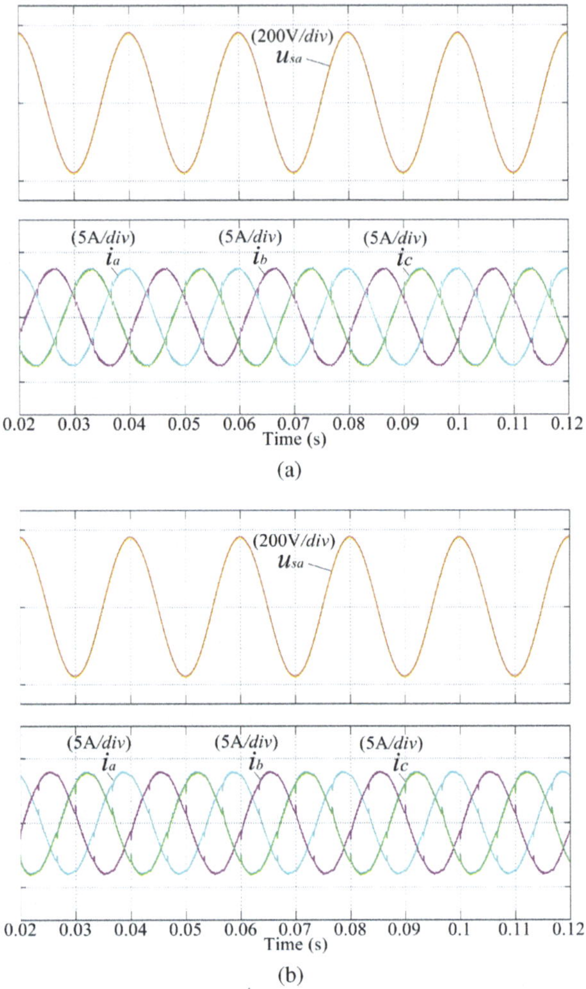

Fig. 7.8 Simulated waveforms of the micro-inverter with different output reactive powers. (**a**) The expected output reactive powers is 268 Var. (**b**) The expected output reactive powers is 577 Var

7.4.2 Experimental Results

In order to validate the theoretical analysis and simulation results, a laboratory prototype of the micro-inverter with the specifications given in Table 7.2 is built, as shown in Fig. 7.9. The control platform of the micro-inverter is based on a floating-point DSP TMS320F28335.

For the flyback stage, considering the 18-55V PV panel voltage, the 220 Vrms grid voltage and a turn ratio of 6.5, the MOSFET IRFS4321 is chosen as the switches of the flyback converters. Besides, Silicon Carbide Schottky Diode C2D05120D is selected as the output rectification diode due to its characteristic of zero reverse-recovery losses. For the flyback transformer, the RM14 ferrite core is adopted, and

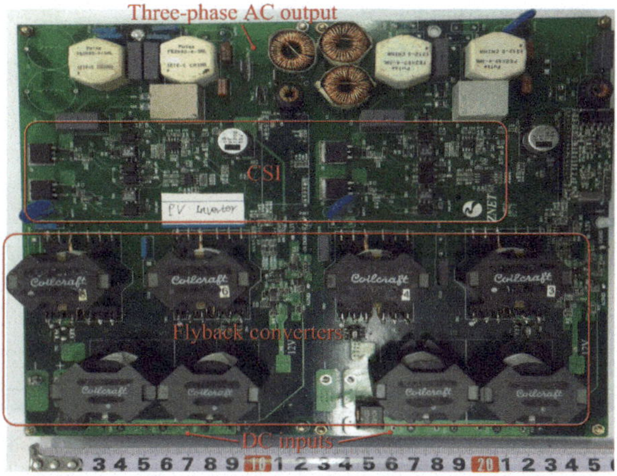

Fig. 7.9 The 1000 W three-phase micro-inverter prototype

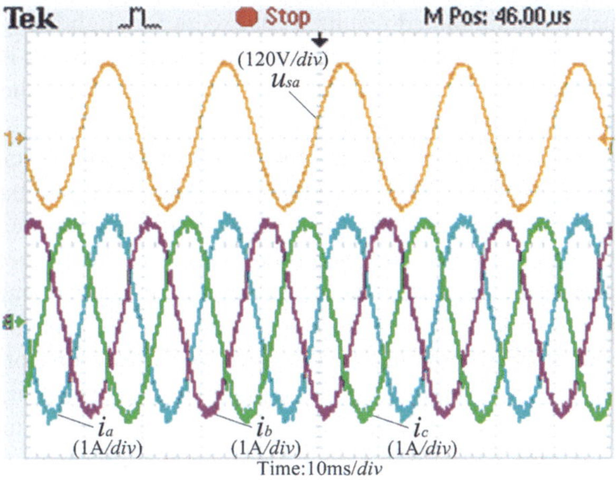

Fig. 7.10 Experimental waveforms of the micro-inverter working with half of the rated power. CH1 is the input phase voltage u_{sa}, CH2 is the output current i_a, CH3 is the output current i_b, and CH4 is the output current i_c

the magnetizing inductance is designed such that the CCM mode is guaranteed within the whole operating range. For the line-commutated switches and the bidirectional switches of the CSI, MOSFET SPW47N60C3 and STD15N50 are chosen due to their low saturation voltages. For the switches of the third-harmonic injection circuit working at high frequency, high speed MOSFET IPB60R190C6 is used.

Figures 7.10 and 7.11 demonstrate the experimental results of the micro-inverter working with half and full of the rated power, respectively; Figure 7.12 shows the experimental results of the micro-inverter with different output displacement angles. It should be noted that the experimental results shown in Figs. 7.10, 7.11, and 7.12 correspond to the simulated results shown in Figs. 7.6, 7.7, and 7.8, and the experimental conditions and commands are exactly the same as these in the simulation. As can be seen from Figs. 7.10, 7.11, and 7.12, the experimental results

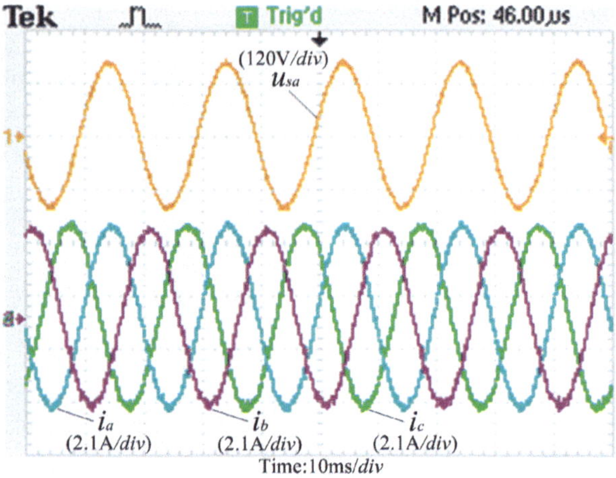

Fig. 7.11 Experimental waveforms of the micro-inverter working in full-load condition. CH1 is the input phase voltage u_{sa}, CH2 is the output current i_a, CH3 is the output current i_b, and CH4 is the output current i_c

match the simulated results well, except for slightly higher distortions of the output currents. The higher distortions of the output currents in the experiments are mainly attributed to the nonidealities of the grid voltage, the nonidealities of the power devices, the dead time effects, etc.

Figure 7.13a–d presents the THDs of the output currents corresponding to the cases in Figs. 7.10, 7.11, and 7.12, where the number of harmonics included in the FFT analysis is 50 (up to 2.5 kHz). It should be noted that the output wires were crossed the current probes twice, in order to improve the signal-to-noise ratio of the measurement. Therefore, the measured amplitudes of the currents were also doubled. As can be seen from the FFT analyses, the THDs of the output current corresponding to the operating conditions of half-load, full-load, output displacement angle of 15 °, and output displacement angle of 30 ° are 6.82%, 4.29%, 3.53%, and 4.2%, respectively, which are acceptable in practice applications.

Figure 7.14 presents the experimental MPPT performance of the micro-inverter. In this experiment, the flyback converter1 corresponding to PV panel1 is connected to a solar array simulator (E4361A, Agilent) whose MPP is set as 250 W, and the other flyback converters are disabled to simulate the partial shading condition of the whole PV system. As can be seen from the result, the practical output power of the solar array simulator tracks the set-point quickly and accurately. Therefore, the results in Fig. 7.14 demonstrate the correctness of the developed MPPT algorithm. Referring to Figs. 7.10, 7.11, 7.12, 7.13, and 7.14, sinusoidal output currents, controllable output power factor, and independent MPPT are achieved in the prototype; thus the validity of the micro-inverter is verified experimentally.

To evaluate the efficiency feature of the presented topology, the power losses and efficiency of the overall system were firstly calculated and then were measured practically. The system power losses can be divided mainly into two categories: semiconductor losses and passive components losses. For the semiconductor devices, the blocking losses can be neglected. Thus only the conduction and switching losses are considered in this chapter.

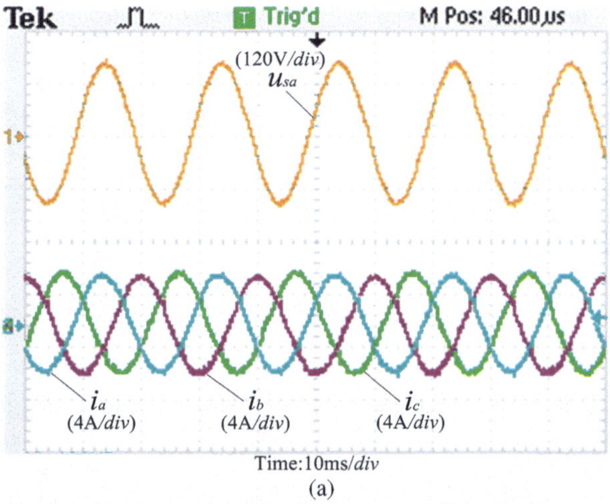

Fig. 7.12 Experimental waveforms of the micro-inverter with different output reactive power. (**a**) The expected output reactive power is 268 Var. (**b**) The expected output reactive powers is 577 Var. CH1 is the input phase voltage u_{sa}, CH2 is the output current i_a, CH3 is the output current i_b, and CH4 is the output current i_c

The conduction losses of the switches and diodes can be calculated by

$$P_{C,S} = I_{S,AVG}U_{CE0} + I_{S,RMS}^2 R_s \tag{7.16}$$

and

$$P_{C,D} = I_{D,AVG}U_{D0} + I_{D,RMS}^2 R_D \tag{7.17}$$

where $P_{C,S}$ and $P_{C,D}$ are the conduction losses of the switch and diode, R_s and R_D are the dynamic on-resistances of the switch and diode, U_{CE0} and U_{D0} are the forward voltages of the switch and diode when the current is zero, $I_{S,AVG}$ and

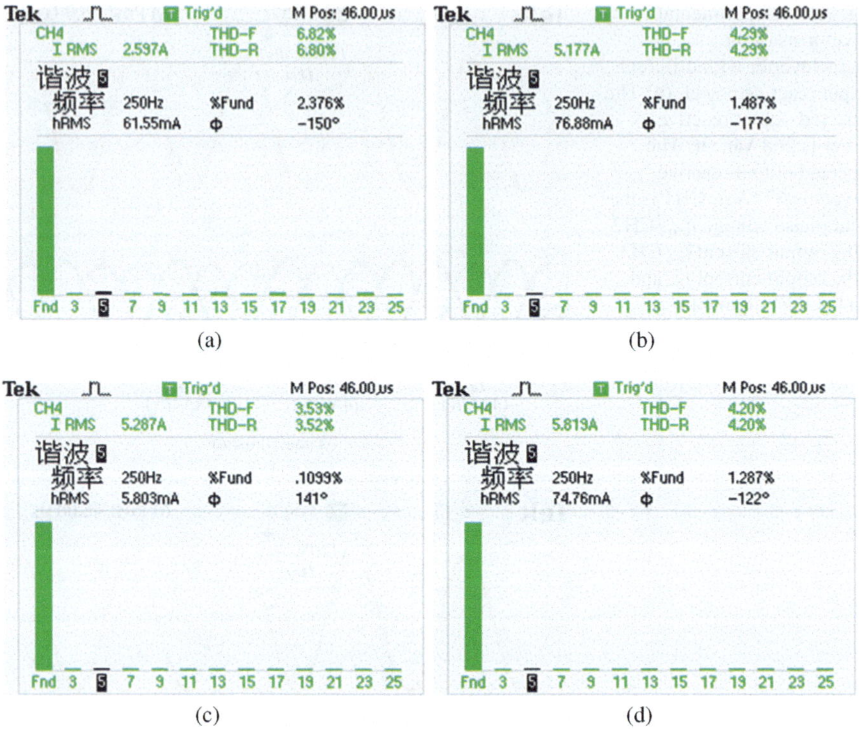

Fig. 7.13 THDs of the output current with (**a**) half of the rated power, (**b**) full-load, (**c**) displacement angle of 15 °, and (**d**) displacement angle of 30 °

Fig. 7.14 Experimental MPPT performance of the micro-inverter. CH1 is the PV voltage u_{PV1}, CH2 is the PV current i_{PV1}, CH3 is the output voltage u_{sa}, CH4 is the output current i_a, and M is the PV output power

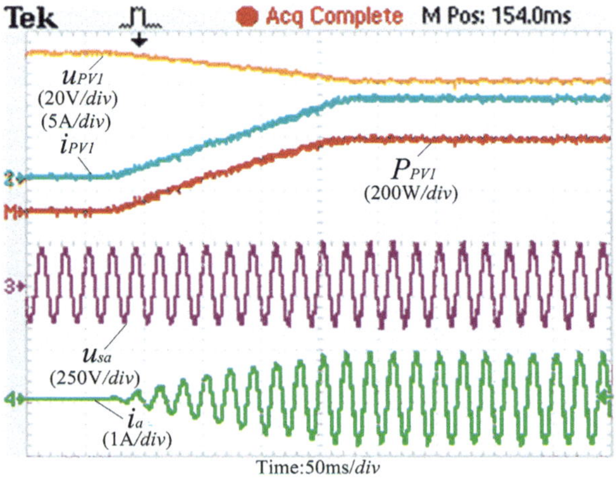

$I_{D,AVG}$ are the average currents of the switch and diode, and $I_{S,RMS}$ and $I_{D,RMS}$ are the RMS currents of the switch and diode.

The switching losses of the switches and diodes are estimated as

$$P_{S,S} = \frac{(E_{on\_R} + E_{off\_R})}{V_{SR} I_{SR}} V_{S,AVG} I_{S,AVG} f_s \tag{7.18}$$

and

$$P_{S,D} = \frac{E_{rr\_R}}{V_{DR} I_{DR}} V_{D,AVG} I_{D,AVG} f_s \tag{7.19}$$

where $P_{S,S}$ and $P_{S,D}$ are the switching losses of the switch and diode, $V_{S,AVG}$ and $V_{D,AVG}$ are the average voltages of the switch and diode, E_{on_R} and E_{off_R} are the switching energies of the switch at the reference voltage V_{SR} and reference current I_{SR}, and E_{rr_R} is the reverse recovery energy of the diode at the reference voltage V_{DR} and reference current I_{DR}.

The losses of the capacitors are obtained as

$$P_C = I_{C,RMS}^2 \frac{\tan\delta}{2\pi C_F f_s} \tag{7.20}$$

where $I_{C,RMS}$ and $\tan\delta$ are the RMS current and loss factor of the input capacitors.

The losses of the inductors can be divided into two parts: core loss and winding resistance loss. Since the inductors in the presented topology operate at low frequency, only low-frequency loss in winding resistance is considered in this chapter. The power loss of the inductor is calculated as

$$P_L = I_{L,RMS}^2 R_L \tag{7.21}$$

where $I_{L,RMS}$ is the RMS current of the inductor and R_L is the DC resistance of the inductor.

According to the above formulas, the overall system power loss distribution is derived as shown in Fig. 7.15. The nominal system efficiency calculated reaches 95.63%.

Figure 7.16 shows the measured efficiency of the micro-inverter with the HIOKI 3390 digital power analyzer, where the voltage of the PV panels is 30 V and the desired output displacement angle is 0. It should be noted that the auxiliary circuits such as the control circuits are all included in the test. As can be seen from Fig. 7.16, the maximum efficiency of 95.52% and California Efficiency of 93.7% are achieved for the proposed micro-inverter.

It is important that the PV inverter should meet the specific EMC standards. The EMI of a PV inverter is determined by many factors, such as the switching frequency. Regarding the influence of the switching frequency, it mainly reflects in two aspects. Firstly, the cut-off frequency of the DM filter is usually designed to

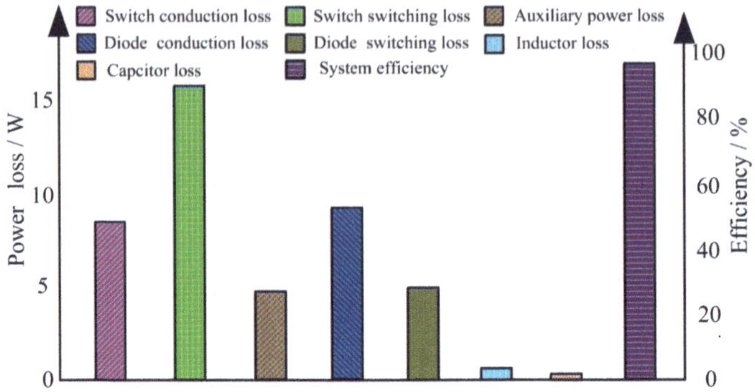

Fig. 7.15 Calculated power loss distribution and efficiency diagram of the micro-inverter

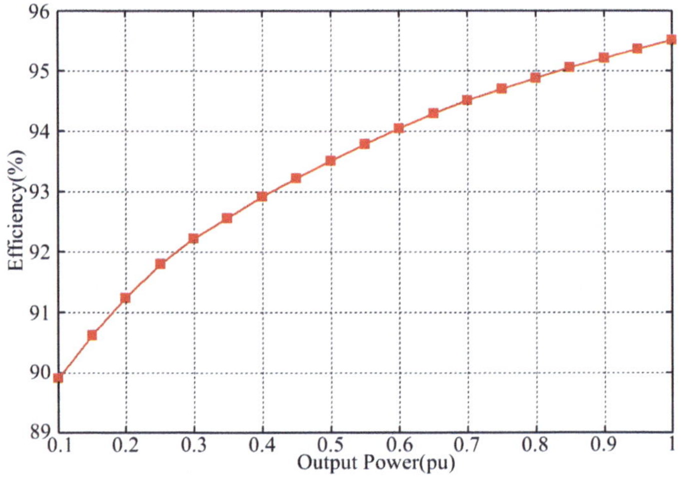

Fig. 7.16 Measured efficiency of the micro-inverter

satisfy the requirement of the output current quality and increases with the switching frequency. Therefore, different switching frequency will result in different cut-off frequency and noise attenuation of the DM filter. Secondly, for the semiconductor power converters switching at high frequency, the EMI noises are mainly generated by the high du/dt and di/dt rates of the power switches. Usually, the du/dt and di/dt rates of the switches, as well as the conduction emission, increase with increasing switching frequency. In Fig. 7.17, the conducted emission of the prototype is measured by an EMI test receiver and an LISN, and the result has been compared with the CISPR 11 standard. As can be seen from Fig. 7.17, the conducted emission levels of the micro-inverter fully meet the standards.

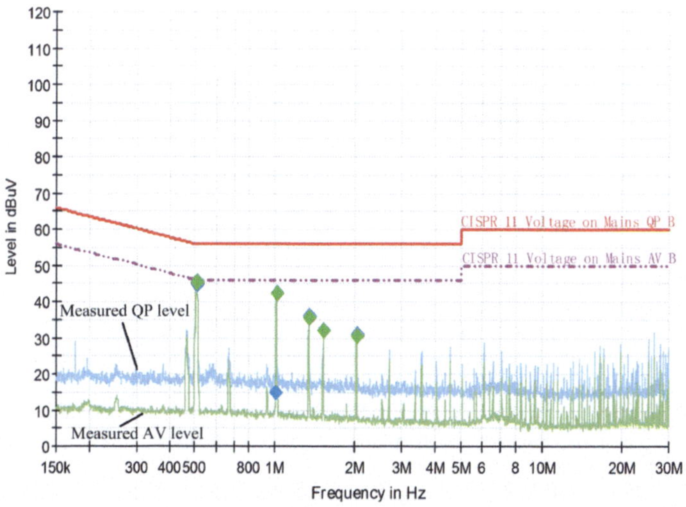

Fig. 7.17 Measured conducted emission of the micro-inverter

7.5 Conclusions

This chapter presents a grid-connected three-phase micro-inverter suitable for AC PV module applications. The proposed micro-inverter uses flyback converters as the DC/DC conversion stage, and MPPT of each PV panel is achieved by proper control of the corresponding flyback converter. A third-harmonic injection circuit is controlled to synthesize the appropriate injection current so as to realize PFC at the output side. A line-commutated CSI is used to convert the DC current to three-phase AC current, which delivers the power generated from the PV panels to the grid. Simulation and experimental results clearly show that the micro-inverter could achieve sinusoidal output currents, controllable output power factor, independent MPPT, and excellent efficiency. By having the potential of achieving high power density/reliability, extended reactive power control range, and three-phase balanced output, the developed micro-inverter is an attractive candidate for AC PV module applications.

References

1. B.K. Bose, Global Warming-Energy, environmental pollution and the impact of power electronics. IEEE Ind. Electron. Magazine **4**(1), 6–17 (2010)
2. European Photovoltaic Industry Association, Global market outlook for photovoltaics until 2016 (2012). www.epia.org
3. Renewables 2016: Global status report. www.ren21.net/status-of-renewables/global-status-report/

4. Bloomberg New Energy Finance (BNEF), Solar panels now so cheap manufacturers probably selling at loss. https://about.bnef.com/blog/solar-panels-now-so-cheap-manufacturers-probably-selling-at-loss/
5. U. S. Department of Energy, $ 1 W photovoltaic systems. www1.eere.energy.gov/solar/sunshot/pdfs/dpw_white_paper.pdf
6. S.B. Kjaer, F. Blaabjerg, Design optimization of a single phase inverter for photovoltaic applications, in *Proceeding of the 34th Annual IEEE Power Electronics Specialist Conference*, vol. 3 (2003), pp. 1183–1190
7. Y. Xue, L. Chang, S. Bakhoj, J. Bordonau, T. Shimizu, Topologies of single-phase inverters for small distributed power generators-An overview. IEEE Trans. Power Electron. **19**(5), 1305–1314 (2004)
8. R. Gonzalez, J. Lopez, P. Sanchis, L. Marroyo, Transformerless inverter for single-phase photovoltaic systems. IEEE Trans. Power Electron. **22**(2), 693–697 (2007)
9. B. Yang, W. Li, Y. Zhao, X. He, Design and analysis of a grid-connected photovoltaic power system energy. IEEE Trans. Power Electron. **25**(4), 992–1000 (2010)
10. W. Li, Y. Gu, H. Luo, W. Cui, X. He, C. Xia, Topology review and derivation methodology of single-phase transformerless photovoltaic inverters for leakage current suppression. IEEE Trans. Ind. Electron. **62**(7), 4537–4551 (2015)
11. Y. Tang, W. Yao, P.C. Loh, F. Blaabjerg, Highly reliable transformerless photovoltaic inverters with leakage current and pulsating power elimination. IEEE Trans. Ind. Electron. **63**(2), 1016–1026 (2016)
12. M. Kasper, D. Bortis, J.W. Kolar, Classification and comparative evaluation of PV panel-integrated DC-DC converter concepts. IEEE Trans. Power Electron. **29**(5), 2511–2526 (2014)
13. S.B. Kjaer, J.K. Pedersen, F. Blaabjerg, A review of single phase grid-connected inverters for photovoltaic modules. IEEE Trans. Ind. Appl. **41**(5), 1292–1306 (2005)
14. M. Calais, J. Myrzik, T. Spooner, V.G. Agelidis, Inverters for single-phase grid connected photovoltaic systems–An overview, in *Proceedings of the IEEE PESC*, vol. 4 (2002), pp. 1995–2000
15. W. Li, X. He, Review of non-isolated high step-up dc/dc converters in photovoltaic grid-connected applications. IEEE Trans. Ind. Electron. **58**(4), 1239–1250 (2011)
16. S.B. Kjær, J.K. Pedersen, F. Blaabjerg, Power inverter topologies for photovoltaic modules–A review, in *Proceedings of the IEEE IAS Conference* (2002), pp. 782–788
17. T. Shimizu, K. Wada, N. Nakamura, Flyback-type single-phase utility interactive inverter with power pulsation decoupling on the DC input for an AC photovoltaic module system. IEEE Trans. Power Electron. **21**(5), 1264–1272 (2006)
18. Q. Li, P. Wolfs, A review of the single phase photovoltaic module integrated converter topologies with three different DC link configurations. IEEE Trans. Power Electron. **23**(3), 1320–1333 (2008)
19. C. Rodriguez, G.A.J. Amaratunga, Long-lifetime power inverter for photovoltaic ac modules. IEEE Trans. Ind. Electron. **55**(7), 2593–2601 (2008)
20. S. Chen, T. Liang, L. Yang, J. Chen, A boost converter with capacitor multiplier and coupled inductor for AC module applications. IEEE Trans. Ind. Electron. **60**(4), 1503–1511 (2013)
21. Y. Li, R. Oruganti, A low cost flyback CCM inverter for AC module application. IEEE Trans. Power Electron. **27**(3), 1295–1303 (2012)
22. A. Nanakos, E. Tatakis, N. Papanikolaou, A weighted-efficiency-oriented design methodology of flyback inverter for AC Photovoltaic modules. IEEE Trans. Power Electron. **27**(7), 3221–3233 (2012)
23. M. Gao, M. Chen, C. Zhang, Z. Qian, A low cost flyback CCM inverter for AC module application. IEEE Trans. Power Electron. **29**(7), 3428–3444 (2014)
24. Z. Zhang, M. Chen, W. Chen, C. Jiang, Z. Qian, Analysis and implementation of phase synchronization control strategies for BCM interleaved flyback microinverters. IEEE Trans. Power Electron. **29**(11), 5921–5932 (2014)

25. Y.-H. Kim, J.-W. Jang, S.-C. Shin, C.-Y. Won, Weighted-efficiency enhancement control for a photovoltaic AC module interleaved flyback inverter using a synchronous rectifier. IEEE Trans. Power Electron. **29**(12), 6481–6493 (2014)
26. M. Rezaei, K. Lee, A. Huang, A high-efficiency flyback micro-inverter with a new adaptive snubber for photovoltaic applications. IEEE Trans. Power Electron. **31**(1), 318–327 (2016)
27. A. Amirahmadi, H. Hu, A. Grishina, L. Chen, Z.J. Shen, I. Batarseh, Improving output current distortion in hybrid BCM current controlled three-phase micro-inverter. in *Proceedings of the IEEE ECCE Conference* (2013), pp. 1319–1323
28. L. Chen, C. Hu, Q. Zhang, K. Zhang, I. Batarseh, Modeling and triple-loop control of ZVS grid-connected DC/AC converters for three-phase balanced micro-inverter application. IEEE Trans. Power Electron. **30**(4), 2010–2023 (2015)
29. F. Chen, Q. Zhang, A. Amirahmadi, I. Batarseh, Modeling and analysis of DC-link voltage for three-phase four-wire two-stage micro-inverter, in *Proceedings of the IEEE APEC Conference* (2014), pp. 3000–3005
30. H. Hu, S. Harb, N. Kutkut, I. Batarseh, Z.J. Shen, A review of power decoupling techniques for microinverters with three different decoupling capacitor locations in PV systems. IEEE Trans. Power Electron. **28**(6), 2711–2726 (2013)
31. H. Hu, S. Harb, N. Kutkut, Z.J. Shen, I. Batarseh, A single-stage microinverter without using electrolytic capacitors. IEEE Trans. Power Electron. **28**(6), 2677–2687 (2013)
32. M. Chen, K. Afridi, D. Perreault, A multilevel energy buffer and voltage modulator for grid-interfaced micro-inverters. IEEE Trans. Power Electron. **30**(3), 1203–1219 (2015)
33. B.J. Pierquet, D.J. Perreault, A single-phase photovoltaic inverter topology with a series-connected energy buffer. IEEE Trans. Power Electron. **28**(10), 4603–4611 (2013)
34. Y. Tang, W. Yao, P.C. Loh, F. Blaabjerg, A single-phase grid-connected photovoltaic Inverter based on a three-switch three-port flyback with series power decoupling circuit. IEEE Trans. Ind. Electron. **64**(3), 2062–2071 (2017)
35. (2011). http://www.vde-verlag.de/standards/0105029/vde-ar-n-4105-anwendungsregel-2011-08-html
36. M. Jantsch, C.W. G. Verhoeve, Inverters with three-phase output and without electrolyte capacitor for improved lifetime, efficiency and costs of grid connected systems, in Proceedings of the 14th European Photovoltaic Solar Energy Conference, Barcelona, Spain (1997)
37. H. Wang, M. Su, Y. Sun, J. Yang, G. Zhang, W. Gui, J. Feng, Two-stage matrix converter based on third-harmonic injection technique. IEEE Trans. Power Electron. **31**(1), 533–547 (2016)
38. H. Wang, M. Su, Y. Sun, Y. Liu, G. Zhang, H. Li, W. Zhang, W. Gui, J. Feng, Active third-harmonic injection indirect matrix converter with dual three-phase outputs. IET Power Electron. **9**(4), 657–668 (2016)
39. L. Wang, H. Wang, M. Su, Y. Sun, J. Yang, M. Dong, X. Li, W. Gui, J. Feng, A three-level T-type indirect matrix converter based on the third-harmonic injection technique. IEEE J. Emerg. Sel. Topics Power Electron **5**(2), 841–852 (2017)
40. H. Wang, M. Su, Y. Sun, G. Zhang, J. Yang, W. Gui, J. Feng, Topology and modulation scheme of a three-level third-harmonic injection indirect matrix converter. IEEE Trans. Ind. Electron **64**(10), 7612–7622 (2017)

Chapter 8
Back-to-Back Matrix Converter with Single-Arm Current Injection

8.1 Introduction

The conventional voltage-source or current-source BBC, which possesses the important capacity to work as a firewall in terms of power quality disturbances, is one of the most popular ac-ac converters and has been applied in different applications such as ac ASDs, renewable energy power generation [1], grid interface [2], WECS, and so on [3–8]. The VBBC works as an active front and has the advantages of sinusoidal inputs and outputs, bidirectional power flow capability, and controllable input power factor [9–11]. However, the energy storage elements of the conventional VBBC such as the bulky electrolytic capacitors usually have a short lifetime and a relatively huge volume compared with the whole converter, which will influence the power density and system reliability [12–15].

As an alternative candidate, MCs realize direct ac-ac power conversion and have been investigated extensively in the past decades owing to the absence of the bulky DC-link energy storage element [16–22]. MCs not only inherit similar advantages to the conventional VBBC but also possess the advantages of compact structure and high reliability [23, 24]. Nevertheless, the drawbacks such as complex commutation strategy, bidirectional power switches configuration, the limited capability of generating the input reactive power, and the non-linearity caused by narrow pulse width hinder the wide application of the MCs.

To overcome the drawbacks of limited input reactive power capability and the nonlinearity caused by narrow pulse width, the 3HIMC has been proposed [25], which has superior characteristics including extended input reactive power capability, a maximum voltage transfer ratio of 0.866 under all operating conditions, and independent control between the rectification and inversion stages and thus flexible modulation strategies implementation [26–28]. However, the extra third-harmonic current injection circuit increases the complexity and cost of the main circuits.

To decrease the complexity and cost of the 3HIMC in [25], an MT-BBC with single-arm current injection is proposed, as shown in Fig. 8.1. The developed MT-

H. Wang, *Advanced Matrix Converters*, Power Systems,
https://doi.org/10.1007/978-3-031-69902-3_8

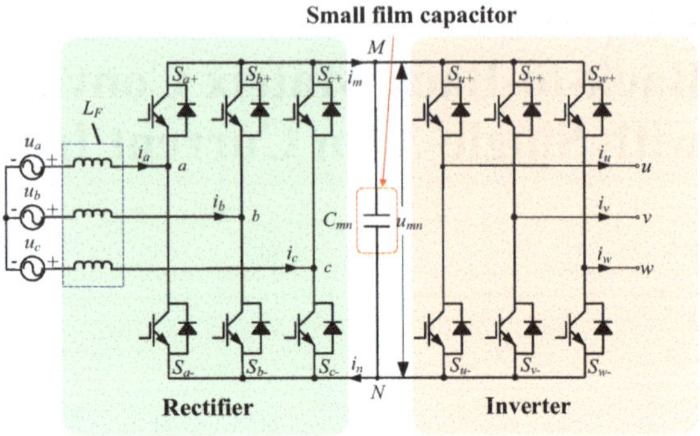

Fig. 8.1 Topology of the proposed MT-BBC

BBC possesses the good features of both the conventional VBBC and the MCs including simple main circuit structure, sinusoidal inputs and outputs, bidirectional power flow capability, high efficiency, direct ac-ac conversion, potentially high power density, and high reliability. Moreover, the input reactive power control range of the MT-BBC is extended significantly, and the input power factor angle varies from $-\pi/2$ to $\pi/2$. The essential properties and operating principles of the MT-BBC are entirely different from those of the conventional VBBC. For the MT-BBC, the bulky DC-link energy storage capacitor in VBBC is replaced with a small film capacitor, and only one bridge arm of the rectifier switches at high frequency during any mains periods. Therefore, the reliability and efficiency of the converter are improved. The rectifier of the MT-BBC can be regarded as the combination of the line-frequency commutated rectifier and the third-harmonic current injection circuit of the 3HIMC. As a result, the inherent functions, and merits of the 3HIMC are retained, and the main power circuit, as well as the control method, is simplified dramatically due to the elimination of the third-harmonic current injection circuit.

The rest of this chapter is organized as follows. In Sect. 8.2, the topology, the operating principles of the MT-BBC, and comparisons with other counterpart structures are introduced in detail. Modulation and control strategies of the rectifier and the inverter are discussed respectively, and the result that three-phase input currents are symmetric and sinusoidal is verified by mathematical proofs in Sect. 8.3. In Sect. 8.4, a 6-kW experimental prototype is built to verify the effectiveness of the proposed methods, and the efficiency of the MT-BBC is evaluated. Finally, the main findings of this chapter are recapitulated and potential applications are outlined in Sect. 8.5.

8.2 Topology and Operating Principles

8.2.1 Topology

The topology of the MT-BBC shown in Fig. 8.1 is a back-to-back converter including a three-phase rectifier, a small film DC-link capacitor, and a three-phase inverter. The rectifier and the inverter are connected by the small film capacitor, and the DC-link voltage presents a time-varying six-pulse shape. $S_{a,b,c.+,-}$, $u_{a,b,c}$, $i_{a,b,c}$, L_F, C_{mn}, u_{mn}, $S_{u,v,w.+,-}$, and $i_{u,v,w}$ represent the active switches of the rectifier, the three-phase mains voltages, the three-phase input currents, the input inductor, the DC-link capacitor, the DC-link voltage, the active switches of the inverter, and the three-phase output currents, respectively.

In terms of the topology, by comparing the MT-BBC and the conventional VBBC, the DC-link capacitor is replaced from a bulky electrolytic capacitor to a small film capacitor. Besides, compared with the MCs, bidirectional switches and consequent complex commutation issues are eliminated. Moreover, the third-harmonic current injection circuit is removed compared with the 3HIMC. In summary, the MT-BBC topology inherits both the advantages of the VBBC and the MCs such as simple main circuit structure, high reliability, potentially high power density, and so on.

8.2.2 *Operating Principles*

The operating principles of the rectifier are discussed as follows. Only one bridge arm switches at high frequency, and other bridge arms commutate at line frequency. The upper switch of the bridge arm connected to the input phase with the highest instantaneous voltage and the lower active switch of the bridge arm connected to the input phase with the lowest instantaneous voltage both are in the on-state. Both active switches of the phase with the medium voltage are kept switching at high frequency with the duty cycle d to inject the quasi-third-harmonic current. The duty cycle d is related to the given steady-state duty cycle D and the compensation duty cycle d_c that is produced by the controller.

For example, when the input voltages satisfy $u_a > u_b > u_c$ (denoted as sector I), active switches S_{a+} and S_{c-} are turned on, and active switches S_{b+} and S_{b-} are kept switching with the duty cycles d and 1-d. The equivalent circuit diagrams of the MT-BBC in sector I are shown in Fig. 8.2, where Fig. 8.2a means the forward power flow and Fig. 8.2b means the reverse power flow. Switching states of the rectifier in other sectors can be obtained by analogy.

The quasi-third-harmonic current i_{ref} in sector I can be expressed as

$$\begin{aligned} i_{ref} &= -I_{pm}\cos(\omega_i t - 2\pi/3) - I_{qm}\sin(\omega_i t - 2\pi/3) \\ &= -GU_m\cos(\omega_i t - 2\pi/3) - GU_m\tan\varphi\sin(\omega_i t - 2\pi/3) \\ &= -\frac{GU_m\cos(\omega_i t - \varphi - 2\pi/3)}{\cos\varphi} \end{aligned} \tag{8.1}$$

where φ and U_m are the input power factor angle and the amplitude of the three-phase mains voltages, I_{pm} and I_{qm} are the amplitudes of the active and reactive components of the three-phase input currents, and G is the equivalent input

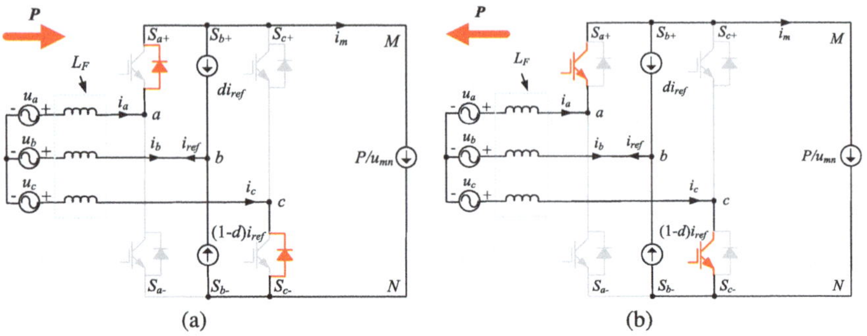

Fig. 8.2 Equivalent circuit diagrams of the MT-BBC in sector I. (**a**) Forward power flow and (**b**) reverse power flow

conductance, which is expressed as

$$G = \frac{2P}{3{U_m}^2} \tag{8.2}$$

where P is the active power. Then the quasi-third-harmonic current i_{ref} in sector I is simplified as

$$i_{ref} = -\frac{2P\cos(\omega_i t - \varphi - 2\pi/3)}{3U_m \cos\varphi} \tag{8.3}$$

The quasi-third-harmonic currents i_{ref} in other sectors can be obtained by the same way.

To derive the steady-state duty cycles D in different sectors, the three-phase mains voltages $u_{a,b,c}$ are sorted into u_{max}, u_{mid}, and u_{min} as follows:

$$\begin{cases} u_{max} = \max(u_a, u_b, u_c) \\ u_{mid} = \text{mid}(u_a, u_b, u_c) \\ u_{min} = \min(u_a, u_b, u_c) \end{cases} \tag{8.4}$$

where u_{max}, u_{mid}, and u_{min} are the maximum one, medium one, and minimum one among the three-phase mains voltages.

According to the voltage-second balance of every input inductor during each switching period, the steady-state duty cycles of the high-frequency switching bridge arm can be obtained as

$$\begin{cases} D = \dfrac{u_{mid} - u_{min}}{u_{max} - u_{min}} \\ \\ D' = \dfrac{u_{max} - u_{mid}}{u_{max} - u_{min}} \end{cases} \tag{8.5}$$

Table 8.1 Sector divisions and main parameters of rectifier

Sector k	$\omega_i t$	S_{a+}	S_{a-}	S_{b+}	S_{b-}	S_{c+}	S_{c-}	D	i_{ref}	u_{mn}
I	0-$\pi/3$	1	0	d	1-d	0	1	$\frac{(u_b-u_c)}{(u_a-u_c)}$	$-2P\cos(\omega_i t-\varphi$ $-2\pi/3))/(3U_m\cos\varphi)$	u_{ac}
II	$\pi/3$-$2\pi/3$	d	1-d	1	0	0	1	$\frac{(u_a-u_c)}{(u_b-u_c)}$	$-2P\cos(\omega_i t-\varphi)$ $/(3U_m\cos\varphi)$	u_{bc}
III	$2\pi/3$-π	0	1	1	0	d	1-d	$\frac{(u_c-u_a)}{(u_b-u_a)}$	$-2P\cos(\omega_i t-\varphi$ $+2\pi/3))/(3U_m\cos\varphi)$	u_{ba}
IV	π-$4\pi/3$	0	1	d	1-d	1	0	$\frac{(u_b-u_a)}{(u_c-u_a)}$	$-2P\cos(\omega_i t-\varphi$ $-2\pi/3))/(3U_m\cos\varphi)$	u_{ca}
V	$4\pi/3$-$5\pi/3$	d	1-d	0	1	1	0	$\frac{(u_a-u_b)}{(u_c-u_b)}$	$-2P\cos(\omega_i t-\varphi)$ $/(3U_m\cos\varphi)$	u_{cb}
VI	$5\pi/3$-2π	1	0	0	1	d	1-d	$\frac{(u_c-u_b)}{(u_a-u_b)}$	$-2P\cos(\omega_i t-\varphi$ $+2\pi/3))/(3U_m\cos\varphi)$	u_{ab}

where D and D' are the steady-state duty cycles of the upper switch and the lower switch, respectively.

The switching states $S_{a,b,c,+,-}$, the steady-state duty cycles D, the quasi-third-harmonic currents i_{ref}, and the DC-link voltages u_{mn} in different sectors are all provided in Table 8.1. In this manner, three-phase sinusoidal input currents and controllable input power factor can be achieved. For the rectifier, two bridge arms commutate at line frequency, while the other one switches at high frequency and thus belongs to the type of current-source rectifier. The operating principle of the inverter is similar to that of the conventional VSI and thus is not elaborated here.

8.2.3 Performance Comparisons

To highlight the superiority of the MT-BBC, this part makes comparisons among the MT-BBC and other counterpart structures including conventional VBBC, 3HIMC, and IMC in terms of the numbers of switches, power losses, and applications. Comparisons between the VBBC and the IMC can be referred to [9], and the final results are listed in Table 8.2.

The results show that the developed MT-BBC requires fewer transistors compared with the 3HIMC and the IMC. Moreover, compared with the VBBC, efficiency, power density, and reliability of the MT-BBC are potentially higher.

Table 8.2 Comparisons among the MT-BBC and other counterpart structures

	MT-BBC	VBBC	3HIMC	IMC
Transistor number	12	12	20	18
Bidirectional switch number	0	0	3	6
Conduction loss	Low	Low	Low	High
Switching loss	Low	Very High	Low	Low
DC-link capacitor	Small	Large	No	No
Input filter components	$3L_f$	$3L_f$	$3L_f$+$3C_f$+$1L_h$	$3L_f$+$3C_f$
Applications	ASDs, WECS	ASDs, WECS	ASDs, WECS	ASDs, WECS

8.3 Modulation and Control Strategies

In this section, the modulation and control strategies of the rectifier and inverter are introduced. The result that the three-phase input currents are symmetric and sinusoidal is verified by mathematical proofs.

8.3.1 *Control of the Rectifier*

A clamped modulation strategy keeping only one-third of the bridge arms switching at high frequency is applied to the rectifier, which can achieve sinusoidal input currents, reduced switching losses, and enhanced input reactive power control range. Based on the operating principles of the rectifier introduced in Sect. 8.2, both active switches of the phase with the medium voltage are kept modulated at high frequency to inject the quasi-third-harmonic current into the corresponding input phase. The duty cycle d is the sum of the compensation duty cycle d_c and the given steady-state duty cycle D. Then, d is compared with the carrier to get the gating signals of the modulated bridge arm. And the gating signals of other two bridge arms can be obtained according to the sector number k. Finally, the three-phase sinusoidal input currents and enhanced input reactive power control range are achieved, and the main waveforms of the rectifier are shown in Fig. 8.3.

To inject the quasi-third-harmonic currents into the high-frequency switching bridge arm of the rectifier well, a proper controller is important. A PR controller is selected to track the ac signal. The dominant frequency of i_{ref} is $3\omega_i$, and ω_i is the angular frequency of the three-phase mains voltages. Then the transfer function of the controller $G_{PR}(s)$ is expressed as

$$G_{PR}(s) = \frac{(k_{PR1}s + k_{PR2})(L_F s + R_F)}{s^2 + (3\omega_i)^2} \tag{8.6}$$

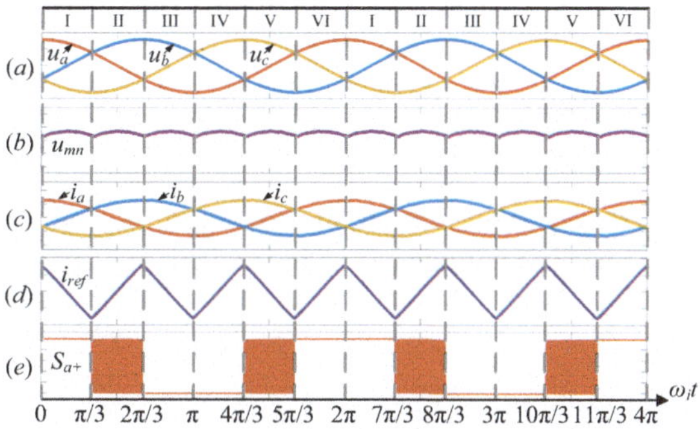

Fig. 8.3 Main waveforms of the rectifier (assuming forward power flow and unity input power factor). (**a**) Three-phase mains voltages $u_{a,b,c}$, (**b**) DC-link voltage u_{mn}, (**c**) three-phase input currents $i_{a,b,c}$, (**d**) quasi-third-harmonic current i_{ref}, and (**e**) schematic view of gating signal S_{a+}

where k_{PR1} and k_{PR2} are parameters for adjusting the bandwidth of the PR controller, and L_F and R_F are the inductance and parasitic resistance of the input inductor.

8.3.2 *Proof of Sinusoidal Input Currents*

The result that the three-phase input currents are symmetric and sinusoidal is verified by mathematical proofs. Sector I is taken as an example for the detailed discussions again.

Three-phase mains voltages $u_{a,b,c}$ can be represented as

$$\begin{cases} u_a = U_m \cos(\omega_i t) \\ u_b = U_m \cos(\omega_i t - 2\pi/3) \\ u_c = U_m \cos(\omega_i t + 2\pi/3) \end{cases} \tag{8.7}$$

where U_m is the amplitude of the three-phase mains voltages.

From the switching states of the rectifier, it is found that the current i_b equals the opposite value of the quasi-third-harmonic current i_{ref}. And the average value of i_b in steady state can be expressed as

$$\overline{i_b} = -i_{\text{ref}} \tag{8.8}$$

The quasi-third-harmonic current i_{ref} in sector I can be obtained from Table 8.1; then the average value of i_b is obtained as

$$\overline{i_b} = \frac{2P\cos(\omega_i t - \varphi - 2\pi/3)}{3U_m \cos\varphi} \tag{8.9}$$

The average value of i_a in steady state can be deduced as

$$\overline{i_a} = i_m + Di_{ref} = \frac{P}{u_{mn}} + Di_{ref} \tag{8.10}$$

In steady state, the DC-link voltage u_{mn} closely follows the peak value of the input line-line voltages u_{l-l}. In sector I, u_{mn} equals u_{ac}, and D is represented as

$$D = \frac{u_{bc}}{u_{ac}} \tag{8.11}$$

The average value of i_a is expressed further as (8.12) by substituting u_{mn}, D, and i_{ref} into (8.10)

$$\overline{i_a} = \frac{P}{u_{ac}} + \frac{u_{bc}}{u_{ac}} \frac{-2P\cos(\omega_i t - \varphi - 2\pi/3)}{3U_m \cos\varphi} \tag{8.12}$$

where

$$u_{bc} = \sqrt{3}U_m \sin(\omega_i t) \tag{8.13}$$

$$u_{ac} = \sqrt{3}U_m \sin(\omega_i t + \pi/3) \tag{8.14}$$

Then (8.12) can be simplified further as

$$\begin{aligned}\overline{i_a} &= \frac{\sqrt{3}P\cos(\varphi) - 2P\sin(\omega_i t)\cos(\omega_i t - \varphi - 2\pi/3)}{3U_m \sin(\omega_i t + \pi/3)\cos\varphi} \\ &= \frac{2P\cos(\omega_i t - \varphi)}{3U_m \cos\varphi} \frac{\sqrt{3}/2\cos(\omega_i t) + 1/2\sin(\omega_i t)}{\sin(\omega_i t + \pi/3)} \\ &= \frac{2P\cos(\omega_i t - \varphi)}{3U_m \cos\varphi}\end{aligned} \tag{8.15}$$

According to Kirchhoff's Current Law, the average current of phase c can be obtained as

$$\overline{i_c} = -\left(\overline{i_a} + \overline{i_b}\right) = \frac{2P\cos(\omega_i t - \varphi + 2\pi/3)}{3U_m \cos\varphi} \tag{8.16}$$

Hence, three-phase input currents are symmetric and sinusoidal with the input power factor angle φ in sector I. Mathematical proofs in other sectors are the same as those in sector I and will not be repeated here. It should be mentioned that, in the aforementioned derivation process, the method is suitable for both directions of power flow. Positive P means the forward power flow operating mode, and negative P means the contrary.

8.3.3 Control of the Inverter

As for the inverter of the MT-BBC, the carrier-based modulation with the variable carrier is applied to achieve three-phase output voltages with desired amplitude and frequency.

Three-phase output reference voltages can be expressed as

$$\begin{cases} u_u^* = U_o \cos(\omega_o t + \psi) \\ u_v^* = U_o \cos(\omega_o t + \psi - 2\pi/3) \\ u_w^* = U_o \cos(\omega_o t + \psi + 2\pi/3) \end{cases} \tag{8.17}$$

where U_o, ω_o, and ψ are the amplitude, the angular frequency, and the initial phase of the three-phase output reference voltages, respectively.

To improve the utilization of the DC-link voltage, a zero-sequence voltage u_{no} is injected. Then the three-phase modulation signals are obtained as

$$\begin{cases} u_{uo}^* = u_u^* + u_{no} \\ u_{vo}^* = u_v^* + u_{no} \\ u_{wo}^* = u_w^* + u_{no} \end{cases} \tag{8.18}$$

where u_{no} is the zero-sequence voltage.

The zero-sequence voltage u_{no} can be represented as

$$u_{no} = -\frac{\max\left(u_u^*, u_v^*, u_w^*\right) + \min\left(u_u^*, u_v^*, u_w^*\right)}{2} \tag{8.19}$$

Then the normalized modulation signals are obtained as

$$\begin{cases} u_{uo} = 2u_{uo}^*/u_{mn} \\ u_{vo} = 2u_{vo}^*/u_{mn} \\ u_{wo} = 2u_{wo}^*/u_{mn} \end{cases} \tag{8.20}$$

where u_{mn} is the six-pulse DC-link voltage. Because u_{mn} is time-varying, real-time voltage compensation of the inverter is realized by adopting (8.20).

The modulation diagram of the inverter is shown in Fig. 8.4. According to the geometry theory, the duty cycles of phases u, v, and w are expressed as

$$\begin{cases} d_u = (1 + u_{uo})/2 \\ d_v = (1 + u_{vo})/2 \\ d_w = (1 + u_{wo})/2 \end{cases} \tag{8.21}$$

Conventional PI controllers are used to control the three-phase output currents, and conventional frequency-domain analysis tools are used to design the controller.

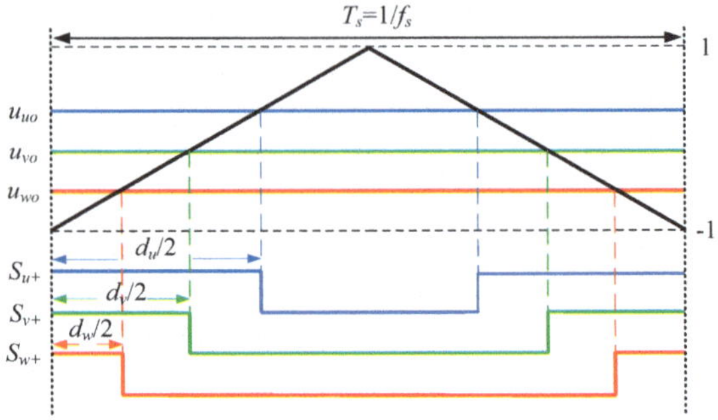

Fig. 8.4 Modulation diagram of the inverter

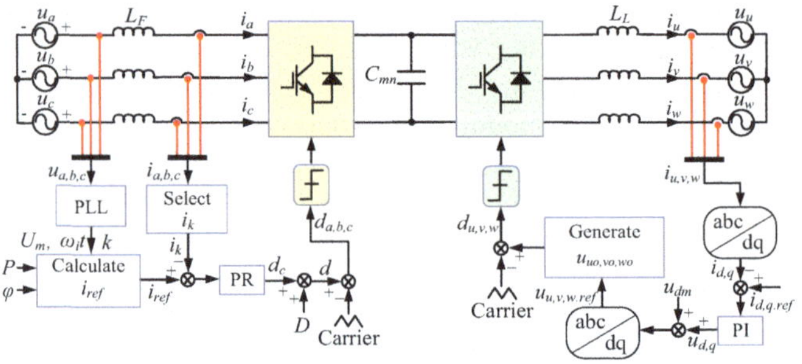

Fig. 8.5 System control schematic diagram

The damping ratio and current loop bandwidth are mainly considered in the design of PI controller, and the damping ratio and current loop bandwidth are selected as 0.707 and 1500 Hz, respectively. The K_p and K_i are designed as 6.36 and 6750, and the detailed design procedure using conventional frequency-domain analysis tools is not given for the sake of brevity. To improve the dynamic response speed, an extra feed-forward term u_{dm} is added.

According to the analysis above, the modulations of the rectifier and inverter in MT-BBC are independent. The clamped modulation strategy is applied to the rectifier, and the carrier modulation strategy with variable carrier is adopted for the inverter. The PR controller and PI controllers are employed in the rectifier and inverter, and the system control schematic diagram of the MT-BBC is shown in Fig. 8.5. Moreover, the power balance between the rectification and inversion is established automatically, and the principle is similar to that of the MCs. The DC-link voltage of the MT-BBC is clamped at the peak value of the input line-line voltages. The three-phase output voltages are synthesized directly with the instantaneous values of the three-phase mains voltages, and the three-phase input currents are synthesized indirectly by the output currents [21].

8.4 Experimental Results

8.4.1 Experimental Results

To verify the effectiveness of the proposed MT-BBC, a 6-kW experimental prototype is built as shown in Fig. 8.6. The main parameters of the MT-BBC are provided in Table 8.3. All of the waveforms are recorded by the DPO 3014 Digital Phosphor Oscilloscope. For the 12 active switches, IGBT FF50R12RT4 is selected.

Figures 8.7 and 8.8 are the experimental results when the MT-BBC is connected with load inductor and load resistor. The load inductance L_L and the load resistance R_L are set as 3 mH and 10 Ω, respectively.

Figure 8.7 shows the experimental waveforms with different input power factor angles. The active power P is set as 6000 W, and the output frequency f_o is set as 50 Hz. The input power factor angle φ is set as 0 in Fig. 8.7a, which means the unity power factor in the rectifier. The input power factor angle φ is set as $-\pi/6$ in Fig. 8.7b, which means a lead power factor of 0.866 in the rectifier. The input power factor angle φ is set as $\pi/6$ in Fig. 8.7c, which means a lag power factor of 0.866 in the rectifier. It is clear that the input and output currents are both sinusoidal, and the displacement angle of input voltage and input current is consistent with the corresponding input power factor angle; thus the good characteristics of sinusoidal

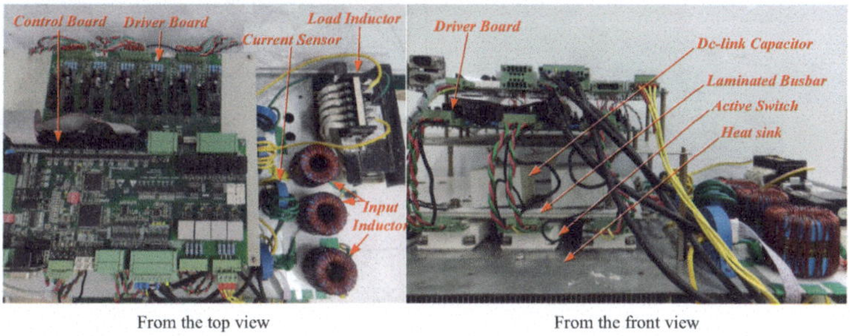

Fig. 8.6 Experimental prototype of the MT-BBC

Table 8.3 Main Parameters of the MT-BBC

Parameters	Value
Active power P	6 kW
Mains voltages $u_{a,b,c}$	220 Vrms
Input inductance L_F	2.2 mH
DC-link capacitance C_{mn}	4 μF
Mains frequency f	50 Hz
Switching frequency f_{sw}	25 kHz
Load resistance R_L	10 Ω
Load inductance L_L	3 mH
Output voltages $u_{u,v,w}$	158 Vrms

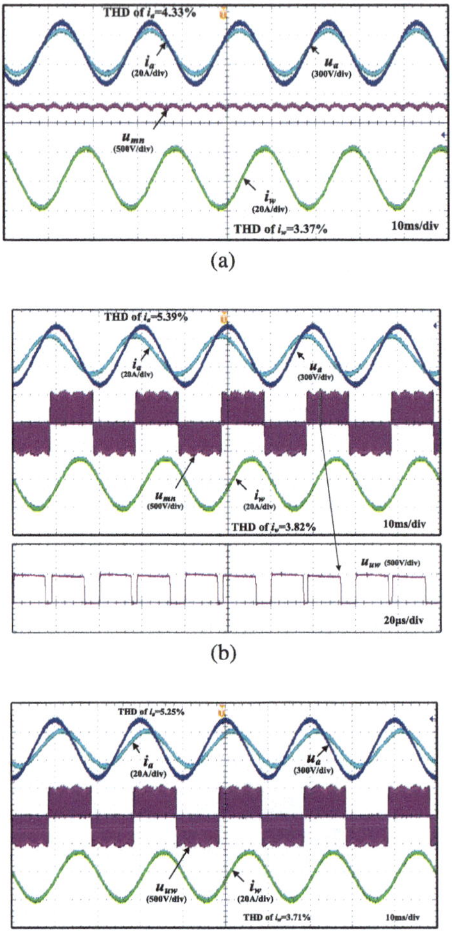

Fig. 8.7 Experimental waveforms of the MT-BBC with different input power factor angles. (**a**) Input power factor angle φ is 0, (**b**) input power factor angle φ is $-\pi/6$, and (**c**) input power factor angle φ is $\pi/6$

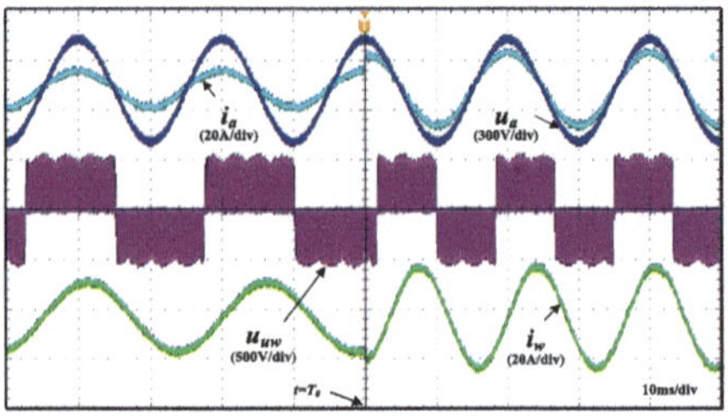

Fig. 8.8 Experimental waveforms of the MT-BBC with the modulation index changing from 0.53 to 0.75 and the output frequency changing from 40 Hz to 60 Hz

inputs and outputs and the controllable input power factor of the MT-BBC are verified.

Figure 8.8 shows the experimental waveforms of the MT-BBC with the modulation index changing from 0.53 to 0.75 and the output frequency changing from 40 Hz to 60 Hz. The result shows that the input current is sinusoidal and in phase with the corresponding mains voltage under different output conditions, which verifies that the MT-BBC possesses the capability of producing three-phase outputs with controllable amplitude and frequency.

Figure 8.9 represents the experimental results when the MT-BBC is in light load, no-load, and STATCOM operating modes. The active power is set as 1000 W by changing the load resistance in Fig. 8.9a. To achieve the STATCOM operating mode, the active power should be guaranteed to be 0. Then the reactive power is set as 2000 Var, and the input power factor angles φ are set as $\pi/2$ and -$\pi/2$, respectively. Figure 8.9 verifies the good feature of the extended input reactive power control range.

Figures 8.10 and 8.11 are the experimental results when the MT-BBC is connected with a three-phase isolated power source. The frequency of the three-phase output voltages is set as 50 Hz. The input power factor angle φ and reactive reference current $i_{q,ref}$ are both set as 0.

Figure 8.10 demonstrates the capability of bidirectional power of the MT-BBC. The experimental waveforms when the MT-BBC is in reverse power flow operating mode are shown, and the active power P is set as $-$ 6000 W. It can be seen that the input current is still sinusoidal and in antiphase with corresponding mains voltage.

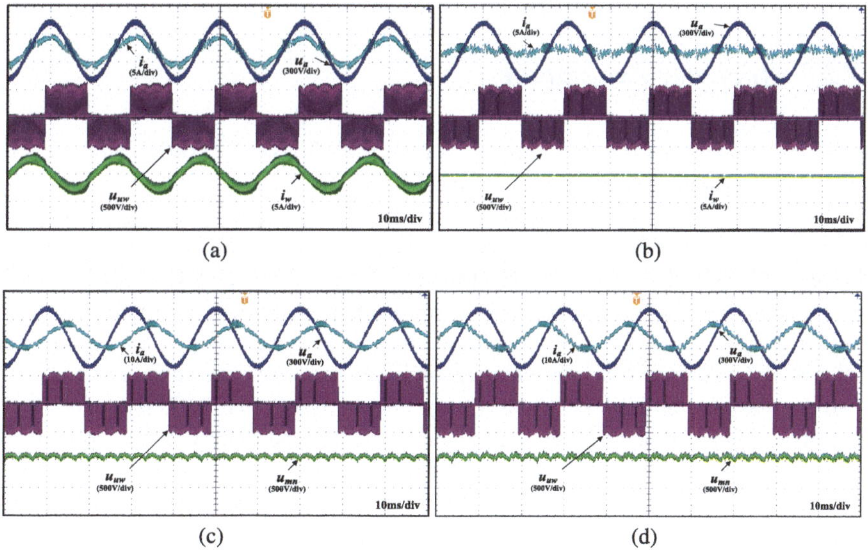

Fig. 8.9 Experimental waveforms when the MT-BBC is in light load, no-load, and STATCOM operating mode. (**a**) Light load operating mode, (**b**) no-load operating mode; (**c**) STATCOM operating mode ($\varphi = \pi/2$), and (**d**) STATCOM operating mode ($\varphi = -\pi/2$)

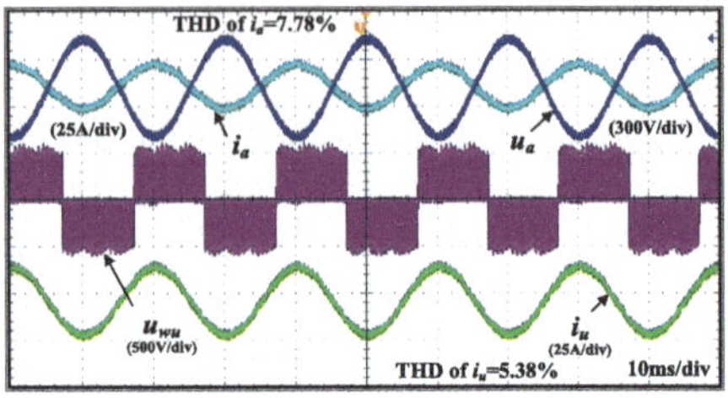

Fig. 8.10 Experimental waveforms when the MT-BBC is in reverse power flow operating mode

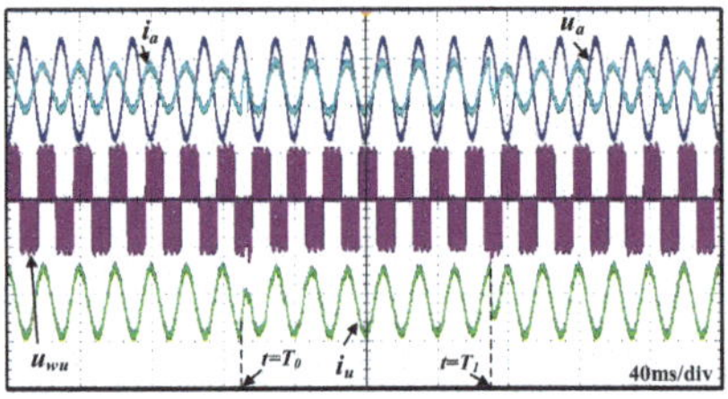

Fig. 8.11 Experimental waveforms of the MT-BBC with the active power P changing from -6000 to 6000 to -6000 W. u_a (300 V/div), i_a (25 A/div), u_{wu} (500 V/div), and i_u (25 A/div)

Figure 8.11 shows the dynamic performance of the proposed MT-BBC. The active power P is set as -6000 W at first. At T_0, P is changing from -6000 W to 6000 W. Then at T_1, P is changing from 6000 W to -6000 W. It can be found from Fig. 8.11 that the output current tracks the reference well, which indicates the good dynamic performance of the MT-BBC.

8.4.2 *Efficiency Evaluation*

To evaluate and compare the efficiency among the MT-BBC, 3HIMC, and conventional VBBC, power losses including IGBT conduction loss, IGBT switching loss, diode loss, inductor loss, and capacitor loss are analyzed. For the sake of fairness, the active power P and the output voltages $u_{u,v,w}$ of the 3HIMC and the VBBC are both set as 6000 W and 158 Vrms. For the parameters of the 3HIMC, the inductance, the capacitance of the input filter, and the inductance of the third-harmonic injection inductor are set as 300 μH, 10 μF, and 2.2 mH, respectively. The main parameters of the VBBC including the inductance of input inductor, the capacitance of the DC-link capacitor, and the DC-link voltage are set as 2.2 mH, 2200 μF, and 700 V, respectively. Due to the DC-link voltage and the clamped modulation keeping only one high-frequency switching bridge arm during any mains periods, the MT-BBC gives better input current harmonic performance for the same switching frequency compared with the VBBC. When the VBBC gives close input current harmonic performance as that of the MT-BBC, the switching frequency of the VBBC is almost 26 kHz, which is adopted to calculate the power losses of the VBBC.

The conduction loss and switching loss of the semiconductor devices are calculated by the values of currents and other parameters as

$$P_{con,s} = I_{S,AVG}U_{CE0} + I_{S,RMS}^2 R_S \tag{8.22}$$

$$P_{sw,s} = k_s I_{S,AVG} f_{sw} \tag{8.23}$$

$$P_{con,D} = I_{D,AVG}U_{D0} + I_{D,RMS}^2 R_D \tag{8.24}$$

where $P_{con,s}$, $P_{sw,s}$, and $P_{con,D}$ are the IGBTs conduction loss, IGBTs switching loss, and diodes loss; $k_s = V_{S,AVG}(E_{on\_R} + E_{off\_R})/(V_R I_R)$, E_{on_R} and E_{off_R} are the turn-on and turn-off energy at the reference voltage V_R and reference current I_R, and $V_{S,AVG}$ is the average switching voltage of the IGBTs; U_{CE0} and U_{D0} are the forward voltages of the IGBTs and diodes when the current is zero; $I_{S,RMS}$ and $I_{D,RMS}$ are the RMS currents of the IGBTs and diodes; R_S and R_D are the dynamic on-resistances of the IGBTs and diodes; $I_{S,AVG}$ and $I_{D,AVG}$ are the average currents of the IGBTs and diodes; and f_{sw} is the switching frequency. Parameters can be obtained from the database of the IGBT FF50R12RT4, and the diode reverse recovery loss is included in $P_{sw,s}$.

The inductor loss is divided into the core loss and winding resistance loss, which can be calculated as

$$P_l = I_{l\_rms}^2 R_l + \frac{1}{\pi}\int_0^{\pi} k\Delta B^{\alpha}(\omega t)\, f_{sw}^{\beta}(\omega t)\, d\omega t \tag{8.25}$$

where I_{l_rms}, R_l, and ΔB are the RMS current of the input inductor, the parasitic resistance of the input inductor, and the peak-to-peak flux ripple, and k, α, and β are constants, which can be obtained from the core manufacturer's datasheet.

Considering the ESR of the DC-link capacitor, capacitor loss can be estimated as

$$P_c = I_{c\_rms}^2 R_c \tag{8.26}$$

where I_{c_rms} is the RMS value of the current flowing through the DC-link capacitor and R_c is the ESR of the DC-link capacitor.

Under the same working conditions, the power losses of the MT-BBC, the 3HIMC, and the conventional VBBC are calculated and compared, which are listed in Fig. 8.12. The DC-link capacitor of the MT-BBC is combined with four small film capacitors in parallel. According to the database, ESR of the 1 μF film capacitor is 3.4 mΩ, and then the total ESR of the DC-link capacitor of the MT-BBC is 0.85 mΩ. The aluminum electrolytic capacitors connected in series and parallel are selected as the DC-link capacitor of the VBBC. Three 2200 μF/400 V aluminum electrolytic capacitors are connected in series as one group, and three same groups are connected in parallel; then the equivalent capacitance is 2200 μF. ESR of the 2200 μF/400 V aluminum electrolytic capacitor is 87 mΩ, then the total ESR of the DC-link capacitor of the VBBC is also 87 mΩ. It can be seen from Fig. 8.12 that

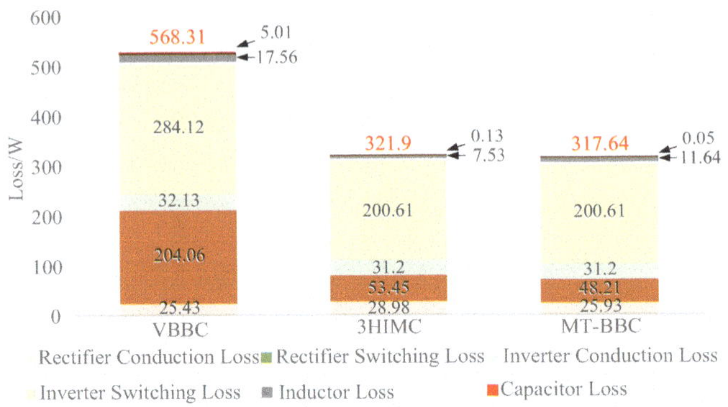

Fig. 8.12 Comparisons of power losses among the conventional VBBC, the 3HIMC, and the MT-BBC

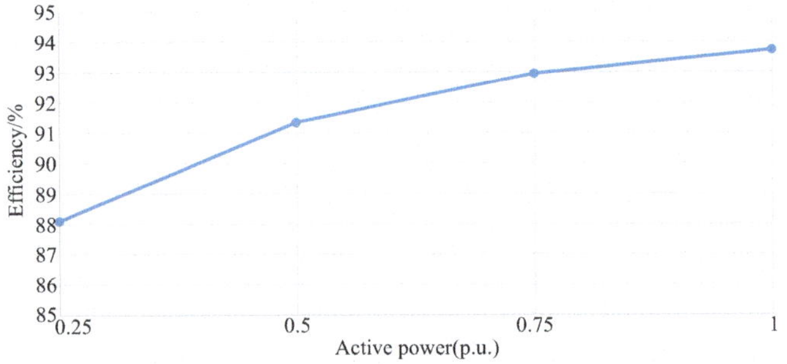

Fig. 8.13 Efficiency curve of the MT-BBC

the switching loss of the rectifier and inverter, the inductor loss, and the capacitor loss of the MT-BBC are all lower compared with the conventional VBBC, and the switching loss of the rectifier is almost decreased more than 76.37%. The calculated efficiency of the MT-BBC is 94.71%, which is 4.18% higher than the conventional VBBC and is almost equal to that of the 3HIMC. The system efficiency is measured by the HIOKI 3390 three-phase wattmeter, and the efficiency curve is shown in Fig. 8.13. The efficiency measured at full power is 93.75%, which is lower than the calculated value. This is because the system losses are related to other factors such as current sensors, which are not considered in the calculation.

8.5 Conclusions

This chapter proposes a matrix-type back-to-back converter with single-arm current injection. The rectifier and the inverter are connected with a small film capacitor rather than a bulky electrolytic capacitor. The clamped modulation strategy is introduced in the rectifier, which achieves sinusoidal input currents, reduced switching losses, and extended input power factor angle control range varying from -$\pi/2$ to $\pi/2$. A 6-kW experimental prototype is built, which verifies that the MT-BBC has essential characteristics such as enhanced input reactive power capability, sinusoidal inputs and outputs, bidirectional power flow capability, direct ac-ac conversion, high conversion efficiency, high reliability, potentially high power density, and good dynamic performance. All the good features make the MT-BBC a good candidate in applications such as ASDs, WECS, and FACTS.

References

1. A.A.A. Radwan, Y.A.-R.I. Mohamed, Grid-Connected wind-solar cogeneration using back-to-back voltage-source converters. IEEE Trans. Sustain Energy **11**(1), 315–325 (2020). https://doi.org/10.1109/TSTE.2019.2890828
2. G.C. Konstantopoulos, A.T. Alexandridis, Full-Scale modeling, control, and analysis of grid-connected wind turbine induction generators with back-to-back AC/DC/AC converters. IEEE J. Emerg. Sel. Topics Power Electron. **2**(4), 739–748 (2014). https://doi.org/10.1109/JESTPE.2014.2325676
3. J.W. Kolar, T. Friedli, J. Rodriguez, P.W. Wheeler, Review of three-phase PWM AC–AC converter topologies. IEEE Trans. Ind. Electron. **58**(11), 4988–5006 (2011). https://doi.org/10.1109/TIE.2011.2159353
4. D. Terazono, J. Liu, Y. Miura, S. Sakabe, H. Bevrani, T. Ise, Grid frequency regulation support from back-to-back motor drive system with virtual-synchronous-generator-based coordinated control. IEEE Trans. Power Electron. **36**(3), 2901–2913 (2021). https://doi.org/10.1109/TPEL.2020.3015806
5. I.D.N. Souza, P.M.d. Almeida, G.A. Fogli, P.G. Barbosa, P.F. Ribeiro, Multivariable optimal control applied to a back-to-back power converter. IEEE Trans. Ind. Electron. **69**(9), 9406–9418 (2022). https://doi.org/10.1109/TIE.2021.3114748
6. Y. Gui, F. Blaabjerg, X. Wang, J.D. Bendtsen, D. Yang, J. Stoustrup, Improved dc-link voltage regulation strategy for grid-connected converters. IEEE Trans. Ind. Electron. **68**(6), 4977–4987 (2021). https://doi.org/10.1109/TIE.2020.2989720
7. L. Shen, S. Bozhko, C.I. Hill, P. Wheeler, DC-Link capacitor second carrier band switching harmonic current reduction in two-level back-to-back converters. IEEE Trans. Power Electron. **33**(4), 3567–3574 (2018). https://doi.org/10.1109/TPEL.2017.2706972
8. H. Zhao, L. Cheng, Open-switch fault-diagnostic method for back-to-back converters of a doubly fed wind power generation system. IEEE Trans. Power Electron. **33**(4), 3452–3461 (2018). https://doi.org/10.1109/TPEL.2017.2705985
9. T. Friedli, J.W. Kolar, J. Rodriguez, P.W. Wheeler, Comparative evaluation of three-phase AC–AC matrix converter and voltage DC-Link back-to-back converter systems. IEEE Trans. Ind. Electron. **59**(12), 4487–4510 (2012) https://doi.org/10.1109/TIE.2011.2179278
10. P. Liu, Z. Wang, Y. Xu, Z. Zou, F. Deng, Y. Li, Improved harmonic profile for high-power PWM current-source converters with modified space-vector modulation schemes. IEEE Trans. Power Electron. **36**(10), 11234–11244 (2021). https://doi.org/10.1109/TPEL.2021.3070885
11. A. Tcai, H.-U. Shin, K.-B. Lee, DC-Link capacitor-current ripple reduction in DPWM-based back-to-back converters. IEEE Trans. Ind. Electron. **65**(3), 1897–1907 (2018). https://doi.org/10.1109/TIE.2017.2745453
12. H. Wang, F. Blaabjerg, Reliability of capacitors for DC-Link applications in power electronic converters–an overview. IEEE Trans. Ind. Appl. **50**(5), 3569–3578 (2014). https://doi.org/10.1109/TIA.2014.2308357

13. M.-H. Wang, Y. He, T. Yang, Y. Jia, Z. Xu, Cascaded voltage control for electric springs with DC-Link film capacitors. IEEE J. Emerg. Sel. Topics in Power Electron. **8**(4), 3982–3994 (2020). https://doi.org/10.1109/JESTPE.2019.2962238
14. M. Ghadrdan, S. Peyghami, H. Mokhtari, F. Blaabjerg, Condition monitoring of DC-Link electrolytic capacitor in back-to-back converters based on dissipation factor. IEEE Trans. Power Electron. **37**(8), 9733–9744 (2022). https://doi.org/10.1109/TPEL.2022.3153842
15. M. Ghadrdan, B. Abdi, S. Peyghami, H. Mokhtari, F. Blaabjerg, On-line condition monitoring system for DC-link capacitor of back-to-back converters using large-signal transients. IEEE J. Emerg. Sel. Topics Power Electron. https://doi.org/10.1109/JESTPE.2022.3163012
16. J. Lei, S. Feng, B. Zhou, H.-N. Nguyen, J. Zhao, W. Chen, A simple modulation scheme with zero common-mode voltage and improved efficiency for direct matrix converter-fed PMSM drives. IEEE J. Emerg. Sel. Topics Power Electron. **8**(4), 3712–3722 (2020). https://doi.org/10.1109/JESTPE.2019.2934730
17. W. Deng, Maximum voltage transfer ratio of matrix converter under DTC with rotating vectors. IEEE Trans. Power Electron. **36**(6), 6137–6141 (2021). https://doi.org/10.1109/TPEL.2020.3033414
18. J. Lei, S. Feng, B. Zhou, W. Cao, J. Zhao, Evaluation and improvement of active stabilization method for matrix converter under input voltage disturbances. IEEE J. of Emerg. Sel. Topics Power Electron. **7**(2), 1116–1125 (2019). https://doi.org/10.1109/JESTPE.2018.2848270
19. M. Siami, M. Amiri, H.K. Savadkoohi, R. Rezavandi, S. Valipour, Simplified predictive torque control for a PMSM drive fed by a matrix converter with imposed input current. IEEE J. Emerging. Sel. Topics Power Electron. **6**(4), 1641–1649 (2018). https://doi.org/10.1109/JESTPE.2018.2837109
20. J. Lei, S. Feng, P. Wheeler, B. Zhou, J. Zhao, Steady-state error suppression and simplified implementation of direct source current control for matrix converter with model predictive control. IEEE Trans. Power Electron. **35**(3), 3183–3194 (2020). https://doi.org/10.1109/TPEL.2019.2928874
21. J. Rodriguez, M. Rivera, J.W. Kolar, P.W. Wheeler, A review of control and modulation methods for matrix converters. IEEE Trans. Ind. Electron. **59**(1), 58–70 (2012). https://doi.org/10.1109/TIE.2011.2165310
22. S. Mondal, D. Kastha, Input reactive power controller with a novel active damping strategy for a matrix converter fed direct torque controlled DFIG for wind power generation. IEEE J. Emerg. Sel. Topics Power Electron. **8**(4), 3700–3711 (2020). https://doi.org/10.1109/JESTPE.2019.2938012
23. L. Ming et al., A SiC-Si hybrid module for direct matrix converter with mitigated current spikes. IEEE J. Emerg. Sel. Topics Power Electron. **10**(4), 3805–3817 (2022). https://doi.org/10.1109/JESTPE.2021.3093627
24. P.P. Patel, M.A. Mulla, A single carrier-based pulsewidth modulation technique for three-to-three phase indirect matrix converter. IEEE Trans on Power Electron. **35**(11), 11589–11601 (2020). https://doi.org/10.1109/TPEL.2020.2988594
25. H. Wang et al., Two-stage matrix converter based on third-harmonic injection technique. IEEE Trans. Power Electron. **31**(1), 533–547 (2016). https://doi.org/10.1109/TPEL.2015.2413452
26. C. Lu, B. Zhou, J. Lei, J. Shan, Mains current distortion suppression for third-harmonic injection two-stage matrix converter. IEEE Trans. Power Electron. **36**(6), 7202–7211 (2021). https://doi.org/10.1109/TPEL.2020.3039596
27. C. Lu, B. Zhou, Q. Chang, J. Lei, Improved third-harmonic injection two-stage matrix converter topology based on decomposition control of injection current. IEEE Trans. Power Electron. **37**(7), 7846–7857 (2022). https://doi.org/10.1109/TPEL.2022.3149484
28. S. Albatran, A.S. Allabadi, A.R. A. Khalaileh, Y. Fu, Improving the performance of a two-level voltage source inverter in the overmodulation region using adaptive optimal third harmonic injection pulsewidth modulation schemes. IEEE Trans. Power Electron. **36**(1), 1092–1103 (2021). https://doi.org/10.1109/TPEL.2020.3001494

Chapter 9
Three-Level Modulation Strategy for Conventional 3 × N Direct Matrix Converters

9.1 Introduction

Due to the potential superiority such as simple and compact structure, sinusoidal input and output currents, and regeneration capability, the DMC has been studied for many years [1]. As shown in Fig. 9.1, the conventional $3 \times N$ ($N \geqslant 3$) DMC is composed of an array of multiple bidirectional power switches. This converter, which directly connects the grid to the N-phase load, synthesizes output voltages with variable frequency and amplitude without using energy storage elements. Thereby, the $3 \times N$ DMC is attractive for many applications such as multiphase wind energy generation and multiphase adjustable speed drives [2, 3].

Due to the high complexity of the DMC system, scholars have made great efforts for modulation and control methods. Classical methods such as Venturini's approach [4, 5], scalar scheme [6], and the double voltage synthesis scheme [7] emerged at the end of the last century and underlined subsequent studies. As an effective and robust strategy, the SVM utilizes basic space vectors to synthesize the expected output voltages and input currents [8–14], and some strong features such as zero common-mode voltage and higher efficiency are achieved via the utilization of rotating vectors [15]. However, the SVM is very complex for multiphase converters, and it is unsuitable, especially for $N > 5$. Thereby, the CBPWM strategy is applied to DMCs as a simple scheme [16–18]. Essentially the same as SVM, CBPWM completes a low complexity modulation for DMCs by selecting the appropriate carrier and offset voltage [19, 20]. Furthermore, the computational burden is effectively reduced through a simplistic modulation algorithm that cancels trigonometry operations and angles [21]. Other attractive schemes such as mathematical construction [22, 23], predictive control [24–26], delta-sigma modulation [27, 28], and the analytic signal PWM [29, 30] provide new ideas for the modulation and control of DMCs. Besides, a generalized modulation algorithm for the different number of inputs and outputs is developed in [3], and the scalar method for 3×3 DMCs is extended to $M \times N$ cases. This attractive algorithm has a simple result and is easy to implement for $M \times N$

H. Wang, *Advanced Matrix Converters*, Power Systems,
https://doi.org/10.1007/978-3-031-69902-3_9

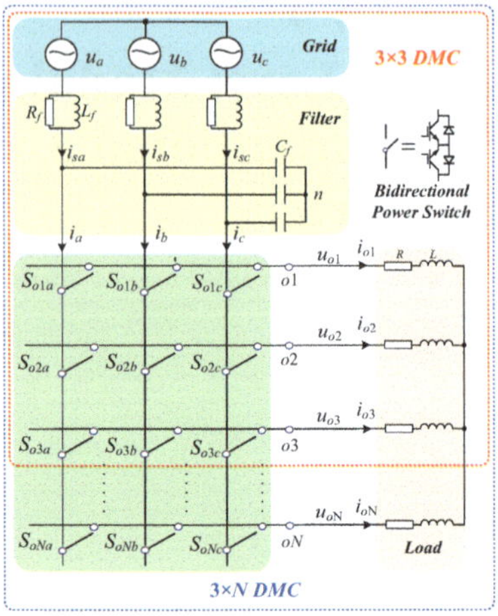

Fig. 9.1 Schematic of a $3 \times N$ DMC system

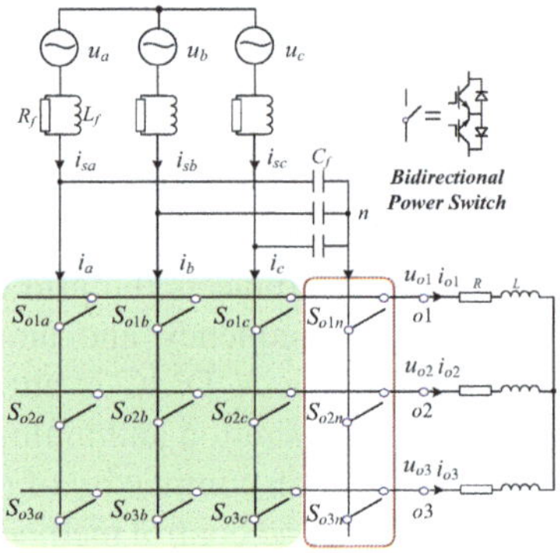

Fig. 9.2 Schematic of a three-level DMC system

DMCs. However, there is still room for improvement in the output waveform quality and efficiency due to the two-level output characteristics in the above strategies [31]. As shown in Fig. 9.2, the derived multilevel topology achieves the three-level output at the cost of a change in the topology and additional power switches [31, 32].

Though DMCs have been studied extensively and remarkable achievements have emerged, the inherent potential of DMCs has not been fully exploited. To achieve three-level outputs for the two-level converter, this chapter presents a 3LPWM strategy for conventional $3 \times N$ DMCs. The DMC is decomposed into a fictitious rectifier and an NPC three-level inverter in this approach. By designing the offset voltage of the modulation signals, sinusoidal grid currents, three-level output voltages, increased efficiency, and controllable input power factors are achieved in conventional $3 \times N$ DMCs without any changes in the topology. The output waveform quality, such as the THD of output voltages and currents, is improved, but the input currents worsen slightly in the developed 3LPWM strategy.

The following parts of this chapter are as follows. Section 9.2 describes the derivation of the developed strategy. The developed 3LPWM strategy is analyzed in detail in Sect. 9.3. Then simulation studies and experimental validation are presented in Sect. 9.4. Finally, Sect. 9.5 draws the conclusion.

9.2 Derivation of Three-Level Output for Conventional DMCs

As an "all-silicon" converter, the DMC realizes direct AC-AC conversion, namely each output phase can be connected to any input phase. Moreover, because the magnitude of each grid voltage is time-varying, the three-phase balanced grid voltages are split into three levels, i.e., high-level u_p, middle-level u_o, and low-level u_n, as shown in Fig. 9.3.

In the developed three-level output strategy, the three-phase grid voltages are assigned to the three voltage levels, which are considered as the required fictitious DC-link voltages. Then the rear-end fictitious part is equivalent to a conventional three-level inverter.

For the conventional $3 \times N$ DMC system shown in Fig. 9.1, three-phase grid voltages are u_a, u_b, and u_c. The N-phase output reference voltages are

$$\begin{cases} u_{o1\_ref} = U_{om} \cos(\omega_o t + \phi) \\ u_{o2\_ref} = U_{om} \cos(\omega_o t - 2\pi/N + \phi) \\ u_{o3\_ref} = U_{om} \cos(\omega_o t - 4\pi/N + \phi) \\ \vdots \\ u_{oN\_ref} = U_{om} \cos(\omega_o t - 2(N-1)\pi/N + \phi) \end{cases} \tag{9.1}$$

where U_{om}, ω_o, and ϕ represent the amplitude, angular frequency, and initial phase angle of the output reference voltages, respectively.

The CBPWM method is a good choice to achieve the three-level output of DMCs due to its low complexity. It compares output references with an imaginary carrier to obtain the switching states. The intersection time between the output reference and the carrier represents the commutation time, and gating signals are

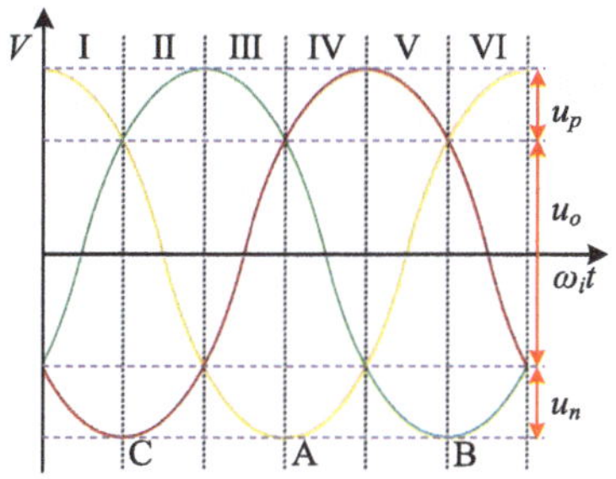

Fig. 9.3 Three levels of grid voltages in the DMC

only determined by the switching states [19]. Thereby, the key to the proposed three-level output modulation strategy lies in the output reference construction and the carrier selection, and details are given in the following sections.

9.3 Three-Level Modulation Strategy

According to Sect. 9.2, the principle of the developed strategy of DMCs is summarized as that three input voltage levels are exploited to synthesize the reference output voltage anytime. To facilitate the analysis, the DMC is split into a fictitious rectifier part and a fictitious inverter part, as shown in Fig. 9.4, where u_{dc} and $u_p/u_o/u_n$ represent the DC-link voltage and the potential of point $p/o/n$, respectively. For convenience, the following analysis is valid in the case of $u_a > u_b > u_c$.

9.3.1 Formation of DC-Link Voltage in the Fictitious Rectifier

To implement sinusoidal input currents and provide the fictitious inverter with three-level voltages, a dual-carrier-based PWM method is developed and explained in detail.

Since the load can be assumed as the current sources in a sampling period T_s, the average upper and lower current (denoted as $\bar{i}_p$ and $\bar{i}_n$) of the fictitious DC-link is synthesized by the output currents. Then the expected average input currents are acquired by adjusting $\bar{i}_p$ and $\bar{i}_n$. The adjustment factors, which are considered as the modulation signals of the fictitious rectifier, are designed as

$$\begin{cases} d_1 = m_1 \cos(\omega_i t - \varphi) \\ d_2 = -m_1 \cos(\omega_i t + 2\pi/3 - \varphi) \\ 0 \le (d_1, d_2) \le 1 \end{cases} \tag{9.2}$$

where ω_i, φ ($-\pi/6 \le \varphi \le \pi/6$), and m_1 ($0 \le m_1 \le 1$) represent the angular frequency of the grid voltages, the given input power factor angle, and the modulation index of the fictitious rectifier, respectively.

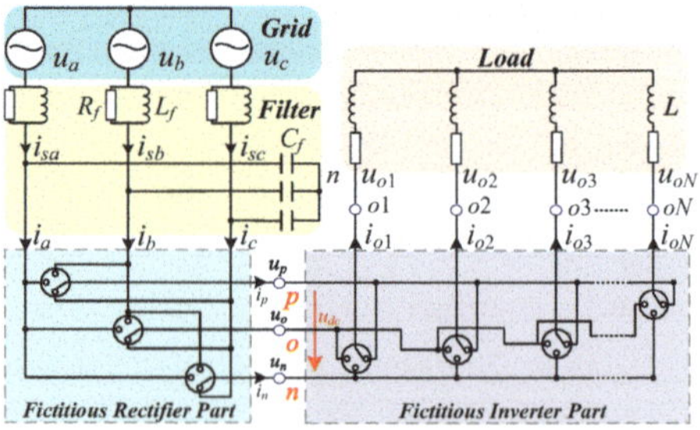

Fig. 9.4 Schematic diagram of the equivalent fictitious topology for the conventional 3 × *N* DMC

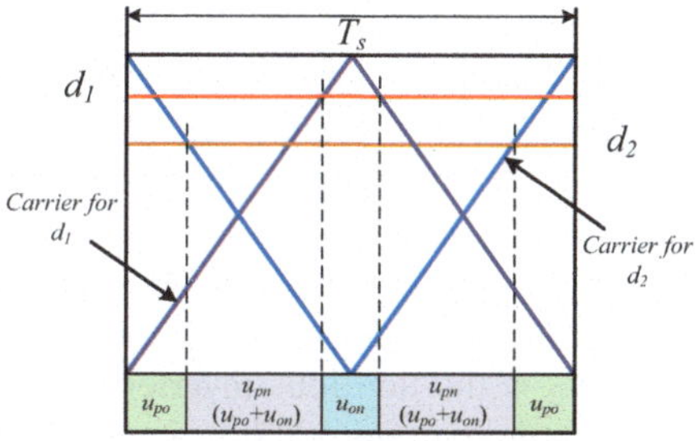

Fig. 9.5 Formation of developed three-level DC-link voltages

By comparing the modulation signals with their corresponding carriers, the fictitious DC-link voltages are determined. As shown in Fig. 9.5, the phase-shifted carriers are adopted to improve the output performance. And the state $u_{pn}(u_{po} + u_{on})$ represents that u_p, u_o, and u_n are the upper, medium, and lower voltages of fictitious DC-link, respectively. Assuming that the grid voltages satisfy $u_a > u_b > u_c$, the average fictitious DC-link voltages in a sampling period are calculated as

$$\begin{cases} \overline{u}_{po} = d_1 \cdot u_{ab} \\ \overline{u}_{on} = d_2 \cdot u_{bc} \\ \overline{u}_{pn} = \overline{u}_{po} + \overline{u}_{on} \end{cases} \tag{9.3}$$

When the grid voltages are balanced, $\overline{u}_{pn} = 1.5m_1U_{im}\cos(\varphi)$, where U_{im} represents the amplitude of the grid voltages. The maximum VTR is $\sqrt{3}m_1\cos(\varphi)/2$ and can reach 0.866 with a unity input power factor. Therefore, m_1 is generally set as 1 to maximize the VTR. Besides, the adjustment factors in other cases are also similar to (9.2), and the general formulas can be described as

$$\begin{cases} d_1 = m_1\cos(\theta_{max} - \varphi) \\ d_2 = -m_1\cos(\theta_{min} - \varphi) \\ 0 \le (d_1, d_2) \le 1 \end{cases} \tag{9.4}$$

where θ_{max} and θ_{min} are the phase angles of the maximum and minimum grid phase voltages, respectively.

9.3.2 Multi-carrier-Based PWM Method for the Fictitious Inverter

Using the above process, three input voltage levels of the fictitious inverter have been determined, and the fictitious inverter can be regarded as a conventional three-level inverter. However, conventional modulation methods of a three-level inverter are not applicable, and a multi-carrier-based DSPWM method is developed to keep the modulation synchronization between the two fictitious parts.

According to (9.1) and Sect. 9.3, the relationship between U_{om} and U_{im} under the balanced grid voltages can be described as

$$U_{om} = \sqrt{3}m_1 m_2 \cos(\varphi) U_{im}/2 \tag{9.5}$$

where m_2 is the fictitious inverter modulation index that ranges from 0 to m_{2max}. Since the maximum VTR is inversely proportional to the number N of output phases, m_{2max} equals 1 in the 3×3 DMC and decreases as N increases.

Then the implementation of the multi-carrier-based DSPWM method is unfolded as given below. To get the utmost out of the DC-link voltage, the concept of offset voltage is also adopted to synthesize the three-phase output voltages. Namely, the fundamental modulation signals in (9.1) are modified by appending the offset voltage as

$$\begin{cases} u'_{o1\_ref} = u_{o1\_ref} + u_0 \\ u'_{o2\_ref} = u_{o2\_ref} + u_0 \\ u'_{o3\_ref} = u_{o3\_ref} + u_0 \\ \vdots \\ u'_{oN\_ref} = u_{oN\_ref} + u_0 \end{cases} \tag{9.6}$$

where u'_{oi_ref} $(i = 1, 2, \cdots, N)$ is the modified modulation signals, and u_0 represents the offset voltage. Furthermore, one degree of freedom must be introduced by separating two parts of the modified modulation signals, which are described as

$$\begin{cases} u'_{o1\_ref} = u_{o1\_ref\_p} + u_{o1\_ref\_n} \\ u'_{o2\_ref} = u_{o2\_ref\_p} + u_{o2\_ref\_n} \\ u'_{o3\_ref} = u_{o3\_ref\_p} + u_{o3\_ref\_n} \\ \vdots \\ u'_{oN\_ref} = u_{oN\_ref\_p} + u_{oN\_ref\_n} \end{cases} \tag{9.7}$$

where $u_{oi\_ref\_p}$ and $u_{oi\_ref\_n}$ $(i = 1, 2, \cdots, N)$ represent the positive modulation signal and the negative modulation signal of the modified modulation signals, respectively.

It is necessary to determine the fundamental modulation signals with the maximum and minimum magnitude by using

$$\begin{cases} u_{max} = \max\left(u_{o1\_ref}, u_{o2\_ref}, \cdots, u_{oN\_ref}\right) \\ u_{min} = \min\left(u_{o1\_ref}, u_{o2\_ref}, \cdots, u_{oN\_ref}\right) \end{cases} \tag{9.8}$$

where u_{max} and u_{min} represent the maximum and minimum of fundamental modulation signals, and "max( )" stands for the operation of selecting the maximum values.

According to (9.6)–(9.8), the positive and negative modulation signals can be designed as

$$\begin{cases} u_{max\_p} = u_{max} + u_0 \\ u_{max\_n} = 0 \\ u_{ox\_ref\_p} = 0.5u_{ox\_ref} + u_{01x} \\ u_{ox\_ref\_n} = 0.5u_{ox\_ref} + u_{02x} \\ u_{min\_p} = 0 \\ u_{min\_n} = u_{min} + u_0 \end{cases} \tag{9.9}$$

where u_{max_p} and u_{min_p} represent the positive modulation signals of the maximum and minimum modified modulation signals and vice versa, x represents the number of the rest, and u_{01x} and u_{02x} are the offset voltages of positive and negative modulation signals, respectively, and satisfy $u_{01x} + u_{02x} = u_0$.

The positive and negative modulation signals still need to be adjusted before a practical application. Firstly, the positive and negative modulation signals are further normalized as $\overline{u}^*_{j\_p}$ and $\overline{u}^*_{j\_n}$ to facilitate a digital implementation.

Based on the formula (9.3), the $\overline{u}^*_{j\_p}$ and $\overline{u}^*_{j\_n}$ are described as

$$\begin{cases} \overline{u}^*_{j\_p} = u_{j\_p}/\overline{u}_{po} \\ \overline{u}^*_{j\_n} = u_{j\_n}/\overline{u}_{on} \end{cases}, j = \{max, ox\_ref, min\} \tag{9.10}$$

The corresponding duty ratios can be expressed by the absolute values of $\overline{u}^*_{j\_p}$ and $\overline{u}^*_{j\_n}$, which are

$$\begin{cases} d_{j\_p} = \overline{u}^*_{j\_p} \\ d_{j\_n} = -\overline{u}^*_{j\_n} \end{cases} \tag{9.11}$$

where d_{j_p} and d_{j_n} are the corresponding upper and lower duty ratios of the switches in the fictitious inverter. Thereby, constraints occur, which can be described as

$$\begin{cases} 0 \le \overline{u}^*_{j\_p} \le 1 \\ -1 \le \overline{u}^*_{j\_n} \le 0 \end{cases} \tag{9.12}$$

The offset voltages with the constraints in (9.12) can then be set as

$$\begin{cases} u_{01x} = \left[-0.5u_{ox\_ref}\left(\overline{u}_{on} - \overline{u}_{po}\right) - \overline{u}_{po}u_{\min}\right]/\overline{u}_{pn} \\ u_{02x} = \left[0.5u_{ox\_ref}\left(\overline{u}_{on} - \overline{u}_{po}\right) - \overline{u}_{on}u_{\max}\right]/\overline{u}_{pn} \\ u_0 = u_{01x} + u_{02x} \end{cases} \tag{9.13}$$

and the proof of sinusoidal grid currents under the balanced grid voltages will come in Sect. 9.3.4.

9.3.3 Implementation of 3LPWM for DMCs

According to Sect. 9.3.1, the average DC-link voltages are governed by d_1 and d_2. To guarantee a modulation synchronization, the sampling period T_s for the fictitious inverter is divided into two time-varying subperiods. In other words, the conventional carrier-based modulation process with a time-varying carrier period is done twice during a sampling period.

Based on the above analysis, the duty ratios, which are regarded as the normalized modulation signals for the DMC, are calculated as

$$\begin{cases} d_{1j\_p} = d_1 d_{j\_p} \\ d_{2j\_n} = d_2 d_{j\_n} \end{cases}, j \in \{\max, ox\_ref, min\} \tag{9.14}$$

Then, the three-level output of the DMC is realized by comparing the modulation signals with their corresponding carriers. Taking the 3×3 DMC surrounded by the red dashed line in Fig. 9.1 as an example, the developed three-level output strategy under the balanced grid voltages is demonstrated in Fig. 9.6. Note that "ox_ref" is replaced by "mid" due to a single x. The state S_{max_p} represents that the switch corresponding to it is on, and so on. As shown in Fig. 9.3, six intervals are divided in the light of the grid voltage phase angle $\omega_i t$ under the balanced grid voltages. The correspondences between the states shown in Fig. 9.6 and the actual switches in the 3×3 DMC system are listed in Table 9.1, when $u_{o1\_ref} > u_{o2\_ref} > u_{o3\_ref}$.

9.3.4 Proof of Sinusoidal Grid Currents

Assuming that the output currents are

$$\begin{cases} i_{o1} = I_{om}\cos\left(\omega_o t + \phi - \varphi_L\right) \\ i_{o2} = I_{om}\cos\left(\omega_o t - 2\pi/N + \phi - \varphi_L\right) \\ \vdots \\ i_{oN} = I_{om}\cos\left(\omega_o t - 2(N-1)\pi/N + \phi - \varphi_L\right) \end{cases} \tag{9.15}$$

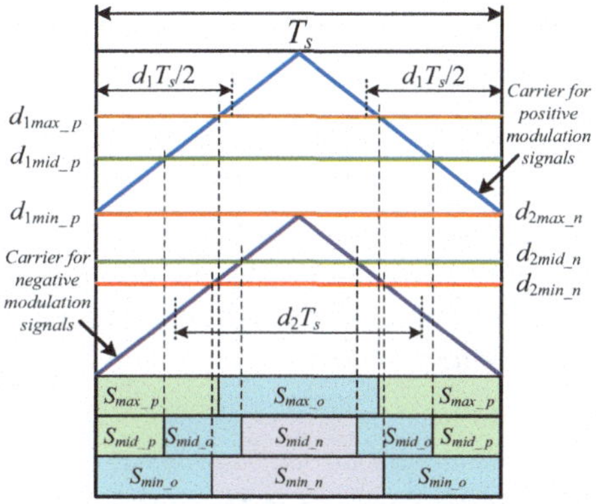

Fig. 9.6 Modulation scheme for 3 × 3 DMC

Table 9.1 Correspondences between states and actual switches for 3 × 3 DMC

State	I	II	III	IV	V	VI
S_{max_p}	S_{o1a}	S_{o1b}	S_{o1b}	S_{o1c}	S_{o1c}	S_{o1a}
S_{max_o}	S_{o1b}	S_{o1a}	S_{o1c}	S_{o1b}	S_{o1a}	S_{o1c}
S_{max_n}	S_{o1c}	S_{o1c}	S_{o1a}	S_{o1a}	S_{o1b}	S_{o1b}
S_{mid_p}	S_{o2a}	S_{o2b}	S_{o2b}	S_{o2c}	S_{o2c}	S_{o2a}
S_{mid_o}	S_{o2b}	S_{o2a}	S_{o2c}	S_{o2b}	S_{o2a}	S_{o2c}
S_{mid_n}	S_{o2c}	S_{o2c}	S_{o2a}	S_{o2a}	S_{o2b}	S_{o2b}
S_{min_p}	S_{o3a}	S_{o3b}	S_{o3b}	S_{o3c}	S_{o3c}	S_{o3a}
S_{min_o}	S_{o3b}	S_{o3a}	S_{o3c}	S_{o3b}	S_{o3a}	S_{o3c}
S_{min_n}	S_{o3c}	S_{o3c}	S_{o3a}	S_{o3a}	S_{o3b}	S_{o3b}

where ϕ_L represents the load power factor angle. In the case of $u_a > u_b > u_c$ and $u_{o1\_ref} > u_{o2\_ref} > \cdots > u_{oN\_ref}$, the average input current corresponding to the maximum phase (denoted as $\bar{i}_a$) and minimum phase (denoted as $\bar{i}_c$) can be obtained as

$$\begin{cases} \bar{i}_a = d_{1max\_p} i_{o1} + \sum d_{1ox\_ref\_p} i_{ox} \\ \bar{i}_c = \sum d_{2ox\_ref\_n} i_{ox} + d_{2min\_n} i_{oN} \end{cases} \tag{9.16}$$

Under the balanced grid voltages, according to (9.1)–(9.15), (9.16) is rewritten as

$$\begin{cases} \bar{i}_a = \dfrac{NU_{om}I_{om}\cos(\varphi_L)\cos(\omega_i t - \varphi)}{3U_{im}cos(\varphi)} \\ \bar{i}_c = \dfrac{NU_{om}I_{om}\cos(\varphi_L)\cos(\omega_i t - \varphi + 2\pi/3)}{3U_{im}\cos(\varphi)} \end{cases} \tag{9.17}$$

Then as three-phase input currents satisfy $\bar{i}_a + \bar{i}_b + \bar{i}_c = 0$, the average input current of the remaining phase (denoted as $\bar{i}_b$) can be calculated as

$$\bar{i}_b = \frac{NU_{om}I_{om}\cos(\varphi_L)\cos(\omega_i t - \varphi - 2\pi/3)}{3U_{im}\cos(\varphi)} \tag{9.18}$$

From (9.20) and (9.21), the grid currents are sinusoidal via the input filter, and $NU_{om}I_{om}\cos(\varphi_L)/3\,(U_{im}\cos(\varphi))$ is the amplitude. In other cases, similar analyses can be done.

9.4 Experimental Results

9.4.1 Experimental Results

To evaluate the effectiveness of the strategy, MATLAB/Simulink simulations of the developed strategy are implemented on the $3 \times N$ DMC system as depicted in Fig. 9.1, and the specific parameters are listed in Table 9.2. The choice of the load inductance is based on a trade-off between the current ripple and the voltage utilization.

To highlight the merits of the developed three-level output strategy, both 3LPWM and the 2LPWM in [19] are executed on the 3×3 DMC system under the balanced grid voltages. Then the output line-line voltage harmonic spectra of two strategies under the condition I ($m = 0.9$, $f_o = 50$ Hz, and $\varphi = 0$) are displayed in Fig. 9.7, where $m = m_1 m_2$ and f_o represent the whole modulation index and output frequency, respectively. Note that $m = 1$ corresponds to the maximum VTR of 0.866. Results show that the realization of three-level output underlies the improved output voltage performance, such as the lower THD. The harmonics concentrate on 8 kHz and its multiples, which are possible to filter out. The current waveforms of the two strategies are shown in Fig. 9.8. 3LPWM improves the output waveform quality while sacrificing input performance to some extent. Figure 9.9 shows the common-mode voltage waveforms of two strategies. Due to the absence

Table 9.2 System parameters

Description	Parameters	Specifications
Grid line-line voltage	$u_{(RMS)}$	380 V
Grid frequency	f_N	50 Hz
Sampling frequency	f_s	8 kHz
Filter inductance	L_f	900 μH
Filter capacitor	C_f	15 μF
Damping resistance	R_f	15 Ω
Load	$L + R$	2.5 mH+40 Ω

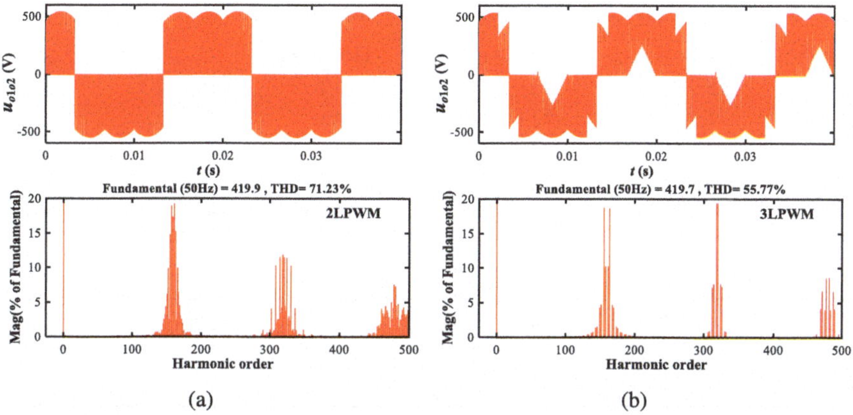

Fig. 9.7 Output line-line voltage harmonic spectra of (**a**) 2LPWM and (**b**) 3LPWM with $m = 0.9$ and $f_o = 50$ Hz for the 3×3 DMC system

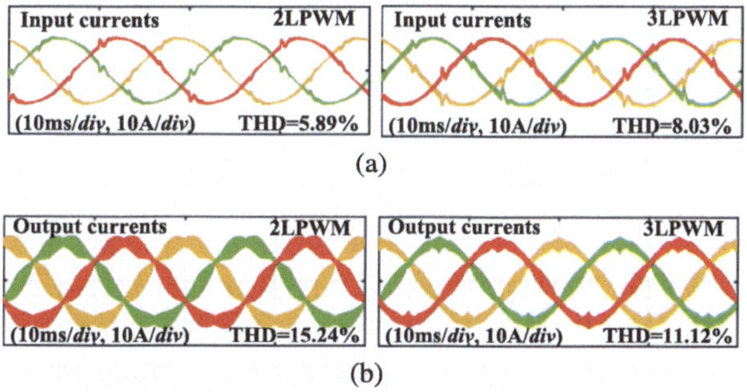

Fig. 9.8 Comparison of current waveforms between 2LPWM and 3LPWM with $m = 0.9$ and $f_o = 50$ Hz for the 3×3 DMC system. (**a**) Input currents. (**b**) Output currents

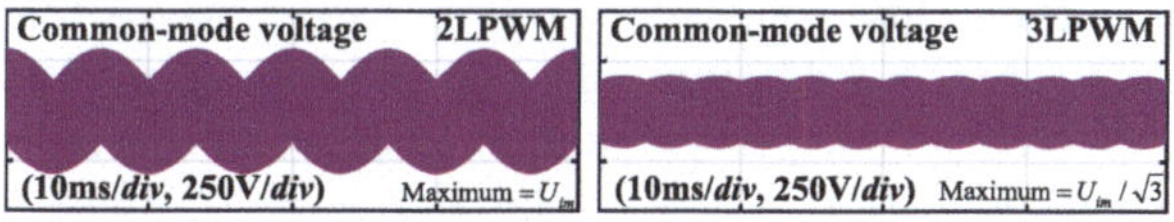

Fig. 9.9 Common-mode voltage waveforms of 2LPWM and 3LPWM

of zero vectors, as shown in Fig. 9.6, the common-mode voltage amplitude U_{cmmax} decreases from U_{im} in 2LPWM to $U_{im}/\sqrt{3}$ in 3LPWM.

For a more comprehensive comparison, some conventional strategies such as the OVM in [5], the NSM in [6], the ISVM strategy in [8], and the DSVM strategy in [9] are also executed under the condition I ($m = 0.9$, $f_o = 50$ Hz, and $\varphi = 0$), and the quantitative comparison is given in Table 9.3. Note that the results yielded by DSVM are very similar to the ISVM [1] and are thus omitted in Table 9.3. The average switching frequency f_{sw} is obtained by counting the total switching jumps X of nine switches over a fixed period of 0.06 s, namely that the average switching frequency is equal to X/9/0.06 [33]. Among the five strategies, the best output waveform quality, the lowest average switching frequency, and the smallest U_{cmmax} appear in 3LPWM, but the input currents of 3LPWM worsen to some extent.

Table 9.3 Simulation quantitative comparison of five strategies

Strategy	THD of u_{o1o2}	THD of i_{o1}	THD of i_a	U_{cmmax}	f_{sw}
3LPWM	55.77%	11.12%	8.03%	$U_{im}/\sqrt{3}$	7.21 kHz
2LPWM	71.23%	15.24%	5.89%	U_{im}	7.58 kHz
OVM	69.98%	17.23%	6.32%	U_{im}	8.00 kHz
NSM	70.50%	17.35%	6.28%	U_{im}	8.00 kHz
ISVM	70.56%	13.61%	4.78%	U_{im}	8.80 kHz

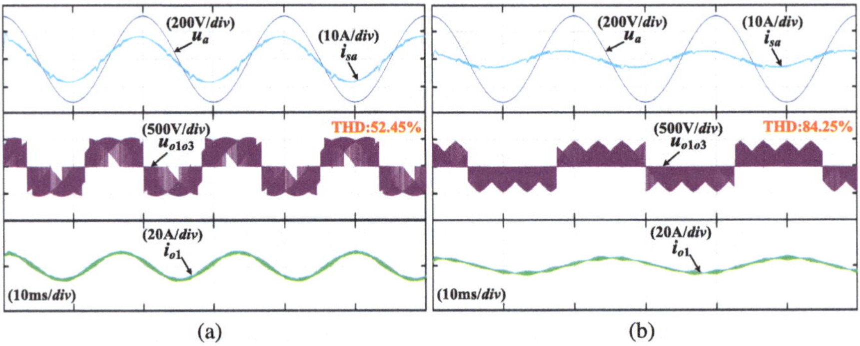

Fig. 9.10 Simulation results with (**a**) $m = 0.9$, $f_o = 60$ Hz and (**b**) $m = 0.5$, $f_o = 40$ Hz

For the 3×5 DMC system, simulation results of the grid voltage u_a and current i_{sa}, output line-line voltage u_{o1o3}, and output current i_{o1} under various operation conditions are provided. Figure 9.10a demonstrates the steady-state waveforms in the case of $m = 0.9$, $f_o = 60$ Hz, and $\varphi = 0$. It serves to show that the output line-line voltage is five-level, which stands for the realization of three-level output. Meanwhile, sinusoidal grid and output currents are also obtained with strong waveforms. As a comparison, another set of waveforms is shown in Fig. 9.10 (b) when $m = 0.5$, $f_o = 40$ Hz, and $\varphi = 0$. Since the absence of large vectors under a low modulation index, the output line-line voltage presents the three-level configuration in Fig. 9.10b. Concerning the ability to regulate the input power factor, results under condition II ($m = 0.9$, $f_o = 50$ Hz, and $\phi = -\pi/6$) and condition III ($m = 0.9$, $f_o = 50$ Hz, and $\phi = \pi/6$) are displayed in Fig. 9.11. The waveforms demonstrate that the developed strategy achieves controllable input power factors.

Power converters need to possess a satisfying dynamic performance. Therefore, the dynamic simulation results with $\varphi = 0$ are demonstrated in Fig. 9.12, where t_1 represents the transition instant from the preceding condition ($m = 0.5$ and $f_o = 40$ Hz) to the following one ($m = 0.9$ and $f_o = 60$ Hz). The dynamic response is fast enough, and there is no fluctuation during the transition, which indicates a strong dynamic performance of the developed strategy. Besides, the two-level strategy for the 3×5 DMC system in [14], which is called 2LPWM3-5, is also implemented and compared. Similar to the 3×3 DMC system, simulation results shown in Figs. 9.13

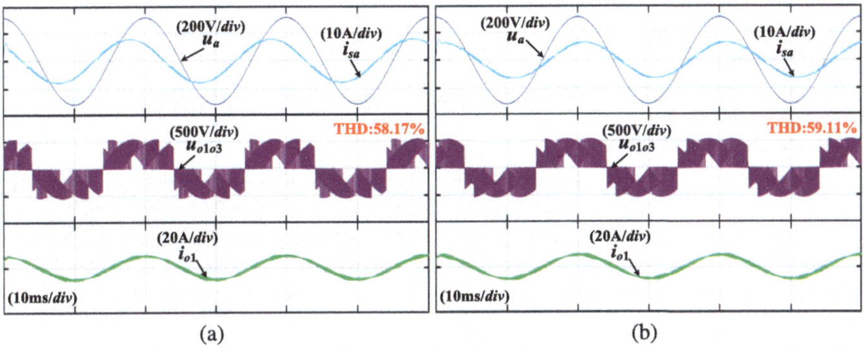

Fig. 9.11 Simulation results with (**a**) $\varphi = -\pi/6$ and (**b**) $\varphi = \pi/6$ ($m = 0.9$, $f_o = 50\,\text{Hz}$)

Fig. 9.12 Dynamic simulation waveforms with preceding condition ($m = 0.9$, $f_o = 40\,\text{Hz}$) and the following one ($m = 0.9$, $f_o = 60\,\text{Hz}$)

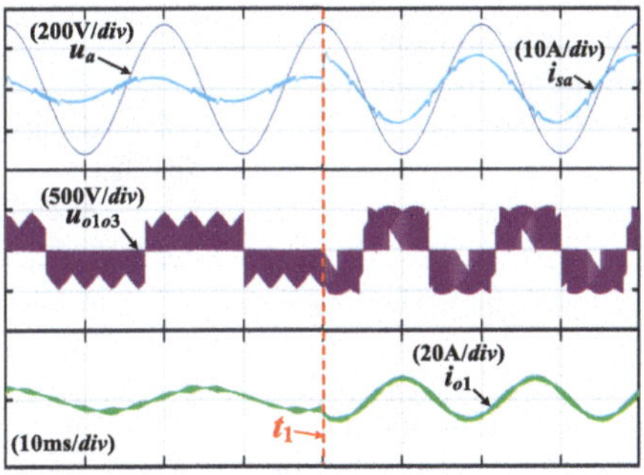

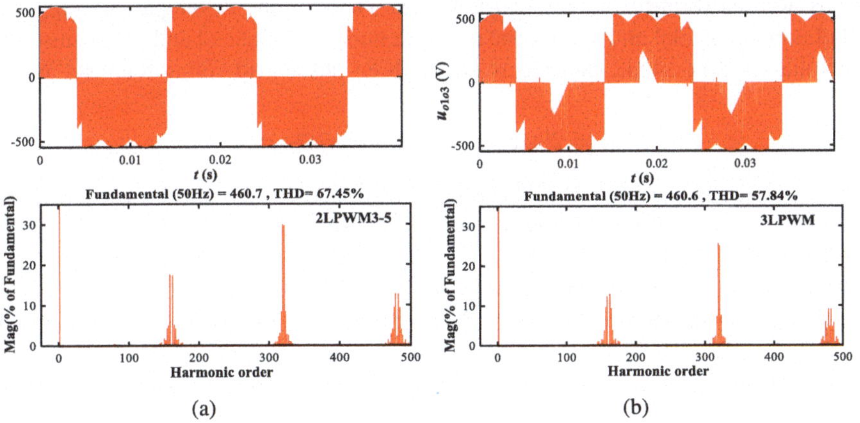

Fig. 9.13 Output line-line voltage harmonic spectra of (**a**) 2LPWM3-5 and (**b**) 3LPWM with $m = 0.9$, $f_o = 50\,\text{Hz}$ for the 3×5 DMC system

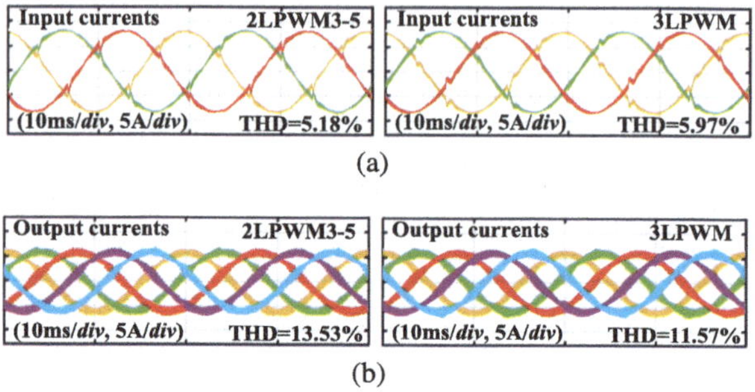

Fig. 9.14 Comparison of current waveforms between 2LPWM and 3LPWM with $m = 0.9$, $f_o = 50$ Hz for the 3×5 DMC system. (**a**) Input currents. (**b**) Output currents

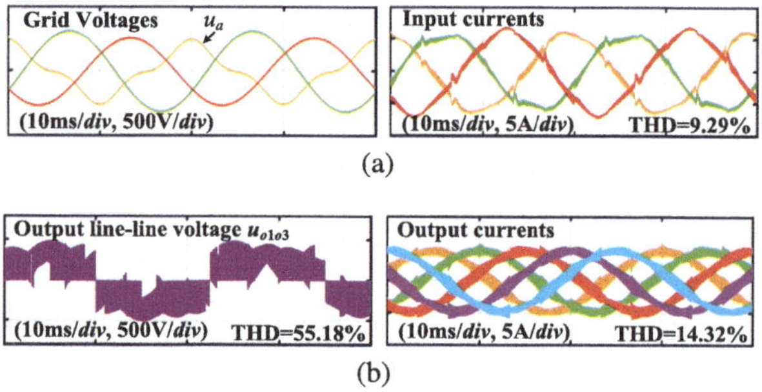

Fig. 9.15 Simulation results under unbalanced and non-sinusoidal grid voltages with $m = 0.7$, $f_o = 40$ Hz, and $\varphi = 0$. (**a**) Input waveforms. (**b**) Output waveforms

and 9.14 demonstrate that the developed three-level output strategy has better output waveform quality but higher THD of input currents.

According to formulas (9.3), (9.9), (9.10), and (9.13), the normalized modulation signals are based on the actual grid voltages. Thereby, the developed strategy can still obtain sinusoidal and symmetrical output under unbalanced and non-sinusoidal grid voltages. To evaluate the effectiveness, three-phase grid voltages are set as

$$\begin{cases} u_a = 200\cos(\omega_i t) + 50\cos(3\omega_i t) \\ u_b = 311\cos(\omega_i t - 2\pi/3) \\ u_c = 260\cos(\omega_i t + 2\pi/3) \end{cases} \tag{9.19}$$

and the simulation waveforms under condition IV ($m = 0.7$, $f_o = 40$ Hz, and $\varphi = 0$) are demonstrated in Fig. 9.15. Results show that sinusoidal and symmetrical output currents and good output voltages are both guaranteed. However, input currents are compromised by distorted grid voltages and thus have poor quality.

Besides, the unbalanced and non-sinusoidal output is also implemented thanks to the real-time adjustment of the modulation signals based on the actual output references in the developed strategy. To evaluate the effectiveness, five-phase output

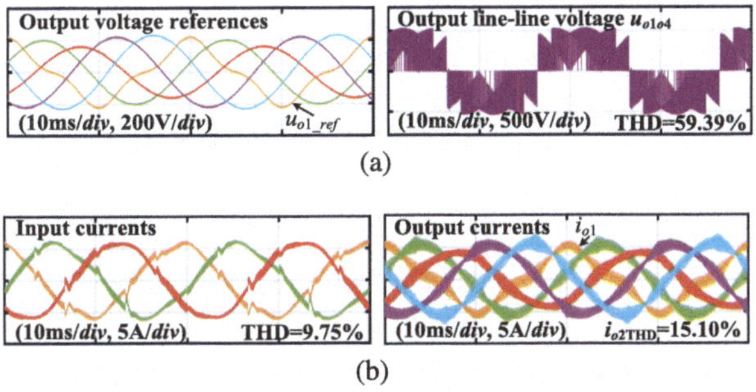

Fig. 9.16 Simulation results for unbalanced and non-sinusoidal output voltage references. (**a**) Voltage waveforms. (**b**) Current waveforms

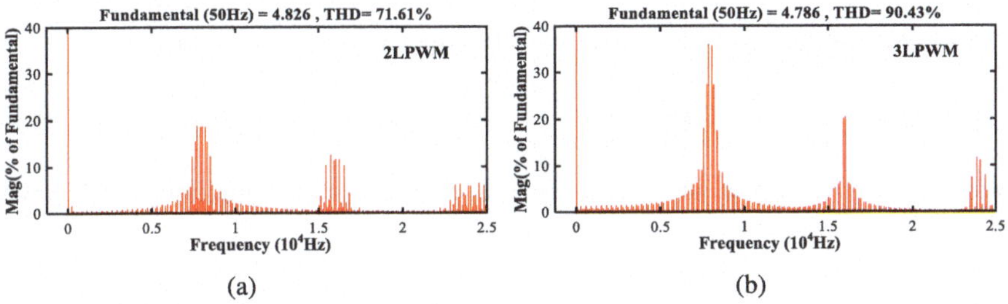

Fig. 9.17 Unfiltered input current harmonic spectra of (**a**) 2LPWM and (**b**) 3LPWM with $m = 0.9$ and $f_o = 50$ Hz for the 3×3 DMC system

voltage references are set as

$$\begin{cases} u_{o1\_ref} = 180\cos(50 \cdot 2\pi \cdot t) + 40\cos(150 \cdot 2\pi \cdot t) \\ u_{o2\_ref} = 200\cos(50 \cdot 2\pi \cdot t - 2\pi/5) \\ u_{o3\_ref} = 160\cos(50 \cdot 2\pi \cdot t - 4\pi/5) \\ u_{o4\_ref} = 220\cos(50 \cdot 2\pi \cdot t - 4\pi/5) \\ u_{o5\_ref} = 230\cos(50 \cdot 2\pi \cdot t - 8\pi/5) \end{cases} \tag{9.20}$$

and the simulation waveforms are demonstrated in Fig. 9.16. Results show that the expected output is generated, but the input currents are slightly distorted. Note that the input currents are formed by modulating the output currents. Thus they have the low tolerance to the output power quality. That is, the distortions of the output currents will be amplified in the input currents.

According to the above simulation results, the input currents deteriorate, and the requirement for the input filter increases in 3LPWM. Next comes the particulars.

Taking the 3×3 DMC case with more serious deterioration of input currents as an example, unfiltered input current harmonic spectra of 2LPWM and 3LPWM under the condition I ($m = 0.9$, $f_o = 50$ Hz, and $\varphi = 0$) are shown in Fig. 9.17. The input current harmonics of both strategies concentrate on 8 kHz and its multiples, and the dominant harmonic components at these frequencies in 2LPWM are about half that in 3LPWM. It means that the necessary attenuation of noise for 3LPWM is about 6 dB more than that for 2LPWM at these frequencies.

According to the model of the input filter, the current transfer function can be derived as

$$G_i(s) = \frac{i_{sa}(s)}{i_a(s)} = \frac{\frac{s}{R_f C_f} + \frac{1}{L_f C_f}}{s^2 + \frac{s}{R_f C_f} + \frac{1}{L_f C_f}} \tag{9.21}$$

where i_{sa} and i_a are the input current and unfiltered input current, respectively.

To obtain similar input currents, the following should be satisfied:

$$20\lg A_{i1}(j\omega) - 20\lg A_{i2}(j\omega) = -6\text{ dB}, \omega = 2\pi \cdot 8 \cdot 10^3 \tag{9.22}$$

where A_{i1} and A_{i2} represent the amplitude of current transfer functions for 3LPWM and 2LPWM, respectively.

According to formulas (9.21) and (9.22), the value of $L_f C_f$ in 3LPWM is about twice that in 2LPWM. To verify the analysis, the simulation of 3LPWM is performed with twice C_f, and the input current THD of 5.77% is similar to 5.89% in 2LPWM. Besides, the difference in input currents between the two strategies and the requirement for the input filter decrease with the increase of N.

9.4.2 *Efficiency Evaluation*

To evaluate the conversion efficiency of the DMC, the power losses, including switching losses, conduction losses, and auxiliary power losses, are calculated. From the topology and modulation strategy, the switching losses are given by

$$P_{sw} = f_{sw}\frac{1}{T} = \int_0^T k_s U(t) I(t)\,dt \tag{9.23}$$

where k_s is the adjustment coefficient acquired from the datasheet, $U(t)$ and $I(t)$ are the real-time voltage and current stresses of the active switch, respectively.

The conduction losses of output phase $o1$ are calculated as

$$P_{cD\_o1} = \frac{1}{2\pi}\int_0^{2\pi} V_F(|i_{o1}|) \cdot |i_{o1}|\,d\theta \tag{9.24}$$

where $V_F(i)$ is acquired from the datasheet.

Besides, the auxiliary power losses P_{aux} consumed by the control electronics and the IGBT drivers are considered. According to the symmetry of the topology, the overall losses are calculated as

$$P_{loss\_all} = 3NP_{sw} + NP_{cD\_o1} + P_{aux} \tag{9.25}$$

Then, the efficiency of the 3×3 DMC for various strategies is calculated at the output power $P_{out} = 2\,k$W. The calculated efficiency for 2LPWM and ISVM is 95.7% and 95.2%, respectively. And it increases to 96.8% for 3LPWM due to the decrease in the average switching frequency and the switching voltage. Note that the literature [8] claims that the measured efficiency for ISVM is 93% at $P_{out} = 2\,k$W. The reasons for the reduction in efficiency may be the additional losses dissipated on the input filter, printed circuit board traces, and current sensors in the experiment and the difference between the power switches.

A 3×5 matrix converter prototype shown in Fig. 9.18 is set up, where the output can be configured from single to five phases. To prove the validity, the developed 3LPWM strategy is tested on the 3×5 matrix converter prototype. The specific parameters are consistent with the simulation, and the dead time is set to 1.5 μs. The IGBT module 18MBI100W-120A from Fuji is used as the power switch. All algorithms are executed on a combination of a floating-point DSP TMS320F28335 that computes duty ratios and an FPGA EP2C8T144C8 that compares the duty ratios with corresponding carriers. Currents and voltages are measured through several Hall-effect current and voltage sensors, transmitted to the DSP through the signal conditioning circuit. To facilitate the analysis, all experimental data are obtained by external probes and an oscilloscope (Tektronix MDO3014) and then imported into a PC. Besides, the clock ticks used by the duty ratio computation modules in various strategies are recorded to evaluate the implementation efforts [21]. Results listed in Table 9.4 show that 3LPWM costs the least clock ticks to compute duty ratios. Note that 2LPWM costs most clock ticks mainly because many inevitable division operations occupy many clock ticks in TMS320F28335. Thereby, the computational

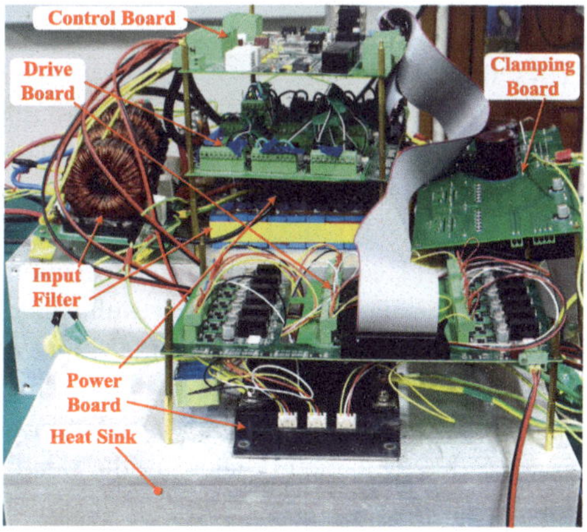

Fig. 9.18 Experimental setup of the 3×5 DMC system

Table 9.4 Clock ticks used by duty ratio computation modules

Strategy	Clock ticks
3LPWM (for 3×3 DMC)	1254
2LPWM	1549
3LPWM (for 3×5 DMC)	1308
2LPWM3-5	1397

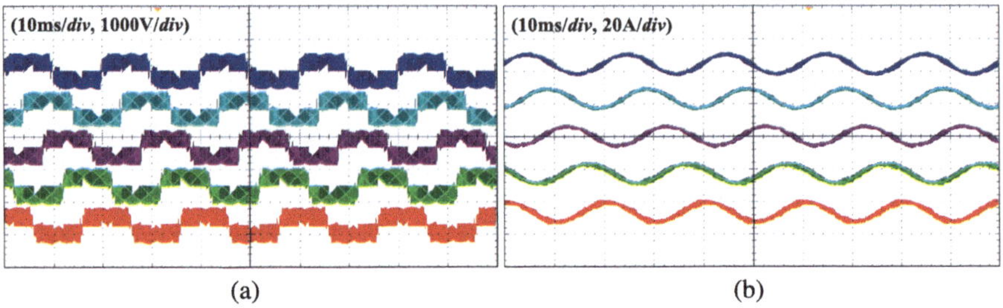

Fig. 9.19 Steady-state experimental waveforms of 3LPWM strategy for 3×5 DMC system with $m = 0.9$, $f_o = 50\,\text{Hz}$, and $\varphi = 0$. (**a**) Output line-line voltages. (**b**) Output currents

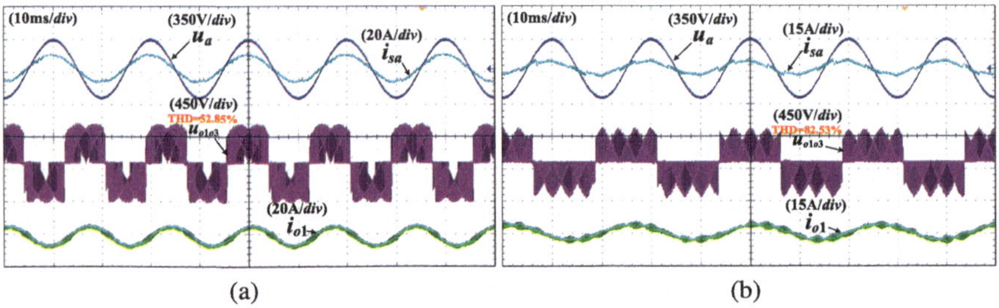

Fig. 9.20 Experimental results with (**a**) $m = 0.9$, $f_o = 60\,\text{Hz}$ and (**b**) $m = 0.9$, $f_o = 40\,\text{Hz}$ ($\varphi = 0$)

burden of 2LPWM may be lower than 3LPWM with a more advanced DSP such as TMS320F28379.

The steady-state output waveforms are displayed in Fig. 9.19, and the advanced calculation of the oscilloscope obtains waveforms of the fifth channel. It shows that good line-line voltages and currents are acquired. The experimental results shown in Figs. 9.20, 9.21, and 9.22 correspond to the simulation results shown in Figs. 9.10, 9.11, and 9.12, and the experimental conditions are the same as those in the simulation. The experimental results match the simulated results well, where the multilevel output voltage, sinusoidal input/output currents, and controllable input power factor are achieved. Thus the validity of the developed strategy is verified experimentally. Besides, slight degradation of the waveform quality may be caused by some practical factors such as the dead time, the control delay of the digital control system, the voltage ripple of the input filtering capacitors, the nonideality of

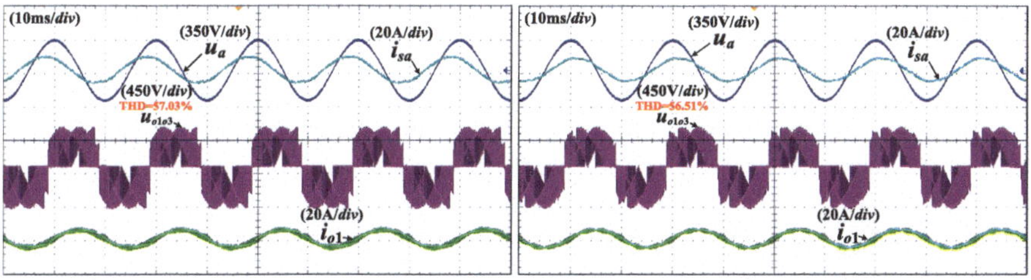

(a) (b)

Fig. 9.21 Experimental results with (**a**) $\varphi = -\pi/6$ and (**b**) $\varphi = \pi/6$ ($m = 0.9$, $f_o = 50\,\text{Hz}$)

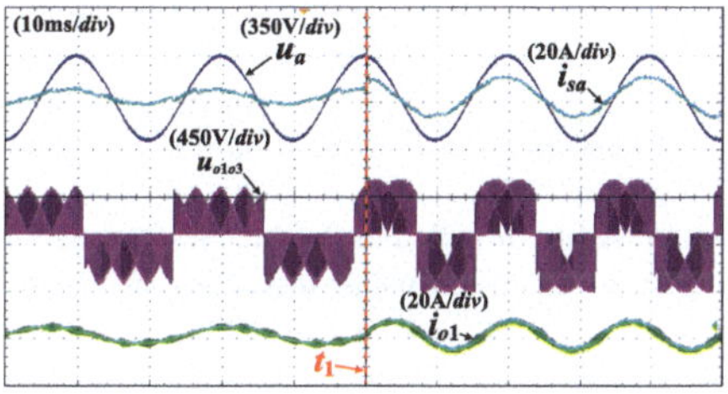

Fig. 9.22 Dynamic experimental waveforms with preceding condition ($m = 0.9$, $f_o = 40\,\text{Hz}$) and the following one ($m = 0.9$, $f_o = 60\,\text{Hz}$)

the power switches, the small distortions or unbalance of the practical grid voltages, and so on.

9.5 Conclusions

In this chapter, a CBPWM strategy is presented to realize the three-level output for conventional $3 \times N$ DMCs. The developed strategy takes full advantage of DMCs by designing the offset voltage and adopting the DSPWM technique. Both simulations and experiments validate the effectiveness and merits of the developed strategy. Results demonstrate that the developed 3LPWM strategy yields better output waveforms and worse input currents than two-level output strategies. For 3×3 DMCs, 3LPWM reduces the THD of the output line-line voltage by about 21.5% and the THD of the output current by about 27.0% with 2LPWM as a benchmark. For 3×5 DMCs, 3LPWM reduces the THD of the output line-line voltage by about 14.2% and the THD of the output current by about 14.4% with 2LPWM3-5 as a benchmark. However, the input current THD of 3LPWM increases by about 25.4% in the 3×3 DMC and 13.2% in the 3×5 DMC with 3LPWM as a benchmark. Meanwhile, high efficiency occurs due to the reduced switching losses, and the developed three-level output strategy can be directly embedded into conventional $3 \times N$ DMCs without any changes in the topology. Thereby, the developed three-level output strategy will find applications where superior output waveform quality and high efficiency are dominated.

References

1. P.W. Wheeler, J. Rodriguez, J.C. Clare, L. Empringham, A. Weinstein, Matrix converters: a technology review. IEEE Trans. Ind. Electron. **49**(2), 276–288 (2002)
2. O. Abdel-Rahim, H. Funato, H. Abu-Rub, O. Ellabban, Multiphase wind energy generation with direct matrix converter, in *2014 IEEE International Conference on Industrial Technology (ICIT)* (2014), pp. 519–523
3. M. Ali, A. Iqbal, M.R. Khan, M. Ayyub, M.A. Anees, Generalized theory and analysis of scalar modulation techniques for a m×n matrix converter. IEEE Trans. Power Electron. **32**(6), 4864–4877 (2017)
4. M. Venturini, A new sine wave in sine wave out, conversion technique which eliminates reactive elements, in *Proceedings of the Powercon*, vol. 7 (1980), pp. E3/1– E3/15
5. A. Alesina, M. Venturini, Analysis and design of optimum-amplitude nine-switch direct AC-AC converters. IEEE Trans. Power Electron. **4**(1), 101–112 (1989)
6. G. Roy, G.-E. April, Cycloconverter operation under a new scalar control algorithm, in *Proceedings of the 20th Annual IEEE Power Electronics Specialists Conference*, vol. 1 (1989), pp. 368–375
7. A. Ishiguro, T. Furuhashi, S. Okuma, A novel control method for forced commutated cycloconverters using instantaneous values of input line-to-line voltages. IEEE Trans. Ind. Electron. **38**(3), 166–172 (1991)
8. L. Huber, D. Borojevic, Space vector modulated three-phase to three-phase matrix converter with input power factor correction. IEEE Trans. Ind. Appl. **31**(6), 1234–1246 (1995)
9. D. Casadei, G. Grandi, G. Serra, A. Tani, Space vector control of matrix converters with unity input power factor and sinusoidal input/output waveforms, in *Proceedings of the EPE Conference*, vol. 7, Brighton (1993), pp. 170–175
10. F. Gao, M. Iravani, Dynamic model of a space vector modulated matrix converter. IEEE Trans. Power Del. **22**(3), 1696–1705 (2007)
11. D. Casadei, G. Serra, A. Tani, L. Zarri, Optimal use of zero vectors for minimizing the output current distortion in matrix converters. IEEE Trans. Ind. Electron. **56**(2), 326–336 (2009)
12. R. Cardenas-Dobson, R. Pena, P. Wheeler, J. Clare, Experimental validation of a space vector modulation algorithm for four-leg matrix converters. IEEE Trans. Ind. Electron. **58**(4), 1282–1293 (2011)
13. W. Cai, Z. Wang, S. Sun, A novel space vector modulation for the 3×5 direct matrix converter, in *Proceedings of the IEEE ICEMS* (2019), pp. 1–6
14. A. Iqbal, S.M. Ahmed, H. Abu-Rub, Space vector PWM technique for a three-to-five-phase matrix converter. IEEE Trans. Ind. Appl. **48**(2), 697–707 (2012)
15. J. Lei, S. Feng, B. Zhou, H. Nguyen, J. Zhao, W. Chen, Simple modulation scheme with zero common-mode voltage and improved efficiency for direct matrix converter Fed PMSM drives. IEEE J. Emerg. Sel. Topics Power Electron. **8**(4), 3712–3722 (2020)
16. P.C. Loh, R. Rong, F. Blaabjerg, P. Wang, Digital carrier modulation and sampling issues of matrix converters. IEEE Trans. Power Electron. **24**(7), 1690–1700 (2009)
17. P. Kiatsookkanatorn, S. Sangwongwanich, A unified PWM method for matrix converters and its carrier-based realization using dipolar modulation technique, IEEE Trans. Ind. Electron. **59**(1), 80–92 (2012)
18. Y. Sun, W. Xiong, M. Su, X. Li, H. Dan, J. Yang, Carrier-based modulation strategies for multimodular matrix converters. IEEE Trans. Ind. Electron. **63**(3), 1350–1361 (2016)
19. Y.-D. Yoon, S.-K. Sul, Carrier-based modulation technique for matrix converter. IEEE Trans. Power Electron. **21**(6), 1691–1703 (2006)
20. Y.-D. Yoon, S.-K. Sul, Carrier-based modulation method for matrix converter with input power factor control and under unbalanced input voltage conditions, in *Proceedings of the 22nd Annual IEEE APEC* (2007), pp. 310–314

21. P. Szczepankowski, W. Sleszynski, T. Bajdecki, A direct modulation for matrix converters based on the one-cycle atomic operation developed in Verilog HDL. IEEE Trans. Ind. Electron. **69**(4), 3303–3312 (2021)
22. W. Gui, Y. Sun, H. Qin, C. Yang, M. Su, A matrix converter modulation based on mathematical construction, in *Proceedings of the IEEE ICIT* (2008), pp. 1–5
23. Y. Sun, W. Xiong, M. Su, H. Dan, X. Li, J. Yang, Modulation strategies based on mathematical construction method for multimodular matrix converter. IEEE Trans. Power Electron. **31**(8), 5423–5434 (2016)
24. M. Rivera, C. Rojas, J. Rodríguez, P. Wheeler, B. Wu, J. Espinoza, Predictive current control with input filter resonance mitigation for a direct matrix converter. IEEE Trans. Power Electron. **26**(10), 2794–2803 (2011)
25. G. Zhang, J. Yang, Y. Sun, M. Su, Q. Zhu, F. Blaabjerg, A predictive- control-based overmodulation method for conventional matrix converters. IEEE Trans. Power Electron. **33**(4), 3631–3643 (2018)
26. W. Xiong et al., A cost-effective and low-complexity predictive control for matrix converters under unbalanced grid voltage conditions. IEEE Access **7**, 43895–43905 (2019)
27. T.N. Mir, B. Singh, A.H. Bhat, Predictive delta sigma modulation for three-phase to three-phase matrix converters. IEEE Trans. Power Electron. **35**(1), 968–976 (2020)
28. T.N. Mir, B. Singh, A.H. Bhat, Delta sigma modulation based common mode voltage elimination in direct matrix converter. IEEE Trans. Ind. Inform. **17**(2), 1048–1057 (2021)
29. P. Szczepankowski, T. Bajdecki, R. Strzelecki, Direct modulation for conventional matrix converters using analytical signals and barycentric coordinates. IEEE Access **8**, 22592–22616 (2020)
30. P. Szczepankowski, P. Wheeler, T. Bajdecki, Application of analytic signal and smooth interpolation in pulse width modulation for conventional matrix converters. IEEE Trans. Ind. Electron. **67**(12), 10011–10023 (2020)
31. R. Rong, P.C. Loh, P. Wang, F. Blaabjerg, A three-level 4×3 conventional matrix converter, in *2007 7th International Conference on Power Electronics and Drive Systems*, Bangkok (2007), pp. 315–319
32. S. Raju, L. Srivatchan, V. Chandrasekaran, N. Mohan, Constant pulse width modulation strategy for direct three-level matrix converter, in *2012 IEEE International Conference on Power Electronics, Drives and Energy Systems (PEDES)*, Bengaluru (2012), pp. 1–5
33. Y. Zhang, J. Zhu, A novel duty cycle control strategy to reduce both torque and flux ripples for DTC of permanent magnet synchronous motor drives with switching frequency reduction. IEEE Trans. Power Electron. **26**(10), 3055–3067 (2011)

Chapter 10
Control Method for Matrix Converter to Enhance Linear Voltage Transfer Ratio

10.1 Introduction

Indirect matrix converter, also known as the TSMC, is a type of direct AC-AC power conversion topology that can generate variable amplitude and frequency output waveforms without the need of bulky energy storage elements [1–6]. Due to the advantages such as sinusoidal input and output currents, controllable input power factor, high power density, zero-current commutation as well as the potential of constructing a cost-effective derived topology with multiple phases and/or multiple output ports [6], TSMCs have attracted an increasing attention in the recent years. However, the relatively low VTR is one of the major drawbacks of the three-phase to three-phase TSMCs, especially when TSMCs are applied in AC ASDs, power supplies, and power systems, and so on. Usually, there are two solutions to improve the VTR: One is altering the topology of TSMCs to boost the output voltage [7–13]; the other is applying modulation strategy to enhance the VTR [14–19].

For the first solution, extra circuits have to be added and merged into the conventional TSMCs. In [7] and [8], two hybrid TSMC topologies with significant improvement of VTR have been proposed, which are formed by integrating an H-bridge inverter and an auxiliary boost converter in the DC-link of TSMC, respectively. In [9], voltage boosting and unity VTR of TSMC were achieved by adding four extra switches to construct an equivalent AC-AC boost topology. References [10–13] focused on the improvement of VTR by introducing Z-source network. Although the VTR has been enhanced greatly, a disadvantage of these methods is the need of extra passive energy storage components, which contradicts the original "all-silicon" feature of TSMCs and leads to a reduced power density.

Regarding the second solution, several modulation schemes have been presented to improve the VTR. In [14], a modulation strategy that improves the VTR of matrix converter operating with non-unity input power factor was suggested, which is able to obtain higher VTR when compared to the traditional strategies. Based on the space vector modulation, reference [15] investigated the derivation about

H. Wang, *Advanced Matrix Converters*, Power Systems,
https://doi.org/10.1007/978-3-031-69902-3_10

the VTR and proved that the theoretical maximum linear VTR can reach 0.928 when the output frequency and initial phase are the same as those of the input side. References [16–19] focused on the improvement of VTR by operating TSMCs in overmodulation region. The overmodulation strategy that can improve the VTR to unity by adding third-harmonic to the modulation signals was presented in [16]. In [17], two overmodulation methods based on square wave modulation and trapezoidal wave modulation were discussed, which can increase the maximum VTR from 0.866 to 0.97 and 0.92, respectively. Reference [18] proposed a simple space vector overmodulation method to increase the VTR of matrix converter. A six-step variable frequency overmodulation strategy that is capable of providing more than unity fundamental component of output voltage was developed in [19]. Although the VTR can be enhanced significantly by these overmodulation strategies without introducing additional hardware circuits, low-order harmonics in the input current and output voltage are inevitable and thus is not desirable. Besides, the VTR is coupled with the input power factor tightly. When the ability of generating the input reactive power is required, the maximum available VTR of TSMC will reduce furthermore [2, 4, 5]. Thus, further research is required to enhance the VTR.

The 3HIMC is a new kind of TSMC [20, 21]. In addition to possess the same essential features as conventional TSMCs such as sinusoidal input/output currents and bidirectional power flow, 3HIMC also has the advantage of enhanced input reactive power capability. The basic TSMC based on the third-harmonic injection technique is presented to perform direct AC-AC power conversion in [20]. To overcome the drawbacks of the conventional topologies with multiple VSIs for the applications that require multiple three-phase outputs such as multiple drives, the active third-harmonic injection indirect matrix converter with multiple VSIs is suggested in [21]. Aiming at overcoming the intrinsic drawback of relatively low linear VTR of TSMCs, the linear VTR capability of 3HIMC is researched systematically in this chapter, and a control method to enhance the linear VTR of 3HIMC from two aspects is proposed. On the one hand, the linear VTR capability of 3HIMC is explored considering the instantaneous six-pulse shape intermediate DC-link voltage, and the maximum linear VTR is extracted by systematically analyzing the SVM strategy of the inversion stage. It is found that the maximum linear VTR can break through the intrinsic limit of 0.866 under certain conditions. Specifically, when the output frequency and initial phase are the same as those of the input side, a theoretical maximum VTR of one can be achieved. On the other hand, the capability of generating input reactive power is utilized as an additional degree of freedom to improve the linear VTR of 3HIMC, and the linear VTR can be increased to infinite in theory at the expense of increasing the voltage and current stresses of the input filter.

After introducing the topology and operating principle briefly in Sect. 10.2, the linear VTR feature of 3HIMC is analyzed, and the relationship between the maximum linear VTR and the output/input frequency ratio, the initial phase of the output voltage and the input reactive power is investigated in Sect. 10.3. In Sect. 10.4, the developed control method to improve the linear VTR of 3HIMC is presented in detail, and the effect of the input filter on the VTR is also discussed. In

Sect. 10.5, the theoretical analysis is verified by simulation and experimental results, and the comparison of 3HIMC and other TSMCs with improved VTR is presented to demonstrate the effectiveness of the proposed method. Finally, the main points of this chapter are summarized in Sect. 10.6.

10.2 Brief Review of the 3HIMC

The topology of 3HIMC is shown in Fig. 10.1, which mainly consists of a line-commutated bidirectional current-source rectifier, a third-harmonic injection circuit, a conventional two-level voltage-source inverter, and an *LC* input filter. The third-harmonic injection circuit is composed of three bidirectional switches, a third-harmonic injection inductor, and a bridge leg.

Assuming that the three-phase input voltages are symmetrical and sinusoidal, the operating principles of 3HIMC are described briefly as follows: For the front-end rectifier, the switches of the rectifier antiparalleled to the conducting diodes are turned on so as to impose the maximum instantaneous input line-line voltage across the intermediate DC-link. Consequently, as with the output voltage of a diode rectifier, the DC-link voltage u_{pn} of 3HIMC exhibits a piecewise six-pulse shape waveform. According to the requirements of the load, the rear-end inverter provides three-phase output voltages with variable frequency and amplitude. The third-harmonic injection circuit is essentially a time-sharing multiplexing single-phase shunt active power filter, and the input power factor correction and the normal operation of 3HIMC depend on injecting the quasi-third-harmonic current into the input side. For the third-harmonic injection circuit, the bidirectional switch that connected to the input phase with the medium instantaneous voltage is turned on, and high frequency switches S_{y+} and S_{y-} of the bridge leg switch complementally to generate the desired quasi-third-harmonic current i_y that is equal to the minus value of the input current corresponding to the input phase with the medium instantaneous voltage [20], as described in (10.1), where I_{rd} and I_{rq} are the amplitudes of the active and reactive components of the input current, and i_{rmid}

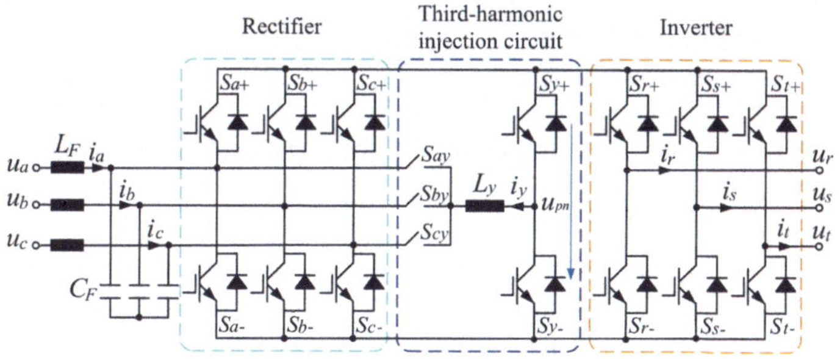

Fig. 10.1 Schematic diagram of 3HIMC

Table 10.1 Switching states of the rectifier and the third-harmonic current injection circuit

θ_i	Sector	S_{ay}	S_{by}	S_{cy}	S_{a+}	S_{a-}	S_{b+}	S_{b-}	S_{c+}	S_{c-}
0-π/3	1	0	1	0	1	0	0	0	0	1
π/3-2π/3	2	1	0	0	0	0	1	0	0	1
2π/3-π	3	0	0	1	0	1	1	0	0	0
π-4π/3	4	0	1	0	0	1	0	0	1	0
4π/3-5π/3	5	1	0	0	0	0	0	1	1	0
5π/3-2π	6	0	0	1	1	0	0	1	0	0

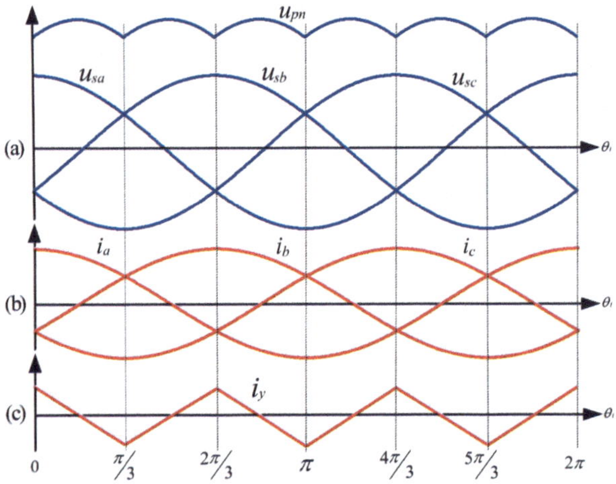

Fig. 10.2 Key waveforms of 3HIMC

and θ_{mid} are the input current before filtering and the phase corresponding to the input phase with the medium instantaneous voltage.

$$i_y = -i_{rmid} = -I_{rd}\cos(\theta_{mid}) - I_{rq}\sin(\theta_{mid}) \tag{10.1}$$

If the switching function of a switch S_{ij} in Fig. 10.1 is defined as

$$S_{ij} = \begin{cases} 0, \ when \ open \\ 1, \ when \ closed \end{cases}, i = a, b, c; \ j = y, +, - \tag{10.2}$$

then the switching states of the rectifier and the third-harmonic current injection circuit can be expressed as shown in Table 10.1, where θ_i is the input phase voltage of the filtering capacitor of phase a.

In this manner, sinusoidal three-phase input currents and controllable input power factor are achieved, as shown in Fig. 10.2. The validity of sinusoidal input currents and controllable power factor has been presented in detail in [20] and thus is not elaborated here.

10.3 Analysis of the Linear VTR of 3HIMC

Suppose that SVM is adopted for the inverter and the reference output voltage vector $\boldsymbol{u}_o$ is given as

$$\boldsymbol{u}_o = U_{om}\mathrm{e}^{j\theta_o} = U_{om}\mathrm{e}^{\mathrm{j}(2\pi f_o t + \varphi_o)} \tag{10.3}$$

Then the duty ratios can be derived as

$$\begin{cases} d_1 = [\sqrt{3}U_{om}\sin(\pi/3 - \theta_o)]/u_{pn} \\ d_2 = [\sqrt{3}U_{om}\sin(\theta_o)]/u_{pn} \\ d_0 = 1 - d_1 - d_2 \end{cases} \tag{10.4}$$

where d_0, d_1, and d_2 are the duty ratios of the zero vector and the two adjacent active vectors, respectively; and U_{om}, θ_o, f_o, and φo are the amplitude, phase, frequency, and initial phase of the rotating reference output voltage vector, respectively.

As can be seen from (10.4), the duty ratios depend on the intermediate DC-link voltage consisting of the maximum input line-line voltage of the input filtering capacitors. Therefore, the linear VTR characteristic of 3HIMC is quite different from that of HIMC, and it can also be inferred that the linear VTR depends on the frequency and initial phase relationship between the inputs and the outputs, as well as the dynamic behavior of the input filter. Assuming that the three-phase inputs and outputs are symmetrical and sinusoidal, the concrete analysis of the linear VTR characteristic of 3HIMC is given as follows.

10.3.1 Ignoring the Effects of the Input Filter

Suppose that the input voltages are given by

$$\begin{cases} u_{sa} = U_{im}\cos(\theta_i) = U_{im}\cos(2\pi f_i t) \\ u_{sb} = U_{im}\cos(\theta_i - 2\pi/3) = U_{im}\cos(2\pi f_i t - 2\pi/3) \\ u_{sc} = U_{im}\cos(\theta_i + 2\pi/3) = U_{im}\cos(2\pi f_i t + 2\pi/3) \end{cases} \tag{10.5}$$

where U_{im} and f_i are the amplitude and frequency of the input voltages.

Assume that the effects of the input filter can be ignored, and then the three-phase input filtering capacitor voltages can be expressed as

$$\begin{cases} u_a = U_{cm}\cos(\theta_{ca}) = u_{sa} \\ u_b = U_{cm}\cos(\theta_{ca} - 2\pi/3) = u_{sb} \\ u_c = U_{cm}\cos(\theta_{ca} + 2\pi/3) = u_{sc} \end{cases} \tag{10.6}$$

where U_{cm} and θ_{ca} are the amplitude and phase of the filtering capacitor voltage of phase a.

According to the operating principles of 3HIMC and (10.6), the DC-link voltage u_{pn} can be expressed as

$$u_{pn} = \sqrt{3}U_{cm}\cos[(\theta_i // (\pi/3)) - \pi/6] \tag{10.7}$$

where "//" is the operator of the remainder value.

For linear modulation of 3HIMC, the duty ratios in each switching period should satisfy

$$\begin{cases} 0 \le d_1 \le 1 \\ 0 \le d_2 \le 1 \\ 0 \le d_1 + d_2 \le 1 \end{cases} \tag{10.8}$$

Substituting (10.4) and (10.7) into (10.8), the constraint of the output voltage can be derived as

$$U_{om} \le \frac{U_{cm}\cos[(\theta_i//(\pi/3)) - \pi/6]}{\cos[(\theta_o//(\pi/3)) - \pi/6]} \tag{10.9}$$

Formula (10.9) contains both input and output phases. To simplify (10.9), define the output/input frequency ratio n_f as

$$n_{\mathrm{f}} = f_{\mathrm{o}}/f_{\mathrm{i}} = 2\pi f_{\mathrm{o}}t/2\pi f_{\mathrm{i}}t = (\theta_{\mathrm{o}} - \varphi_{\mathrm{o}})/\theta_{\mathrm{i}} \tag{10.10}$$

Combining (10.9) and (10.10), we can get the upper limit of the amplitude of the output voltage vector in a switching period as

$$U_{om} \le \frac{U_{cm}\cos[(((\theta_{\mathrm{o}} - \varphi_{\mathrm{o}})/n_{\mathrm{f}})//(\pi/3)) - \pi/6]}{\cos[(\theta_{\mathrm{o}}//(\pi/3)) - \pi/6]} \tag{10.11}$$

Define the VTR of 3HIMC as

$$m = U_{om}/U_{im} \tag{10.12}$$

Then from (10.11), the constraint of the VTR in a switching period can be derived as

$$m \le \frac{\cos[(((\theta_{\mathrm{o}} - \varphi_{\mathrm{o}})/n_{\mathrm{f}})//(\pi/3)) - \pi/6]}{\cos[(\theta_{\mathrm{o}}//(\pi/3)) - \pi/6]} \tag{10.13}$$

It can be found from (10.13) that the maximum VTR in a switching period changes with the phase of the output voltage. Therefore, the maximum linear VTR

m_{max} of 3HIMC can be deduced by searching the minimum value of m in the whole θ_o range:

$$m_{\max} = \min\left(\frac{\cos[(((\theta_o - \varphi_o)/n_f)//(\pi/3)) - \pi/6]}{\cos[(\theta_o//(\pi/3)) - \pi/6]}\right) \tag{10.14}$$

where min() is the operator of the minimum value.

It can be seen from (10.14) that the maximum linear VTR is a nonlinear function of the variables n_f and φ_o. For the given n_f and φ_o parameters, the maximum linear VTR can be obtained by solving (10.14) with respect to the variable θ_o.

Figure 10.3 shows the 3-D plot of the maximum linear VTR of 3HIMC with n_f varying from 0 to 4 and φ_o varying from 0 to 2π. It can be found that, for the given n_f, m_{max} is a periodic function of the variable φ_o, and the maximum value of m_{max} is related to n_f. The minimum value of m_{max} is 0.866, which is irrelevant to n_f. The maximum value of m_{max} depends on n_f, which can achieve 1 when $n_f = 1$.

It can be found from the above analysis that for 3HIMC, the maximum linear VTR can break through the intrinsic limit of 0.866 when the output frequency and initial phase satisfy certain conditions.

10.3.2 Considering the Dynamic Behavior of the Input Filter

For most cases, the effects of the input filter can be neglected when analyzing the VTR characteristic, since usually the voltage drop on the filtering inductor is small. However, for some specific situations such as a relatively high input frequency, a relatively high input inductance, as well as large input reactive power, the dynamic behavior of the input filter should be taken into consideration since the capacitor

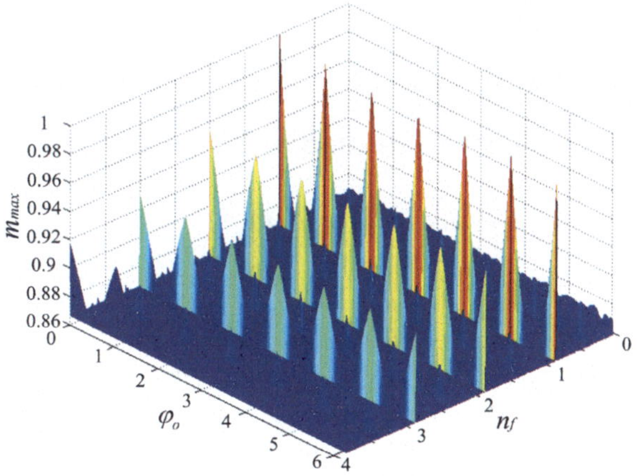

Fig. 10.3 3-D plot of the maximum linear VTR of 3HIMC

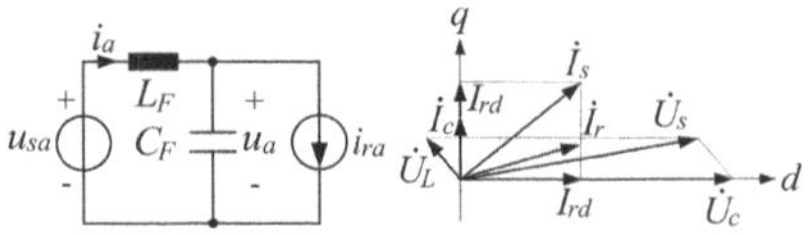

Fig. 10.4 Per-phase equivalent circuit and the steady-state phasor diagram of the input filter. (**a**) Per-phase equivalent circuit. (**b**) Steady-state phasor diagram of the input filter

voltages are used directly for synthesizing the output voltage instead of the input voltages.

The per-phase equivalent circuit and the steady-state phasor diagram of the input filter are illustrated in Fig. 10.4, where $\dot{U}_s$, $\dot{U}_C$, and $\dot{U}_L$ are the input voltage, capacitor voltage, and inductor voltage phasors, respectively; $\dot{I}_s$ and $\dot{I}_r$ are the input current phasor and the input current phasor of the rectifier, respectively.

Referring to Fig. 10.4b, the amplitude of the capacitor voltage phasor can be expressed as

$$|\dot{U}_c| = U_{cm} = \frac{\sqrt{|\dot{U}_s|^2 - (2\pi f_i L_F I_{rd})^2} + 2\pi f_i L_F I_{rq}}{1 - (2\pi f_i L_F C_F)^2} \tag{10.15}$$

It can be found from the operating principles of 3HIMC that the reactive component of the input current in (10.15), labeled as I_{rq}, can be controlled freely, which provides an additional freedom to change the filtering capacitor voltages so as to regulate the VTR of 3HIMC. As can be seen from (10.15), the VTR can be increased by providing positive input reactive current (capacitive). Since the control range of the input reactive power in 3HIMC has no limit in theory, the theoretical linear VTR of 3HIMC is infinite. However, the practical VTR range is limited by some physical factors, such as the voltage rating of the input capacitors and the current rating of the semiconductor switches. This will be discussed further in the next section.

10.4 Control Method to Enhance the Linear VTR

10.4.1 Control Algorithm

As has been analyzed in Sect. 10.3, the linear VTR of 3HIMC can be improved from two aspects. When the frequency and initial phase of the outputs satisfy certain conditions, a VTR greater than 0.866 could be obtained directly without regulating the input reactive power. For other cases, the desired VTR could be obtained by adjusting the input reactive power indirectly. Considering those two aspects, a control method enhancing the linear VTR is presented and analyzed in detail in this section.

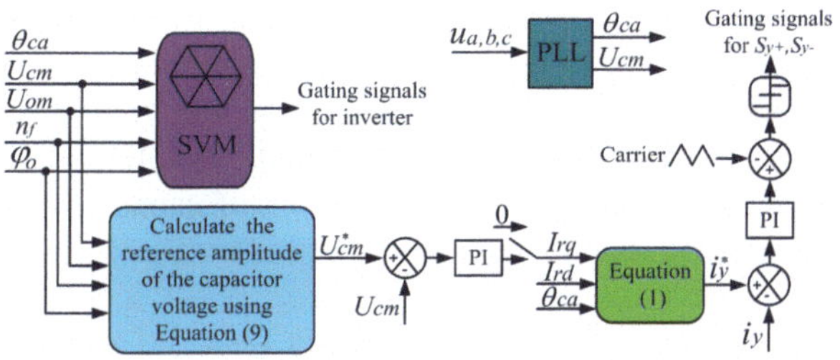

Fig. 10.5 Control algorithm for enhancing the linear VTR of 3HIMC

The detailed control algorithm for improving the linear VTR is depicted in Fig. 10.5. Firstly, the input filtering capacitor voltages are sampled and processed by a PLL to obtain the amplitude and phase of the capacitor voltages. Then based on the output parameters including U_{om}, φ_o, and n_f, the amplitude of the capacitor voltages is examined to verify whether U_{cm} is sufficiently high to synthesize the output voltages according to (10.11) and (10.14). If U_{cm} is high enough and the desired VTR is not greater than the VTR limit, the expected output voltage is synthesized directly by SVM using (10.4), and the input reference reactive current I_{rq} is set as 0. Otherwise, the capacitor voltage has to be boosted by generating input reactive power. In this chapter, a capacitor voltage closed-loop control is adopted. In the implemented control loop, the reference amplitude of the capacitor voltage, denoted as U_{cm}^*, is calculated using (10.11). And U_{cm}^* is compared with the feedback U_{cm} to obtain the voltage error, and then the error is fed into a PI controller to generate the reference reactive current I_{rq}. Finally, the reference third-harmonic injection current i_y^* is calculated according to (10.1), and a PI controller is utilized to track the practical third-harmonic injection current [20]. By doing this, the amplitude of the capacitor voltages can be changed by regulating the input reactive current, and then the expected VTR is obtained.

10.4.2 Effect of the Input Filter and Design Considerations

As has been analyzed in Sect. 10.3.2, the linear VTR of 3HIMC can be improved freely by regulating the input reactive power. In theory, the linear VTR of 3HIMC is infinite. However, this is achieved at the expense of increasing the voltage and current stresses of the input filter. As a result, capacitors and inductors with higher voltage and current rating are needed. Therefore, the practical VTR is limited by the voltage rating of the filtering capacitors and the current rating of the inductors. For most loads, a maximum VTR of 1 is sufficient. That is to say, the voltage rating of the capacitors should be 15% higher than the grid voltage. This will not increase the volume and cost of the capacitors significantly and thus is acceptable.

As can be seen from (10.15), the inductance of the inductor, the capacitance of the capacitor, and the input reactive current component all have effects on VTR. Therefore, the input filter parameters should be properly designed, aiming at obtaining an optimized input filter size. This can be done by performing optimization of the input filter parameters with the optimization objective of minimized total volume. The design procedures can be described briefly as follows. The first step is to select a capacitance of the input capacitor, whose upper and lower limits are decided according to the constraints of capacitor voltage ripple and the reactive current absorbed by the capacitor

$$\frac{I_{om}}{4f_s\delta U_s} \leq C_F \leq \frac{Q_i}{6\pi f_i {U_s}^2} \tag{10.16}$$

where f_s, δ, and Q_i are the switching frequency, voltage ripple index, and input reactive power, respectively. The next step is to choose an inductance of the input inductor

$$L_F = \frac{1}{4\pi^2 f_c^2 C_F}, \; f_c \in (0.1 - 0.5) f_s \tag{10.17}$$

where f_c is the cutoff frequency of the input filter. Then, assuming a maximum VTR of 1 is required in rated condition, the maximum reactive input current I_{rq} can be solved according to (10.15). Finally, the volume models of the capacitor and inductor can be established as

$$V_{LF} = F(L_F, i_{LF}), i_{LF} = \sqrt{I_{rd}^2 + I_{rq}^2} \tag{10.18}$$

$$V_{CF} = F(C_F, U_C) \tag{10.19}$$

And the total volume of the input filter can be expressed as

$$V_F = \Sigma(V_{LF}, V_{CF}) \tag{10.20}$$

By searching and the global minimum V_F, the optimal input filter designs are identified, and one of the input filter designs with the parameters listed in Table 10.2 is selected.

Table 10.2 System parameters of 3HIMC

Parameters	Value
Input line-line voltage	200 Vrms
Input frequency f_i	50 Hz
Switching frequency f_s	16kHz
Capacitor C_F	6.6 μF
Inductor L_F	1 mH
Inductor L_y	1.2 mH
Load inductor	3 mH
Load resistor	25 ohm
Motor	1.1 kW, 4 poles

10.5 Simulation and Experimental Results

To verify the correctness of the developed control method, the VTR characteristic of 3HIMC was firstly verified by simulations using Matlab/Simulink software and then was validated experimentally. The system parameters are listed in Table 10.2.

10.5.1 Simulation Results

Figure 10.6 illustrates the waveforms of 3HIMC with different m_{max} corresponding to different n_f and φ_o. The parameters were set as $n_f = 0.5$, $\varphi_o = \pi/12$, $m_{max} = 0.8967$ in Fig. 10.6a; $n_f = 2$, $\varphi_o = 0$, $m_{max} = 0.9542$ in Fig.10.6b and $n_f = 1$, $\varphi_o = 0$, $m_{max} = 1$ in Fig. 10.6c, respectively. And the reference input reactive current is set as 0 in Fig. 10.6.

The waveforms shown in Fig. 10.6 consist of the input line-line voltage u_{sab}, the input current i_a, the output line-line voltage u_{rs}, and the output current i_r. As can be seen from Fig. 10.6, the input and output currents are almost sinusoidal. Besides, the input current i_a lags behind the input line-line voltage u_{ab} by a phase angle of $\pi/6$. This is reasonable since 3HIMC is in the linear modulation region and the reference input reactive current is 0.

The analyzed fundamental amplitudes of the output line-line voltages are 253.7V, 269.8V, and 282.3V in Fig. 10.6a and c, respectively, which match the expected VTR of 0.8967, 0.9542, and 1 very well. Moreover, the measured input PFs are 0.995, 0.999, and 0.999 in Fig. 10.6a and c, respectively. The results in Fig. 10.6 clearly demonstrate that with the developed method, the linear VTR of 3HIMC can break through 0.866 if the output parameters satisfy certain conditions.

Figure 10.7 shows the waveforms of 3HIMC with the linear VTR greater than 1. In Fig. 10.7, the expected VTR is set as 1.05, and the output frequency is 60 Hz. It is clear that the input and output currents are still sinusoidal, and the phase angle between the input current i_a and the input line-line voltage u_{ab} is no longer $\pi/6$ because the input reactive current has to be regulated to a certain level so as to boost the capacitor voltages and obtain the expected linear VTR. The analyzed fundamental amplitude of the output line-line voltage is 296.3V, which confirms the linear VTR of 1.05 and verifies the effectiveness of the control method.

10.5.2 Experimental Results

To verify the theoretical analysis and simulation results, a 1.5 kW prototype of 3HIMC with the specifications given in Table 10.2 is built. To verify the theoretical analysis sufficiently, both RL load and motor load are tested. The experimental results shown in Fig. 10.8 correspond to the simulation results shown in Fig. 10.6,

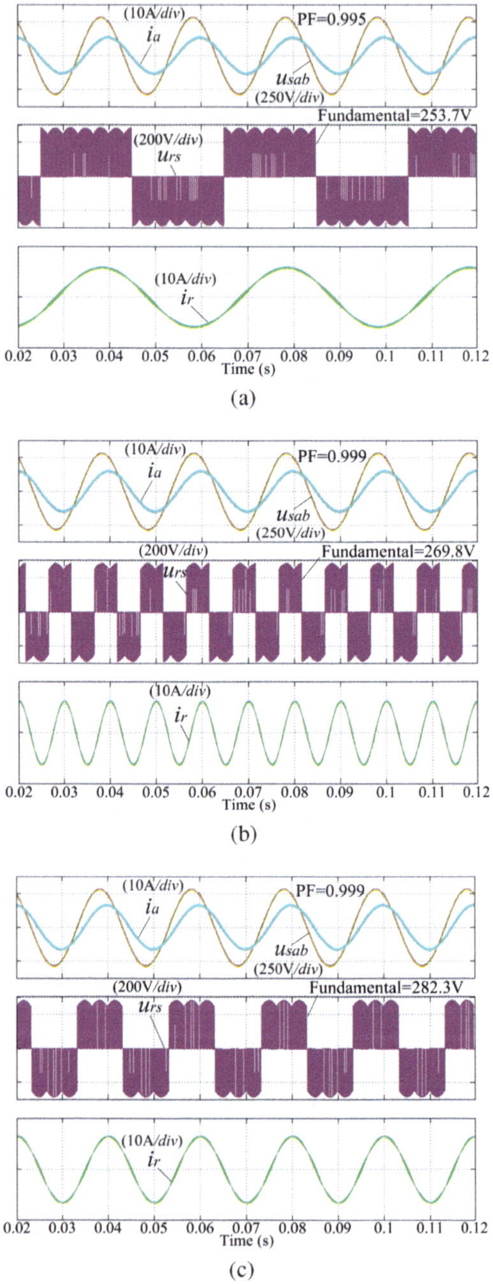

Fig. 10.6 Simulation waveforms of 3HIMC with different m_{max}. (**a**) $n_f = 0.5$, $\varphi_o = \pi/12$, $m_{max} = 0.8967$. (**b**) $n_f = 2$, $\varphi_o = 0$, $m_{max} = 0.9542$. (**c**) $n_f = 1$, $\varphi_o = 0$, $m_{max} = 1$

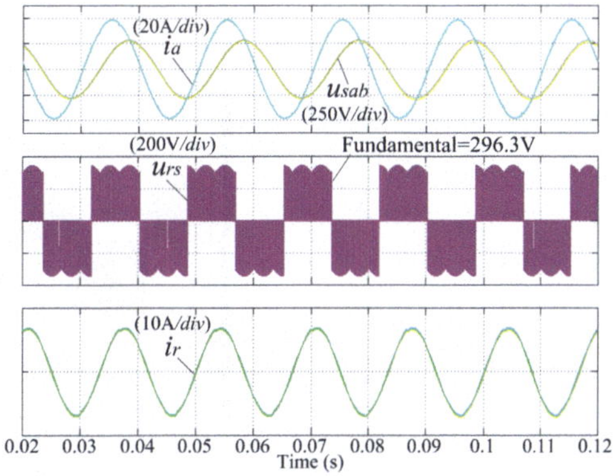

Fig. 10.7 Simulation waveforms with linear VTR of 1.05

and the waveforms of RL load and motor load are depicted on the left and right sides, respectively. As can be seen from Fig. 10.8, sinusoidal input/output currents and unity input power factor are achieved. The measured fundamental amplitudes of the output line-line voltages are 250.6V, 266.7V, and 276.3V for RL load and are 250.1V, 267.5V, and 275.9V for motor load when the desired VTRs are 0.8967, 0.9542, and 1, respectively. Referring to Fig. 10.8, the experimental results match the simulation results well, except for a slight reduction in the fundamental amplitude of the output voltage and distortions of the currents. The differences are primarily attributed to voltage drop of devices and dead time effects.

It can be found from the simulation and experimental results above that with the developed control method, the linear VTR of 3HIMC can break through the intrinsic limit of 0.866. By generating capacitive reactive power at the input side, the linear VTR of 3HIMC can be regulated freely for arbitrary output conditions. Thus the correctness of the method is verified by simulation and experimental results.

10.5.3 Performance Comparisons

Although several solutions have been proposed to enhance the VTR of HIMC, most of them are achieved at the expense of adding extra passive components and/or sacrificing some other performance of the converter, and thus each method has its own limitations. In order to compare the presented method and other HIMCs with improved VTR, several performances including the number of switches, voltage and current rating of the switches, range of the VTR and the waveforms quality, are evaluated and compared. Tables 10.3, 10.4, and 10.5 show the comprehensive performance comparison of proposed method and other methods, where I_{im} and I_{om} are the amplitude of the input and output currents, respectively. In order to enable a fair comparison, the unidirectional rectification stages in [11] and [15] are replaced with the conventional 12 switches current-source rectifier. It should be noted that when calculating the voltage and current stresses, the ripple voltage and

Fig. 10.8 Experimental waveforms of 3HIMC with different m_{max} corresponding to different n_f and φ_o. (**a**) $n_f = 0.5, \varphi_o = \pi/12$, $m_{max} = 0.8967$. (**b**) $n_f = 2$, $\varphi_o = 0, m_{max} = 0.9542$. (**c**) $n_f = 1, \varphi_o = 0, m_{max} = 1$

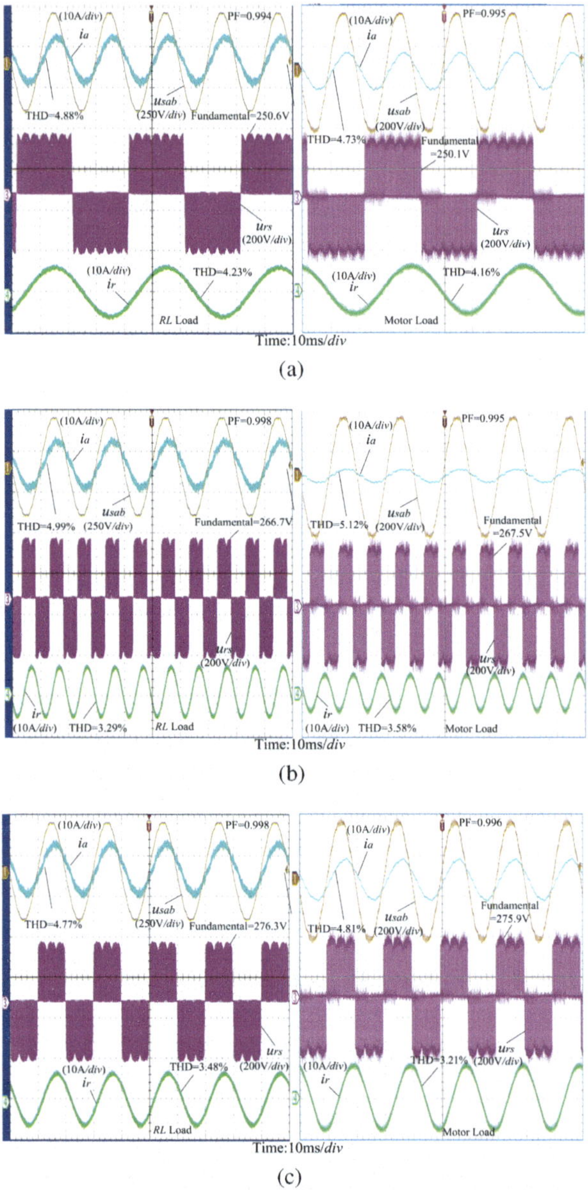

ripple current are ignored for simplicity. The detailed calculation process can refer to [20] and thus is not elaborated here.

As can be seen from Tables 10.3, 10.4, and 10.5, the theoretical maximum VTR in [7, 11, 15, 17] and this chapter are 0.955, ∞, 0.928, 0.97, and ∞, respectively. On the other hand, the required semiconductor switches in [7] and [11, 15, 17] are 22

Table 10.3 Comparison of different methods with improved VTR

Literature	Theoretical VTR Range	Number of Switches	Energy storage components
[7]	0.955	22	1 capacitor
[11]	∞	18	2 capacitors 2 inductors
[15]	0.928	18	0
[17]	0.97	18	0
This work	∞	20	0

Table 10.4 Comparison of different methods with improved VTR

	Rating of the switches (voltage, current)		
Literature	Rectifier	Inverter	Others
[7]	$\sqrt{3}U_{im}, I_{om}$	$[\frac{2\pi\sqrt{3}+6\sqrt{3}-3\pi}{2\pi}]U_{im}, I_{om}$	$[\frac{6\sqrt{3}-3\pi}{2\pi}]U_{im}, I_{om}$
[11]	$\sqrt{3}U_{im}, I_{om}$	$\geq 2\sqrt{3}\text{VTR}U_{im}, I_{om}$	NA
[15]	$\sqrt{3}U_{im}, I_{om}$	$\sqrt{3}U_{im}, I_{om}$	NA
[17]	$\sqrt{3}U_{im}, I_{om}$	$\sqrt{3}U_{im}, I_{om}$	NA
This work	$2U_{im}, I_{im}$	$2U_{im}, I_{om}$	$2U_{im}, \frac{I_{im}}{2}\left(S_{y+}, S_{y-}\right)$ $\sqrt{3}U_{im}, \frac{I_{im}}{2}\left(S_{ay}, S_{by}, S_{cy}\right)$

Table 10.5 Comparison of different methods with improved VTR

Literature	Input current quality	Output current quality
[7]	Sinusoidal (THD = NA)	Sinusoidal (THD = NA)
[11]	Sinusoidal (THD = NA)	Sinusoidal (THD = NA)
[15]	Sinusoidal (THD = 5.11%)	Sinusoidal (THD = 2.8%)
[17]	Distorted (THD = 7%)	Distorted (THD = 5%)
This work	Sinusoidal (THD = 4.77%)	Sinusoidal (THD = 3.48%)

and 18, respectively, while the number of the switches in this chapter is 20. However, it should be noted that the comparison of the number of the switches only gives a first overview of the performance of the topologies, because the current/voltage stress and the losses of the switches for each topology are quite different. For example, although only 18 switches are required in [11], the voltage stress of the inverter switches is high and goes up with the increasing of VTR. As a comparison, the voltage stress of the switches in 3HIMC is relatively low, and the current rating of the switches of the third-harmonic injection circuit is only half of the amplitude of the input current. Besides, the switching losses of the rectifier and the bidirectional switches can be ignored due to their line frequency commutation. Thus, it is possible to make an optimized selection of the switches of 3HIMC.

Regarding the input waveforms quality, 3HIMC has a superior performance than the others in terms of the THD of the input current because of the prefiltering effect of the third-harmonic injection inductor [20], while [7] and [17] have the inferior input current THD performance among the five methods due to the low-order current harmonics caused by the fluctuant input power and overmodulation, respectively. For the comparison of the output waveforms quality, [15] and 3HIMC have the best output voltage THD performance, and the output voltage THD performances in [7] and [11] are a little bit worse because of the higher inverter side DC-link voltage, while the overmodulation in [17] results in the worst output voltage THD performance.

It can be found from the comparison that 3HIMC makes a good trade-off between the VTR and other aspects including the requirements of semiconductor switches/extra energy storage elements, the power density, the efficiency and the input/output power quality when compared with the other methods, although the input capacitors with 15% higher voltage rating have to be selected. Thus, 3HIMC is an attractive choice for the applications where high VTR and excellent input/output power quality are required.

10.6 Conclusions

In this chapter, a control method to enhance the linear VTR of 3HIMC has been presented. Firstly, the maximum VTR under certain output frequency and initial phase is extracted considering the instantaneous intermediate DC-link voltage. Besides, by generating the input capacitive reactive power, the input reactive power capability of 3HIMC is used as an additional degree of freedom to enhance the linear VTR significantly. The proposed method provides a new insight into the improvement of VTR of 3HIMC. The validity and correctness of the theoretical analysis is verified by simulation and experimental results, and the comparison of 3HIMC and other schemes with improved VTR demonstrates the effectiveness of the proposed method. Together with the advantages such as sinusoidal input and output currents, bidirectional power flow, as well as the enhanced input reactive power capability, the presented method makes 3HIMC a potential candidate for power supplies, ASDs, and other applications.

References

1. L. Wei, T.A. Lipo, A novel matrix converter topology with simple commutation, in *Proceedings of the IEEE Industry Applications Society Annual Meeting*, Chicago (2001), pp. 1749–1754
2. J.W. Kolar, F. Schafmeister, S.D. Round et al., Novel three-phase ac-ac sparse matrix converters. IEEE Trans. Power Electron. **22**(5), 1649–1661 (2007)
3. P. Wheeler, J. Rodriguez, J. Clare et al, Matrix converters: a technology review. IEEE Trans. Ind. Electron. **49**(2), 276–288 (2002)
4. J.W. Kolar, T. Friedli, J. Rodriguez, Review of three-phase PWM AC–AC converter topologies. IEEE Trans. Ind. Electron. **58**(11), 4988–5006 (2011)
5. Y. Sun, X. Li, M. Su et al, Indirect matrix converter-based topology and modulation schemes for enhancing input reactive power capability. IEEE Trans. Power Electron. **30**(9), 4669–4681 (2015)
6. T.D. Nguyen, H.-H Lee, Dual three-phase indirect matrix converter with carrier-based PWM method. IEEE Trans. Power Electron. **29**(2), 569–581 (2014)

7. C. Klumpner, A hybrid indirect matrix converter immune to unbalanced voltage supply, with reduced switching losses and improved voltage transfer ratio, in *Proceedings of the IEEE Applications Power Electronics Conference*, Dallas (2006), pp. 133–139
8. T. Wijekoon, C. Klumpner, P. Zanchetta et al, Implementation of a hybrid AC–AC direct power converter with unity voltage transfer. IEEE Trans. Power Electron. **23**(4), 1918–1926 (2008)
9. J. Alireza, R. Ahmad, Indirect matrix converter with unity voltage transfer ratio for AC to AC power conversion. Electr. Power Syst. Res. **96**(96), 157–169 (2013)
10. K. Park, K.B. Lee, F. Blaabjerg, Improving output performance of a Z-source sparse matrix converter under unbalanced input-voltage conditions. IEEE Trans. Power Electron. **27**(4), 2043–2054 (2012)
11. X. Liu, P.C. Loh, P. Wang et al, Improved modulation schemes for indirect Z-source matrix converter with sinusoidal input and output waveforms. IEEE Trans. Power Electron. **27**(9), 4039–4050 (2012)
12. W. Song, Y. Zhong, H. Zhang et al, A study of Z-source dual-bridge matrix converter immune to abnormal input voltage disturbance and with high voltage transfer ratio. IEEE Trans. Ind. Inform. **9**(2), 828–838 (2013)
13. E. Karaman, M. Farasat, A.M. Trzynadlowski, A comparative study of series and cascaded Z-source matrix converters. IEEE Trans. Ind. Electron. **61**(10), 5164–5173 (2014)
14. L. Zarri, M. Mengoni, A. Tani et al., Range of the linear modulation in matrix converters. IEEE Trans. Power Electron. **29**(6), 3166–3178 (2014)
15. C. Xia, S. Li, Y. Yan et al., Research on linear output voltage transfer ratio for ultra sparse matrix converter. IEEE Trans. Power Electron. **31**(3), 1811–1815 (2016)
16. M. Imayavaramban, K. Latha, G. Uma, Analysis of different schemes of matrix converter with maximum voltage conversion ratio, in *Proceedings of the IEEE Mediterranean Electrotechnical Conference*, Dubrovnik (2004), pp. 1137–1140
17. G.T. Chiang, J. Itoh, Comparison of two overmodulation strategies in an indirect matrix converter. IEEE Trans. Ind. Electron. **60**(1), 43–53 (2013)
18. A.M. Bozorgi, M. Monfared, H.R. Mashhadi, Two simple overmodulation algorithms for space vector modulated three-phase to three-phase matrix converter. IET Power Electron. **7**(7), 1915–1924 (2014)
19. B. Wang, G. Venkataramanan, Six step modulation of matrix converter with increased voltage transfer ratio, in Proceedings of IEEE Power Electronics Specialists Conference, Jeju (2006), pp. 1–7
20. H. Wang, M. Su, Y. Sun et al, Two-stage matrix converter based on third-harmonic injection technique. IEEE Trans. Power Electron. **31**(1), 533–547 (2016)
21. H. Wang, M. Su, Y. Sun et al, Active third-harmonic injection indirect matrix converter with dual three-phase outputs. IET Power Electron. **9**(4), 657–668 (2016)

Chapter 11
Cascade PI-SMC Method for Matrix Converter-Fed BDFIM Drives

11.1 Introduction

The BDFIM [1–3] is a type of AC induction machine that has been developed from two cascaded asynchronous motors. Compared to the DFIM, the BDFIM inherits the advantages of the doubly fed machine and has higher reliability as it cancels the slip rings or brushes. Moreover, the BDFIM has a wider range of speed regulation. Therefore, the BDFIM has received extensive attention from researchers [4, 5] and has a broad range of potential applications in the field of variable-speed drive applications [6] and wind power generation [7]. As a promising alternative to the traditional back-to-back converter, matrix converter is employed for the BDFIM drive system in wind power generation application [8, 9]. Since matrix converter [10–12] can realize direct AC-AC conversion without using an intermediate DC-link, this feature of matrix converter will bring several advantages to the matrix converter-fed BDFIM drive systems, such as high power density and long lifetime.

Many control strategies have been proposed for BDFIM drive system, such as scalar control [13–15], DTC [16, 17], MPC [5, 9], and vector control [7, 8, 18–22]. Several scalar control methods have been proposed for the BDFIM system, such as open-loop control [13], closed-loop frequency control [14], and phase angle control [15]. However, the scalar control relies on the static analysis of the BDFM, which makes it only fit for industrial applications with low-performance requirements. In [16, 17], the DTC method for the BDFIG system is proposed. The proper switching vector is chosen by analyzing the effects of different switching vectors on flux and torque. The DTC method has a simple structure and fast dynamic response. However, the torque fluctuation is a disadvantage of this technique. The MPC method has advantages such as conceptual simplicity and easiness of implementation, fast dynamic response, easy inclusion of nonlinearities and constraints of the system, and the flexibility to include other system requirements in the controller [23]. In [9], a predictive torque control strategy has been proposed for controlling the flux and torque of a matrix converter-fed BDFIG. The system

H. Wang, *Advanced Matrix Converters*, Power Systems,
https://doi.org/10.1007/978-3-031-69902-3_11

model is used to predict the system variable and evaluate the switching vector. The optimal switching vector with the minimum cost function value is applied during the next sample period. However, the unmatched model parameters will lead to inaccurate prediction of the motor behavior and deteriorate the performance and stability of the MPC method. The vector control methods based on different flux orientation frames are proposed [8, 18–20, 22]. Wallace et al. developed the dynamic vector model of the BDFIM with an arbitrary pole number referred to the rotor's shaft position [24]. The BDFIM model is built in PW and CW synchronous reference frames related to each pole-pair distribution. The vector control technique implementation is complex for the rotor flux-oriented control method in [18]. The combined magnetizing flux-oriented control for the cascaded doubly fed induction machine is studied in [19]. The rotor current vector is oriented into an artificial magnetizing flux reference frame based on the two stator currents measurements. However, this control method does not lead to natural decoupled active and reactive power control. Thus, the PW flux-oriented vector control strategy [20] is proposed based on a unified reference-frame model in [25]. It is simple to be implemented. Meanwhile, the natural decoupled active and reactive power control is obtained. Thus, the PW flux-oriented vector control method is usually adopted [8, 22]. In [8], the cascaded double closed-loops control method is adopted for a matrix converter-fed BDFM system, including one fast inner current control loop and one outer flux/speed control loop. However, the inner current loop plant of the BDFIM has a complex nonlinear structure and is susceptible to various disturbances. Thus, the robust controller is required for the inner loop.

Since the SMC [26] has the advantages of easy decoupling, disturbance rejection, and parameter variation insensitivity, it has been applied as an alternative robust solution for the induction motor drive applications [27, 28]. For traditional back-to-back inverter-fed BDFIM drive system, the integral SMC method is employed in the outer-loop speed controller [21]. However, the traditional SMC has the phenomenon of "chattering" because the switch does not have the ideal switching characteristics, which will affect the control effect of the system to a certain extent. Therefore, a second-order sliding mode controller based on super twisting is considered a more suitable method for practical applications due to its continuity and robustness. The second-order sliding mode controller eliminates the chattering problem without affecting the tracking accuracy and robustness.

To obtain high robust tracking performance, this chapter proposes a cascade PI-second-order SMC for the matrix converter-fed BDFIM drive system. In this method, the outer-loop PI controller tracks the reference rotor speed, and the inner-loop SMC controller regulates the CW current. The main advantages of this method are that the outer-loop controller provides constraints on the SMC through maximum current and speed limits, and the second-order SMC controller achieves fast and fixed convergence times for the current control loop. Meanwhile, the parameter variations of the BDFIM are considered in the design of the inner-loop SMC controller. The proposed PI-SMC method in this chapter has been described in [29], and more theoretical details and experimental results are added in this chapter.

The rest of this chapter is organized as follows. In Sect. 11.2, the mathematical model and operation principle of matrix converter are described firstly, and the mathematical model of BDFIM drives is presented. Sect. 11.3 shows the sliding mode control of the matrix converter-fed BDFIM. Then in Sect. 11.4, the simulation and experiment for matrix converter-fed BDFIM drives are carried out to verify the proposed PI-SMC method. Finally, Sect. 11.5 concludes the findings.

11.2 Modeling of the Matrix Converter-Fed BDFIM

11.2.1 Mathematical Model of the Matrix Converter

The basic structure of the matrix converter-fed BDFIM is shown in Fig. 11.1. The relationship between the output and input voltages of the matrix converter is shown as follows:

$$\begin{bmatrix} v_{oa} \\ v_{ob} \\ v_{oc} \end{bmatrix} = \begin{bmatrix} S_{Aa} & S_{Ab} & S_{Ac} \\ S_{Ba} & S_{Bb} & S_{Bc} \\ S_{Ca} & S_{Cb} & S_{Cc} \end{bmatrix} \begin{bmatrix} v_{ia} \\ v_{ib} \\ v_{ic} \end{bmatrix} \tag{11.1}$$

where v_{oy} $(y \in \{a, b, c\})$ is the output voltage of the matrix converter and v_{iy} $(y \in \{a, b, c\})$ is the input voltage of the matrix converter. Each group of switching state needs to meet $S_{xa}+S_{xb}+S_{xc} = 1$, $(x \in \{A, B, C\})$. The modulation method based on mathematical construction [30] is used in this chapter.

The switching functions can be replaced by the corresponding duty cycles as the switching frequency is much higher than the input/output frequencies. Thus, Eq. (11.1) can be written as

$$\begin{bmatrix} v_{oa} \\ v_{ob} \\ v_{oc} \end{bmatrix} = M \begin{bmatrix} v_{ia} \\ v_{ib} \\ v_{ic} \end{bmatrix} \tag{11.2}$$

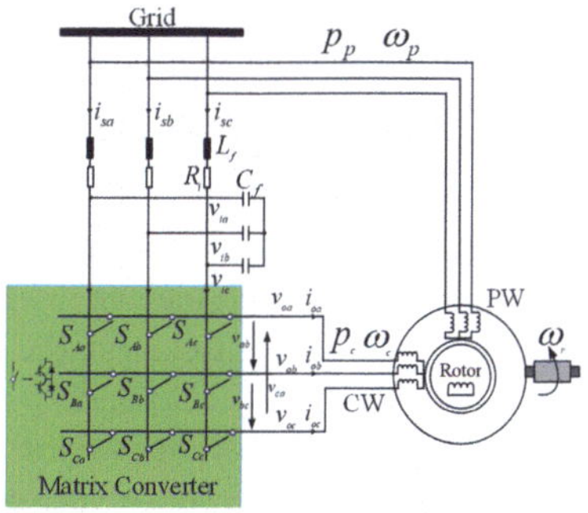

Fig. 11.1 The basic structure of the matrix converter-fed BDFIM drives

where M is the duty cycle matrix and is expressed as

$$M = \begin{bmatrix} m_{11} & m_{12} & m_{13} \\ m_{21} & m_{22} & m_{23} \\ m_{31} & m_{32} & m_{33} \end{bmatrix} = \begin{bmatrix} m''_{11} + x + D & m''_{12} + y + D & m''_{13} + z + D \\ m''_{21} + x + D & m''_{22} + y + D & m''_{23} + z + D \\ m''_{31} + x + D & m''_{32} + y + D & m''_{33} + z + D \end{bmatrix} \tag{11.3}$$

Define the transition modulation matrix M'' as

$$M'' = \begin{bmatrix} m''_{11} & m''_{12} & m''_{13} \\ m''_{21} & m''_{22} & m''_{23} \\ m''_{31} & m''_{32} & m''_{33} \end{bmatrix} = K \begin{bmatrix} \cos(\omega_0 t) \\ \cos(\omega_0 t - 2\pi/3) \\ \cos(\omega_0 t + 2\pi/3) \end{bmatrix} \begin{bmatrix} \cos(\omega_i t - \theta_i) \\ \cos(\omega_i t - \theta_i - 2\pi/3) \\ \cos(\omega_i t - \theta_i + 2\pi/3) \end{bmatrix}^T \tag{11.4}$$

where ω_i, θ_i, ω_0, and K represent input angular frequency, input power factor angle, output angular frequency, and modulation index, respectively. Besides, to meet nonnegative constraints of the duty cycles, x, y, z needs to satisfy (11.5):

$$\begin{cases} x \geq -\min(m''_{11}, m''_{21}, m''_{31}) \\ y \geq -\min(m''_{12}, m''_{22}, m''_{32}) \\ z \geq -\min(m''_{13}, m''_{23}, m''_{33}) \end{cases} \tag{11.5}$$

To make the algorithm as simple as possible, the boundary values in (11.5) are usually adopted. At the same time, the modulation matrix M also needs to satisfy that the sum of elements in each row is 1. Therefore, an offset D is superimposed on each row in M, and D can be selected as $D = [1 - (x + y + z)]/3$, $D \geq 0$.

The mathematical construction-based modulation method can obtain the duty cycles of nine switches directly in the matrix converter, which is simple and can achieve the maximum voltage transfer ratio of 0.866.

11.2.2 Mathematical Model of the BDFIM Drives

The BDFIM consists of three windings: PW, CW, and RW. The power stator side is usually directly connected with the grid side voltage, so the rotating speed of the stator magnetic field on the power side is known and remains unchanged. Therefore, the $d - q$ reference frame based on the PW flux orientation is selected. The voltages and currents vectors of PW, CW, and RW are represented as x_p, x_c, and x_r, respectively. These vectors are transformed from their reference frames to the d-q reference frame using the following corresponding equations:

$$\begin{cases} \boldsymbol{x}_{\boldsymbol{p}}^{\boldsymbol{dq}} = e^{-j\omega_p t}\boldsymbol{x}_{\boldsymbol{p}}^{\boldsymbol{\alpha\beta p}} \\ \boldsymbol{x}_{\boldsymbol{c}}^{\boldsymbol{dq}} = e^{-j(\omega_p-(p_p+p_c)\omega_r)t}(-\boldsymbol{x}_{\boldsymbol{c}}^{\boldsymbol{\alpha\beta p}})^* \\ \boldsymbol{x}_{\boldsymbol{r}}^{\boldsymbol{dq}} = e^{-j(\omega_p-p_p\omega_r)t}\boldsymbol{x}_{\boldsymbol{r}}^{\boldsymbol{\alpha\beta p}} \end{cases} \tag{11.6}$$

where ω_p and ω_c are the angular frequencies of PW and CW, respectively; P_p and P_c are pole pairs of PW and CW, respectively; ω_r is the angular speed of the rotor. To simplify the expressions, the vector superscripts d_q are removed.

Since the rotor resistance is small to be ignored and the rotor flux is nearly zero, using the synchronous rotating coordinate system of power stator as reference coordinate system, the simplified mathematical model of the BDFIM [31] is derived as

$$\begin{cases} \boldsymbol{v_p} = R_p\boldsymbol{i_p} + \dfrac{d\boldsymbol{\psi_p}}{dt} + j\omega_p\boldsymbol{\psi_p} \\ \boldsymbol{v_c} = R_c\boldsymbol{i_c} + \dfrac{d\boldsymbol{\psi_c}}{dt} - j\omega_c\boldsymbol{\psi_c} \\ 0 = R_r\boldsymbol{i_r} + \dfrac{d\boldsymbol{\psi_r}}{dt} + j\omega_c\boldsymbol{\psi_r} \\ \boldsymbol{\psi_r} = L_r\boldsymbol{i_r} + M_p\boldsymbol{i_p} + M_c\boldsymbol{i_c} \\ \boldsymbol{\psi_p} = L_p\boldsymbol{i_p} + M_p\boldsymbol{i_r} \\ \boldsymbol{\psi_c} = L_c\boldsymbol{i_c} + M_c\boldsymbol{i_r} \\ T_e = \dfrac{3}{2}p_p\mathrm{Im}\left[(\boldsymbol{\psi_p})^*\boldsymbol{i_p}\right] + \dfrac{3}{2}p_c\mathrm{Im}\left[(\boldsymbol{\psi_c})^*\boldsymbol{i_c}\right] \\ Q_{pw} = \dfrac{3v_{pq}}{2\left(L_rL_q - M_p^2\right)}M_pM_ci_{cd} + \dfrac{3L_rv_{pq}^2}{2\left(L_rL_q - M_p^2\right)\omega_p} \\ J\dfrac{d\omega_r}{dt} = T_e - T_L - B_g\omega_r \end{cases} \tag{11.7}$$

where v_p, v_c, i_p, i_c, i_r, ψ_c, and ψ_r are the power winding stator voltage vector, CW stator voltage vector, PW stator current vector, CW stator current vector, rotor current vector, PW stator flux vector, CW stator flux vector and rotor flux vector in the d-q rotating coordinate system; R_p, R_c, R_r, L_p, L_c, and L_r are the PW stator resistance, CW stator resistance, rotor resistance, PW stator inductance, CW stator inductance, and rotor inductance; M_p is the coupling inductance between the PW stator and rotor winding; M_c is the coupling inductance between the CW stator and rotor winding; T_e is the electromagnetic torque; ()* is the conjugate symbol; Im[] is a symbol to obtain the imaginary part of vector; J and B_g are the moments of inertia and friction coefficient; T_L is the load torque; Q_{pw} is the reactive power. When the rotor resistance and the rotor flux are not considered, according to (11.7), the voltage equation of the BDFIM can be expressed as

$$\begin{cases} \dfrac{di_{cd}}{dt} = -\dfrac{\sigma_1 R_c}{\sigma_2} i_{cd} + \dfrac{\sigma_1}{\sigma_2} v_{cd} - \omega_c i_{cq} \\ \dfrac{di_{cq}}{dt} = -\dfrac{\sigma_1 R_c}{\sigma_2} i_{cq} + \dfrac{\sigma_1}{\sigma_2} v_{cq} + \omega_c i_{cd} + D_1 \end{cases} \tag{11.8}$$

where ψ_{pm} is the amplitude of the PWstator flux vector, $\sigma_1 = L_p L_r - M_p^2$, $\sigma_2 = L_c L_p L_r - M_c^2 L_p - M_p^2 L_c$, and $D_1 = M_p M_c \psi_{pm}/\sigma_2$.

11.3 Sliding Model Control for Matrix Converter-Fed BDFIM

The SMC diagram for the matrix converter-fed BDFIM is shown in Fig. 11.2. The control method adopts a double closed-loop control structure. The outer loops use the conventional PI control strategy to track the reactive power of the PW and the reference rotor speed. The output of the outer-loop controller is the reference stator current of the inner loop control. To improve the robustness and interference immunity ability of the nonlinear inner-loop dynamics, a second-order SMC method is introduced as the current loop to obtain the reference voltage of the CW and the duty cycles of the matrix converter.

11.3.1 Current Loop Design

According to the nonlinear system shown in (11.8), in order to achieve zero steady-state error tracking of i_{cd}^* and increase the degree of freedom in system bandwidth adjustment, the sliding mode surface is selected as

$$S_{sl} = X_1 + \lambda_1 X_2 \tag{11.9}$$

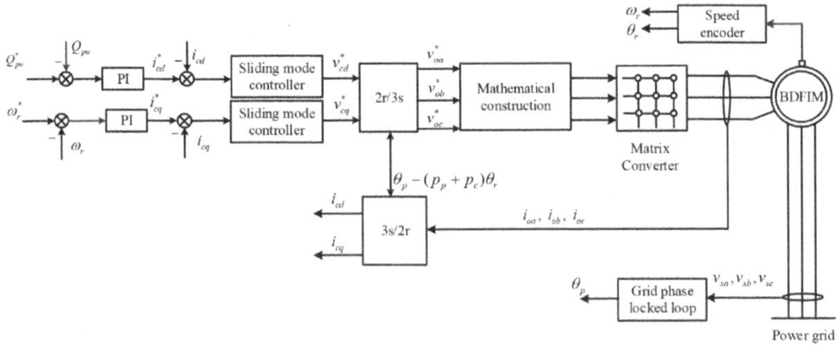

Fig. 11.2 Sliding mode control diagram of the matrix converter-fed BDFIM drives

where $X_1 = i_{cd} - i_{cd}^*$, $X_2 = \int i_{cd} - i_{cd}^* dt$, and λ_1 is the sliding mode coefficient to adjust the current loop bandwidth. When the sliding occurs, i.e., $S_{sl} = 0$, $\dot{S}_{sl} = 0$, the steady-state error i_{cd} can be expressed as

$$X_1(t) = X_1(0)\, e^{\lambda_1 t} \tag{11.10}$$

The control law of the sliding mode controller is selected as shown in (11.11):

$$\begin{cases} v_{cd} = v_{cd}^{eq} + v_{cd}^{st} \\ v_{cd}^{eq} = \left(\dfrac{\sigma_1 R_c}{\sigma_2} i_{cd} + \omega_c i_{cq} + \dfrac{di_{cd}^*}{dt} - \lambda_1 X_1\right)\dfrac{\sigma_2}{\sigma_1} \\ v_{cd}^{st} = -c_1 \left|S_{sl}\right|^{\frac{1}{2}} \operatorname{sign}(S_{sl}) - c_2 \int \operatorname{sign}(S_{sl})\, dt \end{cases} \tag{11.11}$$

where v_{cd}^{eq} is the equivalent control, v_{cd}^{st} is a super-twist control item, and c_1 and c_2 are the positive constants, respectively.

Considering that the high-frequency noise can be presented in the system, it is not easy to obtain the derivative of i_{cd}^*. In order to solve this problem, this solution uses a second-order sliding mode differentiator as shown in (11.12) to obtain the derivative of i_{cd}^*. Supposing i_{cd}^* contains Langberg measurement noise, and $\left(di_{cd}^*/dt\right) < L$, then:

$$\begin{cases} \dot{z}_0 = v_0,\ v_0 = -\beta_0 L^{1/3} \left|z_0 - i_{cd}^*\right|^{2/3} \operatorname{sign}\left(z_0 - i_{cd}^*\right) + z_1 \\ \dot{z}_1 = v_1,\ v_1 = -\beta_1 L^{1/2} \left|z_0 - v_0\right|^{1/2} \operatorname{sign}(z_0 - v_0) + z_2 \\ \dot{z}_2 = -\beta_2 L \operatorname{sign}(z_2 - v_1) \end{cases} \tag{11.12}$$

In (11.12), the parameter selection method in terms of β_0, β_1, β_2 can refer to the literature [32]. In the absence of input noise, the following equation is true after a finite time transient process.

$$\begin{cases} z_0 = i_{cd}^* \\ v_0 = z_1 = \dfrac{di_{cd}^*}{dt} \end{cases} \tag{11.13}$$

11.3.2 Speed Loop Design

The outer speed loop adopts the conventional PI controller. According to (11.8), the mathematical model of the BDFIM can be expressed as

$$J\frac{d\omega_r}{dt} = \frac{3M_p M_c (p_p + p_c)\psi_{pm}}{2\left(L_p L_r - M_p^2\right)} i_{cq} - T_L - B_g \omega_r \tag{11.14}$$

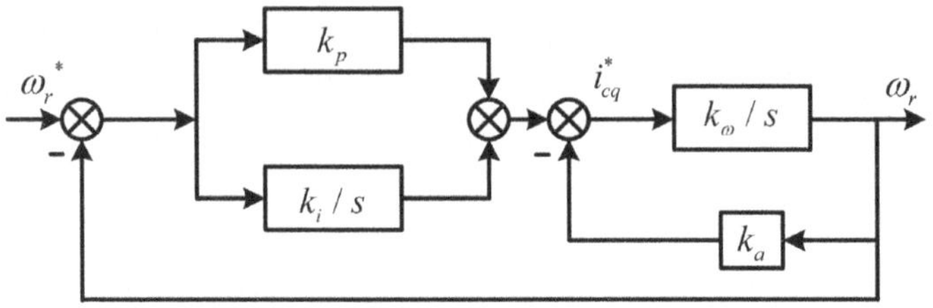

Fig. 11.3 The block diagram of speed outer loop with active damping item

and the item $B_g\omega_r$ can usually be treated as the disturbance part. The factor of i_{cq} is represented by k_ω, and then Eq. (11.14) can be simplified as

$$\frac{d\omega_r}{dt} = k_\omega i_{cq} \tag{11.15}$$

Assuming the inner loop's bandwidth is 10 times the bandwidth of the outer loop, the speed equation (11.15) is regarded as a first-order speed system. The control block diagram of the outer loop is shown in Fig. 11.3, and the PI controller with active damping item [33] is applied in this manuscript.

$$i_{cq}^* = k_p\left(\omega_r^* - \omega_r\right) + k_i\int_0^t\left(\omega_r^* - \omega_r\right)dt - k_a\omega_r \tag{11.16}$$

The closed-loop transfer function of the system is obtained as

$$\Phi(s) = \frac{k_\omega\left(k_p s + k_i\right)}{s^2 + k_\omega\left(k_p + k_a\right)s + k_\omega k_i} \tag{11.17}$$

where V_{dc} represents the DC side voltage, and α is the angle between the reference voltage vector $\boldsymbol{V}_{ref}$ and the voltage basis vector $\boldsymbol{V}_1$.

In this manuscript, the parameter of the active damp item is selected as $k_a = k_p$. The closed-loop transfer function of the speed control loop is designed as an inertial link in a first-order system with ω_{ac} as the bandwidth.

$$\Phi(s) = \frac{\omega_{ac}}{s + \omega_{ac}} \tag{11.18}$$

According to Eqs. (11.17) and (11.18), the parameter of the PI controller is calculated as $k_p = \omega_{ac}/k_\omega$, $k_i = \omega_{ac}^2/k_\omega$, $k_a = \omega_{ac}/k_\omega$.

11.3.3 Robust Analysis and Proof of Stability

Considering the physical limitation of the system in practice and its behavior during the operation, such as eddy current loss and parameter uncertainties, the second-order sliding mode control method is applied to the current inner loop to improve the system stability and anti-disturbance ability [32].

According to (11.8) and (11.11), considering the parameters variation, the unmodeled dynamics, and any other external disturbances, the current differential equation of the BDFIM with disturbance is shown in (11.19):

$$\begin{aligned}\frac{di_{cd}}{dt} &= -\frac{\hat{\sigma}_1\hat{R}_c}{\hat{\sigma}_2}i_{cd} + \frac{\hat{\sigma}_1}{\hat{\sigma}_2}v_{cd} - \omega_c i_{cq}\\ &+\frac{\hat{\sigma}_1\sigma_2\hat{R}_c - \sigma_1\hat{\sigma}_2R_c}{\sigma_2\hat{\sigma}_2}i_{cd} + \frac{\sigma_1\hat{\sigma}_2 - \hat{\sigma}_1\sigma_2}{\sigma_2\hat{\sigma}_2}v_{cd} + \xi\end{aligned} \tag{11.19}$$

where ξ includes the unmodeled dynamics and any other disturbances. The upper notation " $\wedge$ 0" represents the nominal value, after that $\hat{\sigma}_1$ and $\hat{\sigma}_2$ are shown as

$$\begin{cases}\hat{\sigma}_1 = \sigma_1 + \varDelta\sigma_1 = \hat{L}_p\hat{L}_r - \hat{M}_p^2\\ \hat{\sigma}_2 = \sigma_2 + \varDelta\sigma_2 = \hat{L}_c\hat{L}_p\hat{L}_r - \hat{M}_c^2\hat{L}_p - \hat{M}_p^2\hat{L}_c\end{cases} \tag{11.20}$$

The parameter variation is shown in (11.21):

$$\begin{cases}\hat{L}_r = L_r + \varDelta L_r,\ \hat{L}_p = L_p + \varDelta L_p,\ \hat{L}_c = L_c + \varDelta L_c\\ \hat{R}_c = R_c + \varDelta R_c,\ \hat{M}_p = M_p + \varDelta M_p,\ \hat{M}_c = M_c + \varDelta M_c\end{cases} \tag{11.21}$$

where $\Delta L_r, \Delta L_p, \Delta L_c, \Delta M_p, \Delta M_c$ are the parametric errors, and $\Delta\sigma_1$ and $\Delta\sigma_2$ in (11.20) denote the parameter errors caused by the variation of parameters L_r, L_p, L_c, M_p, M_c and are shown as

$$\begin{aligned}\varDelta\sigma_1 &= \varDelta L_rL_p + L_r\varDelta L_p + \varDelta L_r\varDelta L_p - 2M_p\varDelta M_p - \varDelta M_p^2\\ \varDelta\sigma_2 &= \varDelta L_cL_rL_p + \varDelta L_rL_cL_p + \varDelta L_pL_cL_r + \varDelta L_c\varDelta L_rL_p\\ &\quad +\varDelta L_c\varDelta L_pL_r + \varDelta L_r\varDelta L_pL_c + \varDelta L_c\varDelta L_r\varDelta L_p\\ &\quad -2L_pM_c\varDelta M_c - L_p\varDelta M_c^2 - M_c^2\varDelta L_p\\ &\quad -2L_cM_p\varDelta M_p - L_c\varDelta M_p^2 - M_p^2\varDelta L_c\end{aligned} \tag{11.22}$$

For ease of analysis, the errors in (11.19) can be integrated into the form shown in (11.23):

$$d = \frac{\hat{\sigma}_1\sigma_2\hat{R}_c - \sigma_1\hat{\sigma}_2 R_c}{\sigma_2\hat{\sigma}_2} i_{cd} + \frac{\sigma_1\hat{\sigma}_2 - \hat{\sigma}_1\sigma_2}{\sigma_2\hat{\sigma}_2} v_{cd} + \xi \tag{11.23}$$

In practice, the matrix converter-fed BDFIM needs to meet certain physical constraints in operation. The following assumptions can be made:

(1) The capacity of the matrix converter is limited, that is, the output current and voltage are limited.
(2) The disturbance of the brushless doubly fed induction machine is bounded. The parameter variations ΔL_r, ΔL_p, ΔL_c, ΔM_p and ΔM_c are bounded.

According to the above assumption, the total disturbance is bounded as $|d| \leq \varepsilon \varepsilon$ is a known positive constant, which represents the upper limit of the known current inner-loop error. According to (11.9) and (11.19), the first-order derivative of the sliding surface S_{sl} is derived as

$$S_{sl} = \dot{X}_1 + \lambda_1 X_1 = \hat{X}_1 + \lambda_1 X_1 + d \tag{11.24}$$

where $\hat{X}_1$ is the nominal value of X_1. Substituting (11.11) into (11.24), the closed-loop system is derived as

$$\begin{cases} \dot{S}_{sl} = -c_1 |S_{sl}|^{\frac{1}{2}} \operatorname{sign}(S_{sl}) + v + d \\ \dot{v} = -c_2 \operatorname{sign}(S) \end{cases} \tag{11.25}$$

To prove the stability of the system, the Lyapunov function is selected as [32]

$$\begin{aligned} V = 2c_2 |S_{sl}| + \frac{1}{2}v^2 + \frac{1}{2}\left(c_1 |S_{sl}|^{1/2} \operatorname{sign}(S_{sl}) - v^2\right) \\ = \zeta^T P_{sl} \zeta \end{aligned} \tag{11.26}$$

where ζ^T and P_{sl} are expressed as

$$\zeta^T = \left[\operatorname{sign}(S_{sl}) |S_{sl}|^{1/2} \, v\right] \tag{11.27}$$

$$P_{sl} = \frac{1}{2}\begin{bmatrix} 4c_2 + c_1^2 & -c_1 \\ -c_1 & 2 \end{bmatrix} \tag{11.28}$$

From (11.26), it can be derived that:

$$\dot{V} = -\frac{1}{|S_{sl}|^{1/2}}\zeta^T Q \zeta + \frac{d}{|S_{sl}|^{1/2}} q^T \zeta \tag{11.29}$$

where Q and q^T can be expressed as

$$Q = \frac{c_1}{2}\begin{bmatrix} 2c_1 + c_1^2 & -c_1 \\ -c_1 & 1 \end{bmatrix}, q^T = \begin{bmatrix} \left(2c_2 + \frac{c_1^2}{2}\right) & -\frac{c_1}{2} \end{bmatrix} \tag{11.30}$$

Since the disturbance of the current inner loop is bounded, it can be shown that [34]:

$$\dot{V} \leq -\frac{1}{|S|^{(1/2)}}\zeta^T \tilde{Q}\zeta \tag{11.31}$$

where

$$\tilde{Q} = \frac{c_1}{2}\begin{bmatrix} 2c_2 + c_1^2 - \left(\frac{4c_2}{c_1} + c_1\right)\varepsilon & -(c_1 + 2\varepsilon) \\ -(c_1 + 2\varepsilon) & 1 \end{bmatrix} \tag{11.32}$$

When V is positive and $\dot{V}$ is negative, the system satisfies the asymptotic stability condition, if $\tilde{Q} > 0$. Therefore, by setting the values of c_1 and c_2, the system is stable under bounded noise interference. The choice of c_1 and c_2 is shown in (11.33):

$$\begin{cases} c_1 > 2\varepsilon \\ c_2 > c_1 \dfrac{5c_1\varepsilon + 4\varepsilon^2}{2(c_1 - 2\varepsilon)} \end{cases} \tag{11.33}$$

Therefore, the second-order sliding mode controller of i_{cd} is designed according to the theory of sliding mode control, and its robustness is proved by mathematical method. In the same theory, the second-order sliding mode controller of i_{cq} can be designed in the same way.

11.4 Simulation and Experimental Results

11.4.1 Simulation Results

To verify the correction of the proposed method, numerical simulations are performed on MATLAB/Simulink. The parameters of BDFIM are shown in Table 11.1.

Figure 11.4 shows the simulation results of the PI-SMC method. It can be seen that the actual speed of the BDFIM reaches a steady state within 1s when the reference rotor speed changes. The steady-state error of speed is approximately zero, and the reactive power fluctuates slightly and keeps to zero rapidly. When the load torque changes abruptly from 0 N · m to 30 N · m, the rotor speed has a drop but recovers in 1s. The reactive power is almost maintained at zero.

Table 11.1 Simulation parameters

Symbol	Parameter	Value
P_p	The number of pole pairs of the PW	1
P_c	The number of pole pairs of the CW	3
R_p	PW stator resistance (Ω)	0.40355
R_c	CW stator resistance (Ω)	0.5470
R_r	Rotor resistance (Ω)	0.785244
L_p	PW stator inductance (mH)	253.3
L_c	CW stator inductance (mH)	129.6
M_p	PW excitation inductance (mH)	247.4
M_c	CW excitation inductance (mH)	123.7

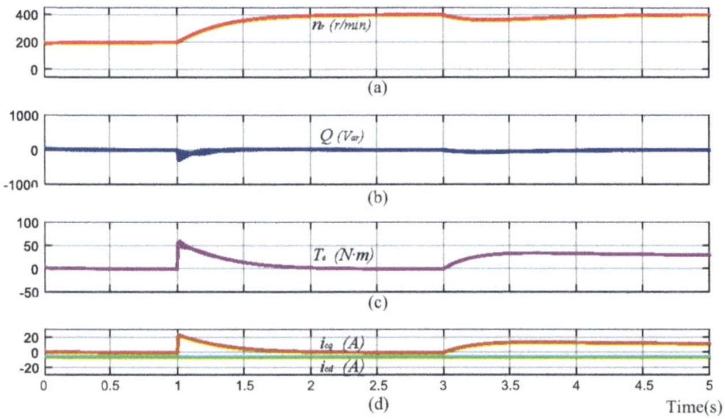

Fig. 11.4 Simulation results of the PI-SMC method: The reference rotor speed is changed from 200 r/min to 400 r/min at $t = 1$s; the load torque is changed from 0 N · m to 30 N · m at $t = 3$s. (**a**) The rotor speed, (**b**) the reactive power, (**c**) the electromagnetic torque, (**d**) CW dq-axis stator current

To simplify the analysis, the variations of parameters M_p and M_c are considered. The simulation results under parametric error $\pm 20\% M_p$ and $\pm 20\% M_c$ are presented in Figs. 11.5 and 11.6 to validate the robustness of the proposed method. The steady and dynamic control performances of the proposed method are different when the parameter M_p has $\pm 20\%$ error. Nonetheless, the tracking performances of the rotor speed and CW stator current are both satisfied with quick tracking processes. When the parameter M_c has $\pm 20\%$ error, the tracking performances of the rotor speed and CW stator current are both acceptable as shown in Fig. 11.6. In conclusion, when the parametric error exists, the waveforms in steady and dynamic states are affected to some extent. However, the system stability under variable parameters is guaranteed with suitable parameters c_1 and c_2.

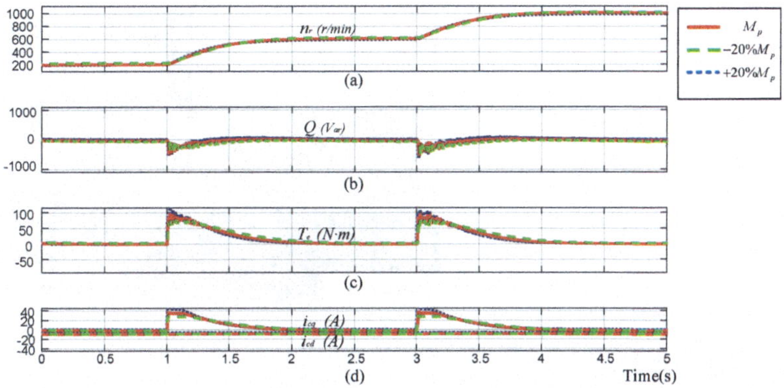

Fig. 11.5 Simulation results with parameter errors: $\pm 20\% M_p$: The reference rotor speed is changed from 200 r/min to 600 r/min and 1000 r/min at t = 1s and 4s; (**a**) the rotor speed, (**b**) the reactive power, (**c**) the electromagnetic torque, (**d**) CW dq-axis stator current

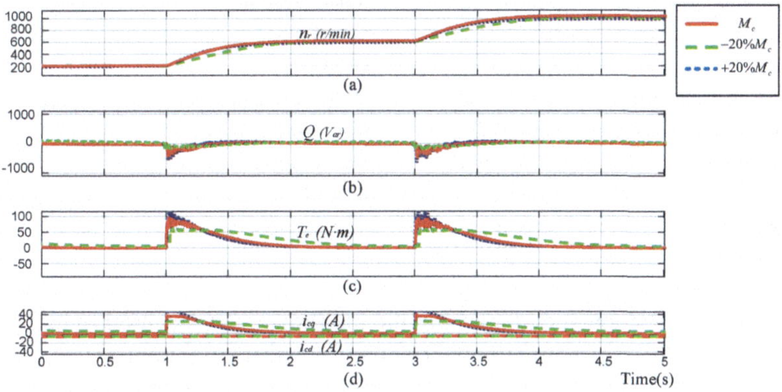

Fig. 11.6 Simulation results with parameter errors: $20 \pm \% M_c$: The reference rotor speed is changed from 200 r/min to 600 r/min and 1000 r/min at t = 1s and 4s; (**a**) the rotor speed, (**b**) the reactive power, (**c**) the electromagnetic torque, (**d**) CW dq-axis stator current

11.4.2 Experimental Results

To verify the correctness of the proposed cascade PI-SMC method for the matrix converter-fed BDFIM drive, an experimental prototype is built. The experimental system arrangement is shown in Fig. 11.7. The root means square value of input line-to-line voltage is 200V. An 11 kW three-phase IM is mechanically coupled to the 15 kVA BDFIM and driven by a matrix converter. The matrix converter-fed BDFIM drive system consists of four parts: control platform, clamp circuit, drive system, and IGBT module. The control platform mainly consists of a DSP (TMS320F28335) and FPGA (EP2C8T144C8N). The drive system is utilized to convert the acquired duty cycle into high or low levels to drive the IGBTs of the matrix converter. The clamp

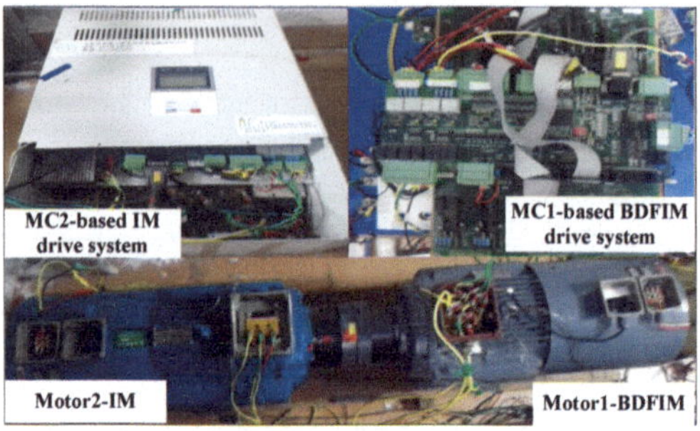

Fig. 11.7 Experimental setup

Table 11.2 Motor specification of BDFIM and IM

Motor1-BDFIM		Motor2-IM	
Type	ZWSF250L-8	Type	YZPFM180L-8
Rated power	15kVA	Rated power	11kW
Rated voltage	400V	Rated voltage	380V
Rated current	17.3A	Rated current	24A
Speed range	0–1450 r/min	Speed range	0–1450 r/min
Pole pairs	3	Polar pairs	4
PW stator resistance R_p	0.40355 Ω	Stator resistance R_s	0.3 Ω
CW stator resistance R_c	0.5470 Ω	Rotor resistance R_r	0.317065 Ω
Rotor resistance R_r	0.785244 Ω	Stator inductance L_s	0.055165 H
$(L_p+L_c+L_r)$	0.0178 H	Rotor Inductance L_r	0.055165 H
$(M_p//M_c)$	0.0824 H	Mutual inductance L_{sr}	0.0498 H

circuit is equipped to prevent damage to the matrix converter caused by issues such as commutation failure or shut down. The precise values of parameters L_p, L_c, L_r, M_p, and M_c are unknown. The data relationship between the inductors is obtained through a simple measurement. The motor specifications of BDFIM and IM are shown in Table 11.2.

To verify the steady-state performance and dynamic performance of the proposed PI-SMC method under the condition of unknown parameters, the following tests are carried out.

Figure 11.8a shows the experimental waveforms of PW stator phase voltage v_{pa}, PW stator current i_{pa}, and CW stator current i_{oa} when the reference speed is set as 400 r/min for the matrix converter-fed BDFIM. The PW stator phase current is approximately sinusoidal and in phase with the PW stator phase voltage. Thus, the reactive power is nearly zero. The data in Fig.11.8b–c is obtained through the digital-to-analog converter of the control platform. The rotor speed tracks the reference speed well, as shown in Fig. 11.8b. The steady-state error of rotor speed is ±3 r/min after filtering sampling noise. The real values of dq-axis CW stator current are stable with small fluctuations and track the reference value well, as shown in Fig. 11.8c. The proposed PI-SMC method still has a perfect steady-state performance in the case of uncertain motor specific parameters.

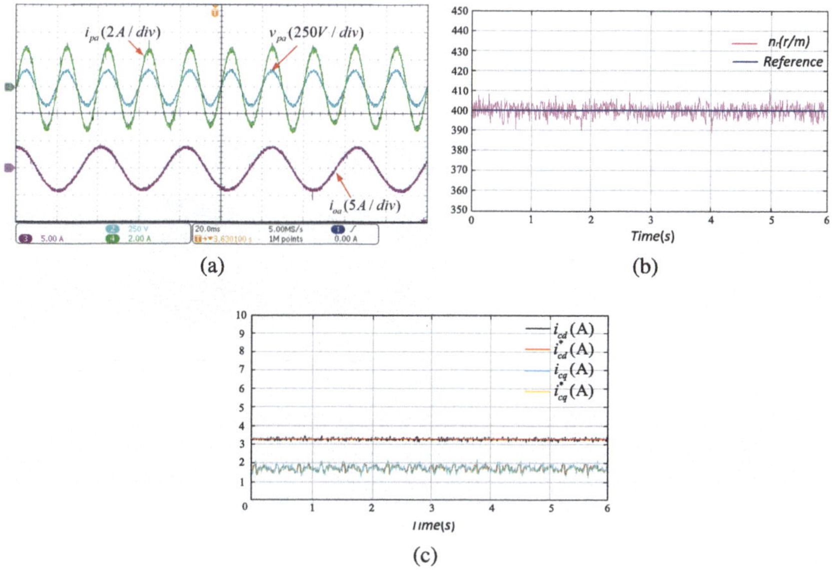

Fig. 11.8 Experimental results of the cascade PI-SMC method when the reference rotor speed is 400 r/min. (**a**) Experimental waveforms of PI-SMC method, (**b**) the reference and real values of rotor speed, (**c**) the reference and real values of dq-axis CW stator current

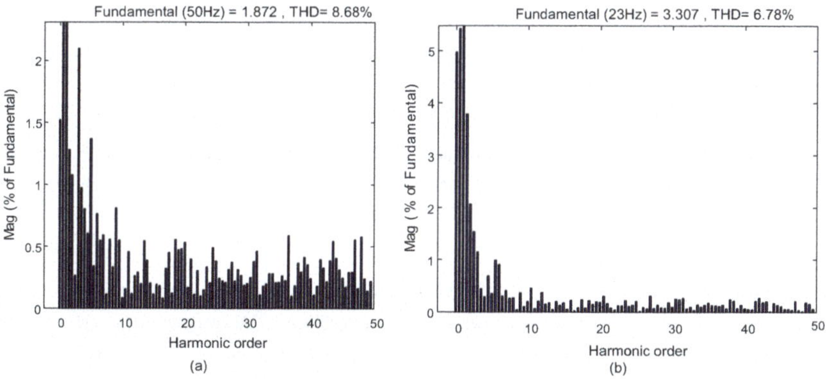

Fig. 11.9 The harmonic spectra of PW and CW current when the reference rotor speed is 400 r/min. (**a**) PW stator current i_{pa}. (**b**) CW stator current i_{oa}

The harmonic spectra of PW and CW current when the reference rotor speed is 400 r/min are shown in Fig. 11.9. The THD values of PW and CW current are 8.68% and 6.78%, respectively. The distortion of PW and CW current are generated due to the device voltage drop, commutation dead time, and other nonlinear factors.

Figure 11.10 shows the experimental waveforms of PW stator phase voltage v_{pa}, PW stator current i_{pa}, and CW stator current i_{oa} when the reference rotor

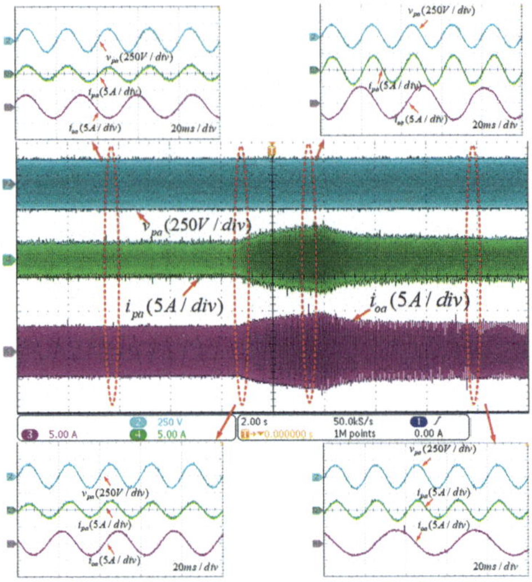

Fig. 11.10 Experimental waveforms of cascade PI-SMC method when the reference rotor speed changes from 200 r/min to 400 r/min

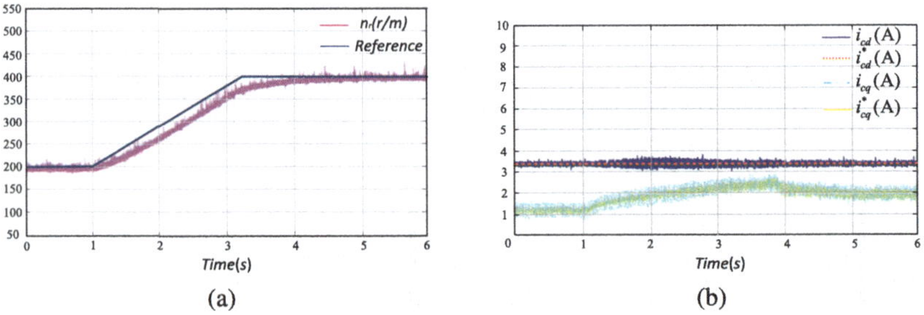

Fig. 11.11 Experimental waveforms of cascade PI-SMC method when the reference rotor speed changes from 200 r/min to 400 r/min. (**a**) The reference and real values of rotor speed. (**b**) The reference and real values of dq-axis CW stator current

speed changes from 200 r/min to 400 r/min. The enlarged drawings of the red dotted boxes in four different states are given. The PW stator current and CW stator current increase to provide larger electromagnetic torque when the reference rotor speed increases. Then PW stator current and CW stator current are stable after the reference rotor speed keeps constant. As shown in Fig. 11.11a, when the reference rotor speed changes, the real rotor speed can track the reference value accurately with the proposed cascade PI-SMC method. The real values of the dq-axis CW stator current track the reference value in the dynamic process, as shown in Fig. 11.11b, which shows a good dynamic performance.

Figure 11.12 shows the experimental waveforms of PW stator phase voltage v_{pa}, PW stator current i_{pa}, and CW stator current i_{oa} when the load torque changes from 0 N·m to 5 N·m and changes back to 0 N·m with a constant reference rotor speed. The enlarged drawings of the red dotted boxes in four different states are given. The PW stator current and CW stator current increase or decrease when the load torque

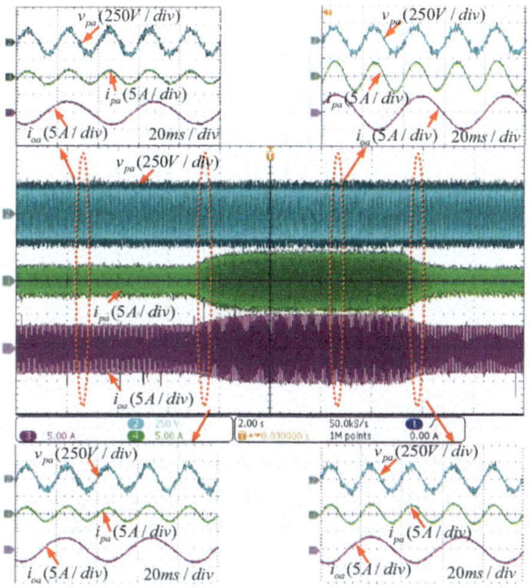

Fig. 11.12 Experimental waveforms of cascade PI-SMC method: The load torque changes from 0 N·m to 5 N·m and changed back to 0 N·m when the reference rotor speed is 400 r/min

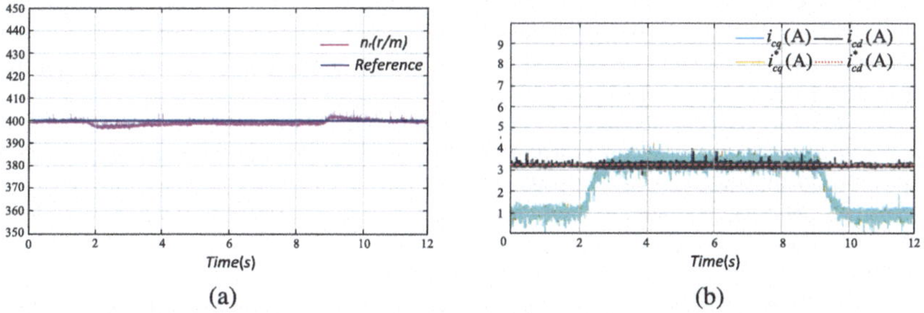

Fig. 11.13 Experimental results of the cascade PI-SMC method: The load torque changes from 0 N·m to 5 N·m and changed back to 0 N·m when the reference rotor speed is 400 r/min. (**a**) The reference and real values of rotor speed. (**b**) The reference and real dq-axis values of CW stator current

increases or decreases. As shown in Fig. 11.13, a few adjustments occur in the real rotor speed and dq-axis CW stator current when the load torque changes. Then the rotor speed recovers to its original value in a short time. The proposed cascade PI-SMC method has a good anti-disturbance interference ability and achieves good dynamic performance.

11.5 Conclusions

In this chapter, a cascade PI-SMC method is proposed for a matrix converter-fed BDFIM drive system. The proposed method consists of a PI speed controller and an inner-loop second-order SMC. The outer loop adopts the PI controller with the active damping to track the reference value of the rotor speed and reactive power.

The inner loop uses the continuous super-twisting sliding mode controller to control the CW current of the brushless doubly fed motor, which eliminates the chattering problem. The proposed cascade PI-SMC method has a good anti-disturbance interference ability. Meanwhile, it has excellent robust tracking performance. The experimental results verify the effectiveness and correction of the proposed cascade PI-SMC method.

References

1. M.B. Field, H.W. Wilson, L.J. Hunt, Discussion on 'A new type of induction motor'. J. Inst. Electr. Eng. **39**(186), 668–677 (1907)
2. P. Han, M. Cheng, S. Ademi, M.G. Jovanovic, Brushless doubly-fed machines: opportunities and challenges. Chinese J. Electr. Eng. **4**(2), 1–17 (2018)
3. M.E. Mathekga, S. Ademi, R.A. McMahon, Brushless doubly fed machine magnetic field distribution characteristics and their impact on the analysis and design. IEEE Trans. Energy Conversion **34**(4), 2180–2188 (2019)
4. Y. Liu, W. Xu, J. Zhu, F. Blaabjerg, Sensorless control of standalone brushless doubly fed induction generator feeding unbalanced loads in a ship shaft power generation system. IEEE Trans. Ind. Electron. **66**(1), 739–749 (2019)
5. X. Li, T. Peng, H. Dan, G. Zhang, W. Tang, W. Jin, P. Wheeler, M. Rivera, A modulated model predictive control scheme for the brushless doubly-fed induction machine. IEEE J. Emerg. Sel. Topics Power Electron. **6**(4), 1681–1691 (2018)
6. S. Shao, E. Abdi, R. McMahon, Low-cost variable speed drive based on a brushless doubly-fed motor and a fractional unidirectional converter. IEEE Trans. Ind. Electron. **59**(1), 317–325 (2012)
7. J. Chen, W. Zhang, B. Chen, Y. Ma, Improved vector control of brushless doubly fed induction generator under unbalanced grid conditions for offshore wind power generation. IEEE Trans. Energy Conversion **31**(1), 293–302 (2016)
8. R. Cárdenas, R. Peña, P. Wheeler, J. Clare, A. Muñoz, A. Sureda, Control of a wind generation system based on a brushless doubly-fed induction generator fed by a matrix converter. Electr. Power Syst. Res. **103**, 49–60 (2013)
9. S. Bayhan, P. Kakosimos, M. Rivera, Predictive torque control of brushless doubly fed induction generator fed by a matrix converter, in *IEEE 12th International Conference on Compatibility, Power Electronics and Power Engineering* (2018)
10. P.W. Wheeler, J. Rodriguez, J.C. Clare, L. Empringham, A. Wein-stein, Matrix converters: a technology review. IEEE Trans. Ind. Electron. **49**(2), 276–288 (2002)
11. M. Khosravi, M. Amirbande, D.A. Khaburi, M. Rivera, J. Riveros, J. Rodriguez, A. Vahedi, P. Wheeler, Review of model predictive control strategies for matrix converters. IET Power Electron. **12**(12), 3021–3032 (2019)
12. T. Friedli, J.W. Kolar, J. Rodriguez et al., Comparative evaluation of three-phase AC–AC matrix converter and voltage DC-link back-to-back converter systems. IEEE Trans. Ind. Electron. **59**(12), 4487–4510 (2011)
13. D. Zhou, R. Spee, A.K. Wallace, Laboratory control implementations for doubly-fed machines, in *Proceedings of IECON '93 - 19th Annual Conference of IEEE Industrial Electronics*, Maui (1993), pp. 1181–1186
14. R. Li, R. Spee, A.K. Wallace, G.C. Alexander, Synchronous drive performance of brushless doubly-fed motors. IEEE Trans. Ind. Appl. **30**(4), 963–970 (1994)
15. S.Y. Shao, E. Abdi, R.A. McMahon, Low-cost variable speed drive based on a brushless doubly-fed motor and a fractional unidirectional converter. IEEE Trans. Ind. Electron. **59**(1), 317–325 (2012)

16. W.R. Brassfield, R. Spee, T.G. Habetler, Direct torque control for brushless doubly-fed machines. IEEE Trans. Ind. Appl. **32**(5), 1098–1104 (1996)
17. R. Sadeghi, S.M. Madani, M. Ataei, A new smooth synchronization of brushless doubly-fed induction generator by applying a proposed machine model. IEEE Trans. Sustain. Energy **9**(1), 371–380 (2018)
18. D. Zhou, R. Spee, G.C. Alexander, Experimental evaluation of a rotor flux oriented control algorithm for brushless doubly-fed machines. IEEE Trans. Power Electron. **12**(1), 72–78 (1997)
19. B. Hopfensperger, D.J. Atkinson, R.A. Lakin, Combined magnetizing flux oriented control of the cascaded doubly-fed induction machine. IEE Proc. Electr. Power Appl. **148**(4), 354–362 (2001)
20. J. Poza, E. Oyarbide, I. Sarasola, M. Rodriguez, Vector control design and experimental evaluation for the brushless doubly fed machine. IET Elect. Power Appl. **3**(4), 247–256 (2009)
21. G. Zhang, J. Yang, Y. Sun, M. Su, W. Tang, Q. Zhu, H. Wang, A robust control scheme based on ISMC for the brushless doubly fed induction machine. IEEE Trans. Power Electron. **33**(4), 3129–3140 (2018)
22. W. Xu, A.K. Ebraheem, Y. Liu, J. Zhu, M.G. Hussien, O.M. Elbabo Mohammed, An MRAS speed observer based on control winding flux for sensorless control of stand-alone BDFIGs. IEEE Trans. Power Electron. **35**(7), 7271–7281 (2020)
23. J. Rodriguez et al., State of the art of finite control set model predictive control in power electronics. IEEE Trans. Ind. Inform. **9**(2), 1003–1016 (2013)
24. M.S. Boger, A.K. Wallace, R. Spee, Ruqi Li, General pole number model of the brushless doubly-fed machine. IEEE Trans. Ind. Appl. **31**(5), 1022–1028 (1995)
25. J. Poza, E. Oyarbide, D. Roye, M. Rodriguez, Unified reference frame DQ model of the brushless doubly fed machine. IEE Proc. Electr. Power Appl. **153**(5), 726–734 (2006)
26. Y. Shtessel, C. Edwards, L. Fridman, A. Levant, *Sliding Mode Control and Observation* (Birkhauser, New York, 2014)
27. J. Mishra, L. Wang, Y. Zhu, X. Yu, M. Jalili, A novel mixed cascade finite-time switching control design for induction motor. IEEE Trans. Ind. Electron. **66**(2), 1172–1181 (2019)
28. C. Lascu, A. Argeseanu, F. Blaabjerg, Super twisting sliding-mode direct torque and flux control of induction machine drives. IEEE Trans. Power Electron. **35**(5), 5057–5065 (2020)
29. X. Zhao, H. Wang, H. Dan, Z. Di, Y. Sun, M. Su, M. Rivera, P. Wheeler, A cascade PI-SMC method for brushless doubly-fed induction machine with matrix converter, in Presented at the 35th Annual IEEE Applied Power Electronics Conference & Exposition, New Orleans, Louisiana, 2020
30. G. Weihua, S. Yao, Q. Hengsi, Y. Chunhua, S. Mei, A matrix converter modulation based on mathematical construction, in *2008 IEEE International Conference on Industrial Technology*, Chengdu, pp. 1–5
31. S. Tohidi, Analysis and simplified modelling of brushless doubly-fed induction machine in synchronous mode of operation. IET Electr. Power Appl. **10**(2), 110–116 (2016)
32. A. Levant, Higher-order sliding modes, differentiation and output-feedback control. Int. J. Control **76**, 924–941 (2003)
33. L. Harnefors, K. Pietilainen, L. Gertmar, Torque-maximizing field-weakening control: design, analysis, and parameter selection. IEEE Trans. Ind. Electron. **48**(1), 161–168 (2001)
34. J.A. Moreno, M. Osorio, Strict Lyapunov functions for the super-twisting algorithm. IEEE Trans. Autom. Control **57**(4), 1035–1040 (2012)

Appendix A
Conclusions and Perspective for Future Work

As an "all-silicon" solution, matrix converter is a type of "green" AC-AC converter that eliminates the bulky passive energy storage components compared to the traditional back-to-back PWM converters. It has the characteristics of high efficiency, high reliability, and high power density. Therefore, it has great potential for volume/weight-critical, efficiency-critical, and lifetime-critical applications such as WECS, MEA, military electrified ship, and the occasion that require continuous low-speed operation and bidirectional power flows.

Perfect thing is nothing, behind of the advantages of matrix converters, there are also some disadvantages, such as complicated main circuit and commutations, limited voltage transfer ratio, and limited input reactive power capacity. In order to meet the demand on enhanced performance for emerging fields and promote the industrial application of matrix converters, this book proposes a series of solutions to these problems. The content of this book mainly revolves around the topology, modulation strategy, and control method of matrix converters. Usually, each chapter starts with a brief introduction, followed by specific methods to solve the problem, and finally concludes with a chapter summary. The core of each chapter includes problem description, method details, as well as simulation and experimental verification.

Chapter 1 introduces the research background, historical development trajectory, and industrial application status of matrix converters. Chapter 2 provides a detailed introduction to the basic third-harmonic injection matrix converter, including the topology configuration, operating principles, modulations, and designs. Chapters 3 to 6 present the derived third-harmonic injection matrix converter topology family, including the third-harmonic injection matrix converter with multiple-outputs and several three-level third-harmonic injection matrix converters for different applications. The main purpose of Chap. 7 is to promote the engineering application of third-harmonic injection concept by developing a third-harmonic injection micro-inverter topology for AC photovoltaic module applications. Chapter 8 mainly elaborates on the topology of back-to-back matrix converter with single-arm current

H. Wang, *Advanced Matrix Converters*, Power Systems,
https://doi.org/10.1007/978-3-031-69902-3

injection, which fully utilizes the low cost, low complexity of the main circuit, and low switching loss characteristics of the back-to-back converter with passive front-end converter. Chapter 9 mainly focuses on the advanced three-level output modulation strategy for conventional matrix converter, which improves the output waveforms quality extremely without increasing the complexity of the main circuit. Chapters 10 and 11 mainly introduce the advanced control methods of the matrix converters, including the control method to enhance linear voltage transfer ratio and the high-performance cascade PI-SMC method for matrix converter-fed BDFIM drives.

This book elaborates on the theory of advanced matrix converters technology comprehensively, from the topology constructions to modulation strategies and control methods. Although some researches have been done on advanced matrix converters and part achievements have been made, this is just a small step in the industrialization process of matrix converters, and there is still a lot of work, such as controls under abnormal operation conditions and common-mode voltage suppression to be completed urgently in the future.

Index

H. Wang, *Advanced Matrix Converters*, Power Systems, https://doi.org/10.1007/978-3-031-69902-3

The manufacturer's authorised representative in the EU is Springer
Nature Customer Service Centre GmbH, Europaplatz 3, 69115 Heidelberg,
Germany. If you have any concerns regarding our products, please
contact ProductSafety@springernature.com

Printed and bound by CPI Group (UK) Ltd, Croydon, CR0 4YY
01/10/2026
02225932-0002